Advanced Biotechnologies for Water and Wastewater Treatment

Advanced Biotechnologies for Water and Wastewater Treatment

Guest Editors

Yung-Tse Hung
Tsuyoshi Imai
Rehab O. Abdel Rahman
Issam A. Al-Khatib

MDPI

Basel • Beijing • Wuhan • Barcelona • Belgrade • Novi Sad • Cluj • Manchester

Guest Editors

Yung-Tse Hung
Civil and Environmental Engineering
Cleveland State University
Cleveland
United States

Tsuyoshi Imai
Environmental Science and Engineering
Yamaguchi University
Ube City
Japan

Rehab O. Abdel Rahman
Radioactive Waste Management Dep.
Atomic Energy Authority of Egypt
Cairo
Egypt

Issam A. Al-Khatib
Institute of Environmental and Water Studies
Birzeit University
Birzeit
West Bank

Editorial Office
MDPI AG
Grosspeteranlage 5
4052 Basel, Switzerland

This is a reprint of the Special Issue, published open access by the journal *Water* (ISSN 2073-4441), freely accessible at: www.mdpi.com/journal/water/special_issues/34BS041DY6.

For citation purposes, cite each article independently as indicated on the article page online and using the guide below:

Lastname, A.A.; Lastname, B.B. Article Title. *Journal Name* **Year**, *Volume Number*, Page Range.

ISBN 978-3-7258-3476-1 (Hbk)
ISBN 978-3-7258-3475-4 (PDF)
https://doi.org/10.3390/books978-3-7258-3475-4

Contents

About the Editors

Yung-Tse Hung

Professor Yung-Tse Hung, Ph.D., P.E., DEE, Fellow of the ASCE, was a Professor of Civil Engineering (retired) at Cleveland State University, Cleveland, Ohio, USA, for 43 years from 1981 to 2024. He obtained his B.S. and M.S. degrees in Civil Engineering from Cheng Kung University in Taiwan in 1964 and 1966 and his Ph.D. degree in Civil Engineering from the University of Texas in Austin in 1970. He has taught at 16 universities in eight countries. He started the public health engineering program at the Department of Civil Engineering, University of Canterbury, Christchurch, New Zealand, in 1972. His primary research interests and publications have been involved with biological wastewater treatment, industrial water pollution control and industrial waste treatment, and municipal wastewater treatment. He is a Fellow of the American Society of Civil Engineers (ASCE), a Diplomate of the American Academy of Environmental Engineers (AAEE), a Fellow of the Ohio Academy of Science, a member of the Association of Environmental Engineering Professors, a life member of the Water Environment Federation, and a life member of the American Association of Engineering Education. He is the Editor-in-Chief of the *International Journal of Environment and Waste Management* (IJEWM) and of the *International Journal of Environmental Engineering* (IJEE). He is the co-editor of several environmental engineering books by Springer and CRC. He is the Editor-in-Chief of three environmental engineering books by World Scientific Publishers. He is a registered professional engineer in Ohio and North Dakota.

Tsuyoshi Imai

Prof. Tsuyoshi Imai is a professor of the Graduate School of Sciences and Technology for Innovation, Yamaguchi University. He graduated from the Graduate School of Engineering (doctoral course), Kyushu University, Japan, in 1995. His research field is environmental/sanitary engineering including aerobic/anaerobic wastewater treatment and resource recovery, solid waste treatment, and resource recovery and the development of water environment conservation technology.

Rehab O. Abdel Rahman

Prof. Dr. Rehab O. Abdel Rahman is an executive project manager, NLSD, and Prof. and Head of the Radioactive Waste Management Department, EAEA, Egypt. She received her Ph.D. degree in Nuclear Engineering and has participated in international projects and meetings on safety cases and safety assessments, especially for radioactive waste disposal facilities. Her published research focuses on the different aspects of safe radioactive and hazardous waste management, including liquid wastewater treatment, the immobilization and stabilization of hazardous and radioactive wastes, and assessing the performance of disposal barriers. She supports national capacity building, where she supervises post graduate students; she has also taught undergraduate courses and mentored the safety assessment team. She serves as a member of international scientific committees, an editor for international journals, a (co)-editor for several books, a reviewer for scientific papers and grant applications, and has contributed to the publication of several book chapters. She was awarded the State Encouragement Award in Engineering sciences, Egypt, 2011.

Issam A. Al-Khatib

Prof. Dr. Issam A. Al-Khatib is a faculty member at the Institute of Environmental and Water Studies, Birzeit University, Palestine. His expertise spans across water resource management, environmental assessment, wastewater management, and climate change, with a focus on environmental health and sustainable development. Dr. Al-Khatib has led numerous research and evaluation projects on topics such as medical waste management, solid waste, water and sanitation, and urban environments. He is dedicated to promoting public environmental awareness through training, workshops, and community campaigns. He has contributed to national committees on environmental standards, including swimming pool water quality regulations. Dr. Al-Khatib has supervised over 40 MSc theses focusing on these critical environmental issues, and he is an expert on academic quality evaluation. Currently, he is working on a project to evaluate the attitudes, perceptions, and behaviors of Birzeit University students toward the management of single-use plastic waste. Prof. Al-Khatib is ranked among the top 0.5% of researchers globally according to ScholarGPS in 2025, recognized for his exceptional contributions to environmental science, monitoring, and waste management.

water

Editorial

Advanced Biotechnologies for Water and Wastewater Treatment

Yung-Tse Hung [1], Rehab O. Abdel Rahman [2,*], Issam A. Al-Khatib [3] and Tsuyoshi Imai [4]

[1] Department of Civil and Environmental Engineering, Cleveland State University, Cleveland, OH 44115, USA; y.hung@csuohio.edu
[2] Hot Laboratory Center, Atomic Energy Authority of Egypt, Inshas, Cairo P.O. Box 13759, Egypt
[3] Institute of Environmental and Water Studies, Faculty of Graduate Studies, Birzeit University, Birzeit P.O. Box 14, Palestine; ikhatib@birzeit.edu
[4] Graduate School of Sciences and Technology for Innovation, Yamaguchi University, 2-16-1, Tokiwadai, Ube City 755-8611, Yamaguchi, Japan; imai@yamaguchi-u.ac.jp
* Correspondence: alaarehab@yahoo.com

1. Introduction

The use of biotechnology to control and prevent the contamination of water resources has a long track record that goes back to the beginning of the 20th century. In that early stage of development, activated sludge technology was developed and implemented for the treatment of wastewater [1]. This technological breakthrough, at that time, was a result of several efforts to enhance the performance of the innovative biological filter [1,2]. Since then, scientists and engineers have started a quest to improve the effective implementation of this technology in partial or full stabilization of biodegradable contaminants, i.e., organic contaminants [3]. This led to the development of several technologies that use varied modes of contact between the wastewater and the biological mass in the presence or absence of air [3]. The wide implementation of biological treatment technologies in the management of several industrial waste streams can be attributed to the efficiency of these technologies, the competitive construction and operational costs, the low energy requirements, the operational flexibility, and the reduced environmental impacts.

Recently, the advancements in the applications of biotechnologies for water and wastewater treatment are being investigated as a part of the quest to improve the sustainability of integrated water and wastewater management systems. These technologies are directed not only to improve the conventional secondary treatment of wastewater, i.e., biological treatment, but also extended to address the tertiary treatment, i.e., sorption [4] and photodegradation [5]. The advances in this field cover not only the use of bacteria but also algae and fungi [6] and advances in improving sludge management and contact technologies, e.g., microbial fuel cells [7], but is also extended to cover the removal of toxic elements, e.g., heavy metals. This Special Issue is directed to present the recent advances in the implementation of biotechnology in water and wastewater treatment with a focus on aspects related to the sustainability of the technologies, i.e., improved efficiency, reduced costs, and minimized energy consumption.

check for updates

Received: 2 February 2025
Accepted: 6 February 2025
Published: 11 February 2025

Citation: Hung, Y.-T.; Abdel Rahman, R.O.; Al-Khatib, I.A.; Imai, T. Advanced Biotechnologies for Water and Wastewater Treatment. *Water* **2025**, *17*, 509. https://doi.org/10.3390/w17040509

2. Summary of Contributions

This Special Issue includes 11 papers (Contributions 1 to 11) and this editorial. Improving biomass management by using thermal degradation to valorize the biomass or sludge separation to enhance recyclability was addressed in two research papers {C2, C4}. In the first paper, the integration of the pyrolysis and gasification technologies was proposed to improve the management process {C2}. In that context, the results of pilot-scale trials were presented to valorize the biomass via the production of biochar. The feasibility of the presented technology was investigated by coupling the energy analysis and preliminary techno-economic assessments. The work concludes on the efficiency of the proposed system to remove emerging contaminants, e.g., PFAS, from the product, scrubber water and gaseous emissions. Moreover, the optimization conditions for operating self-sufficient energy were presented {C2}. In the second paper, the improvement of the granulation of the activated sludge with poor settle ability was investigated for variable high-strength effluent {C4}. In this respect, a microbial selection strategy to form aerobic granules was implemented via the use of an anaerobic feast/aerobic famine strategy in a sequencing batch reactor. The work reveals that the granulated sludge has significantly lower resistance compared to the seed sludge due to pore blockage and has significantly higher sustainable flux. The research work recommends the implementation of adequate influent pretreatment technology to improve the performance of the adopted strategy in the granulation of the activated sludge {C4}.

Heavy metal removal from water and wastewater is one of the challenging problems that affect the applicability of biotechnologies in this field. In this Special Issue, five research papers addressed the effect of the presence of these metals on the biomass and the improvement in the removal performance and the optimization of the process {C3, C5, C6, C10, C11}. Insights into the effect of the presence of lead in municipal wastewater on the symbiosis of algal–bacterial granular activated sludge were provided in a research paper {C10}. This effect was studied for initial lead concentration in the range of 2.5–10 ppm. Within that range, the study reveals that the presence of lead has little effect on the performance of the algal–bacterial granular sludge {C10}. The optimization of the Cr removal performance in microbial fuel cells was investigated in a research paper by using an adaptive neuro-fuzzy inference system (ANFIS) and artificial ecosystem optimization (AEO) {C11}. The study considered the effect of the Cu(II)/Cr(VI) ratio, the substrate concentration, and the external resistance during the optimization process and concluded that the integration between ANFIS and AEO to model and optimize the process led to increasing both the power density and the Cr removal considerably {C11}. Another research paper addressed the process modeling for the treatment of water contaminated by Cr via using bacterial cellulose biomass {C3}. Lab-scale and pilot-scale experiments were conducted in that research work to study the removal of chromium using the biomass. In addition, the reusability of the biomass was investigated by studying biomass elution using EDTA. The work concluded on the applicability of the used biomass to remove heavy metals from aqueous solutions {C3}. The fourth paper investigated the use of biotechnology in the removal of Cr from aqueous solutions using transformed *E. crassipes* biomass with sodium Tripolyphosphate (TPP){C6}. In that respect, the sorption capacity and the cost were investigated, and the elution performance to reuse the used biomass was determined. The results suggest that *E. crassipes* biomass, enhanced with PTT and through EDTA elutions, could be proposed as a suitable water remediation technology based on sorption technology. The last research paper that covered the removal of heavy metals investigated the construction of a genetically engineered *Escherichia coli* cell factory for the removal of Cd {C5}. In this respect, molecular biology techniques were adopted to fuse the recombinant human ferritin (rHF) gene and the synthetic phytochelatin (EC) gene. The work shows the successful soluble expression of the recombinant fusion protein in *E. Coli* cells that can remove 9.2 μmol of Cd^{2+} in vitro over a wide range of temperatures (16–45 °C) and in slightly acidic–slightly neutral pHs (5–9). The main emphasis of that work is on the synthesis of multivalent chelating peptides in E. coli cells to enhance Cd removal {C5}.

Finally, improving the degradation of organic contaminants in the tertiary treatment was presented in four papers {C1, C7–C9}. The first paper addressed the effect of the preparation conditions of the ZnO nano-particles on their photo-catalytic activity {C1}.

The work addresses the effect of the calcination temperature during the preparation of ZnO nanoparticles on the structural, morphological, and optical properties of the prepared material. The study demonstrates that the calcination temperature affects photo-catalytic activity, and the performance of the optimized material in the degradation of RhodamineB was investigated under different operating conditions, either solely or combined with H_2O_2oxidation. The study concluded on the optimized conditions to fully degrade the Rhodamine B {C1}. Another study assessed the potential use of the immobilized enzymes, i.e., Horseradish Peroxidase (HRP) and Myoglobin (MB), in sodium alginate for the treatment of organic pollutants {C7}. The prepared material was tested for its potential use in the degradation of aniline, phenol, and p-nitrophenol. The paper concludes on the resilience of the enzymes to the changes in the temperature and pH and on the promising reusability performance {C7}. The third paper in this category studied the removal of Xenobiotics by Trametes hirsute LE-BIN 072 activated carbon-based Mycelial pellets {C8}. In that context, the roles of fungal mycelium and fungal enzymes in the sorption and biodegradation of the dye were evaluated. The work reveals that laccases were proposed as the main contributing enzymes in the degradation process, where RT-qPCR measurements demonstrated an increase in transcription for the two laccase genes—*lacA* and *lacB*. It has been proven that the composite Mycelial pellets of *T. hirsute* show improved sorption ability, which emphasizes the role of fungal mycelium in improving dye removal {8}. Finally, the last paper investigated phenol degradation using immobilized *Alcaligenes faecalis* strain JH1 in Fe_3O_4-modified biochar from pharmaceutical residues{C9}. In that work, four types of biochar were used as carriers to immobilize different *Alcaligenes faecalis* strains to have insights into the removal mechanism of phenol and the reusability of the material was investigated. The study indicated that the immobilization process improves the tolerance of the bacteria to the operating conditions. The durability of the material, its stability, and reproducibility were proven, and the role of Fe_3O_4nano-particle in enhancing the removal was concluded.

3. Conclusions

Despite the use of biotechnology in water and wastewater treatment having a long operational track, innovation in this field is directed to improve the overall performance of the technologies and ensure their sustainability. The presented research papers in this Special Issue cover these motivations, where biomass management, improving the performance of heavy metal removal, and organic contaminant degradation were experimentally and theoretically investigated. The presented papers highlighted the possibility of improving the practice of treatment and balancing the sustainability measures, e.g., cost-effectiveness, energy efficiency, and reduced environmental impacts. Yet efforts are still needed to synergize the sustainability performance indicators with the technical assessment during the feasibility assessment of the use of different biotechnologies in water and wastewater treatment.

Author Contributions: Conceptualization and writing—original draft preparation, Y.-T.H. and R.O.A.R.; writing—review and editing, all authors. All authors have read and agreed to the published version of the manuscript.

Conflicts of Interest: The authors declare no conflicts of interest.

List of Contributions:

1. Saidani, A.; Boudraa, R.; Fendi, K.; Benouadah, L.; Benabbas, A.; Djermoune, A.; Salvestrini, S.; Bollinger, J.-C.; Alayyaf, A.A.; Mouni, L. Effect of Calcination Temperature on the Photocatalytic Activity of Precipitated ZnO Nanoparticles for the Degradation of Rhodamine B Under Different Light Sources. *Water* **2025**, *17*, 32. https://doi.org/10.3390/w17010032.

2. Rathnayake, N.; Patel, S.; Hakeem, I.G.; Veluswamy, G.; Al-Waili, I.; Agnihotri, S.; Vuppaladadiyam, A.K.; Surapaneni, A.; Bergmann, D.; Shah, K. The Pyrolysis of Biosolids in a Novel Closed Coupled Pyrolysis and Gasification Technology: Pilot Plant Trials, Aspen Plus Modelling, and a Techno-Economic Analysis. *Water* **2024**, *16*, 3399.
3. Sayago, U.F.C.; Ballesteros, V.B.; Aguilar, A.M.L. Designing, Modeling and Developing Scale Models for the Treatment of Water Contaminated with Cr (VI) through Bacterial Cellulose Biomass. *Water* **2024**, *16*, 2524. https://doi.org/10.3390/w16172524.
4. Ahmed, M.; Goettert, D.; Vanherck, C.; Goossens, K.; Dries, J. Microbial Selection for the Densification of Activated Sludge Treating Variable and High-Strength Industrial Wastewater. *Water* **2024**, *16*, 2087. https://doi.org/10.3390/w16152087.
5. Tian, L.; Wang, D.; Liu, Y.; Wei, M.; Han, X.; Sun, X.; Yin, L.; Luo, G. Construction of Genetically Engineered Escherichia coli Cell Factory for Enhanced Cadmium Bioaccumulation in Wastewater. *Water* **2024**, *16*, 1759.
6. Sayago, U.F.C.; Ballesteros, V.A.B. The Design of a Process for Adsorbing and Eluting Chromium (VI) Using Fixed-Bed Columns of *E. crassipes* with Sodium Tripolyphosphate (TPP). *Water* **2024**, *16*, 952. https://doi.org/10.3390/w16070952.
7. Wang, X.; Ghanizadeh, H.; Khan, S.; Wu, X.; Li, H.; Sadiq, S.; Liu, J.; Liu, H.; Yue, Q. Immobilization of Horseradish Peroxidase and Myoglobin Using Sodium Alginate for Treating Organic Pollutants. *Water* **2024**, *16*, 848. https://doi.org/10.3390/w16060848.
8. Glazunova, O.A.; Moiseenko, K.V.; Fedorova, T.V. Xenobiotic Removal by Trametes hirsuta LE-BIN 072 Activated Carbon-Based Mycelial Pellets: Remazol Brilliant Blue R Case Study. *Water* **2024**, *16*, 133. https://doi.org/10.3390/w16010133.
9. Zeng, Z.; Xiao, J.; Li, M.; Wu, J.; Zhang, T. Degradation of Phenol by Immobilized Alcaligenes faecalis Strain JH1 in Fe3O4-Modified Biochar from Pharmaceutical Residues. *Water* **2023**, *15*, 4084. https://doi.org/10.3390/w15234084.
10. Yang, J.; Zhang, Y.; Wang, S. Over-Produced Extracellular Polymeric Substances and Activated Antioxidant Enzymes Attribute to Resistance of Pb(II) for Algal–Bacterial Granular Sludge in Municipal Wastewater Treatment. *Water* **2023**, *15*, 3833. https://doi.org/10.3390/w15213833.
11. Abdelkareem, M.A.; Alshathri, S.I.; Masdar, M.S.; Olabi, A.G. Adaptive Neuro-Fuzzy Inference System Modeling and Optimization of Microbial Fuel Cells for Wastewater Treatment. *Water* **2023**, *15*, 3564. https://doi.org/10.3390/w15203564.

References

1. Park, J.H.; Park, H.D. Advanced biological wastewater treatment. In *Current Developments in Biotechnology and Bioengineering*; Elsevier: Amsterdam, The Netherlands, 2021; pp. 107–123.
2. Herschy, R.W. Activated sludge process: Historical. In *Encyclopedia of Hydrology and Lakes. Encyclopedia of Earth Science*; Springer: Dordrecht, The Netherlands, 1998. [CrossRef]
3. Cheremisinoff, N.P. *Biotechnology for Waste and Wastewater Treatment*; Elsevier: Amsterdam, The Netherlands, 1997.
4. Maldonado-Carmona, N.; Piccirillo, G.; Godard, J.; Heuzé, K.; Genin, E.; Villandier, N.; Calvete, M.J.; Leroy-Lhez, S. Bio-based matrix photocatalysts for photodegradation of antibiotics. *Photochem. Photobiol. Sci.* **2024**, *23*, 587–627. [CrossRef] [PubMed]
5. Alhalabi, A.M.; Meetani, M.A.; Shabib, A.; Maraqa, M.A. Sorption of pharmaceutically active compounds to soils: A review. *Environ. Sci. Eur.* **2024**, *36*, 161. [CrossRef]
6. Chan, S.S.; Khoo, K.S.; Chew, K.W.; Ling, T.C.; Show, P.L. Recent advances biodegradation and biosorption of organic compounds from wastewater: Microalgae-bacteria consortium—A review. *Bioresour. Technol.* **2022**, *344*, 126159. [CrossRef]
7. Jalili, P.; Ala, A.; Nazari, P.; Jalili, B.; Ganji, D.D. A comprehensive review of microbial fuel cells considering materials, methods, structures, and microorganisms. *Heliyon* **2024**, *10*, e25439. [CrossRef] [PubMed]

water

MDPI

Article

Effect of Calcination Temperature on the Photocatalytic Activity of Precipitated ZnO Nanoparticles for the Degradation of Rhodamine B Under Different Light Sources

Amira Saidani [1], Reguia Boudraa [2], Karim Fendi [1], Lamia Benouadah [1,3], Abderrahim Benabbas [1], Atmane Djermoune [4], Stefano Salvestrini [5], Jean-Claude Bollinger [6], Abdulmajeed Abdullah Alayyaf [7] and Lotfi Mouni [3,*]

1 Laboratory Processes for Materials, Energy, Water and Environment, Faculty of Science and Technology, University of Bouira, Bouira 10000, Algeria; a.saidani@univ-bouira.dz (A.S.); k.fendi@univ-bouira.dz (K.F.); l.benouadah@univ-bouira.dz (L.B.); a.benabbas@univ-bouira.dz (A.B.)
2 Laboratory of Materials and Durable Development (LM2D), University of Bouira, Bouira 10000, Algeria; r.boudraa@univ-bouira.dz
3 Laboratoire de Gestion et Valorisation des Ressources Naturelles et Assurance Qualité, Faculté SNVST, Université de Bouira, Bouira 10000, Algeria
4 Centre de Recherche Scientifique et Technique en Analyses Physico-chimiques (CRAPC), Bou-Ismaïl, Tipaza 42004, Algeria; atmane.djermoune@univ-bejaia.dz
5 Department of Environmental, Biological and Pharmaceutical Sciences and Technologies, University of Campania "Luigi Vanvitelli", via Vivaldi 43, 81100 Caserta, Italy; stefano.salvestrini@unicampania.it
6 Laboratoire E2Lim (Eau Environnement Limoges), Université de Limoges, 87060 Limoges, France; jean-claude.bollinger@unilim.fr
7 Department of Chemistry, College of Sciences, King Saud University, Riyadh 11451, Saudi Arabia; mayyaf@ksu.edu.sa
* Correspondence: l.mouni@univ-bouira.dz; Tel.: +213-784-42-62-28

check for updates

Academic Editor: Chengyun Zhou

Received: 2 November 2024
Revised: 23 December 2024
Accepted: 23 December 2024
Published: 26 December 2024

Citation: Saidani, A.; Boudraa, R.; Fendi, K.; Benouadah, L.; Benabbas, A.; Djermoune, A.; Salvestrini, S.; Bollinger, J.-C.; Alayyaf, A.A.; Mouni, L. Effect of Calcination Temperature on the Photocatalytic Activity of Precipitated ZnO Nanoparticles for the Degradation of Rhodamine B Under Different Light Sources. *Water* **2025**, *17*, 32. https://doi.org/10.3390/w17010032

Abstract: This research provides valuable insights into the application of ZnO nanoparticles in photocatalytic wastewater treatment. Process optimization was carried out by determining the ratio of the surface area to the energy band gap (S/E) in the photocatalysis rate under different sources of light (UV light, visible light, sunlight). The nanoparticles were synthesized using the precipitation technique, and the calcination process was carried out within a temperature range of 400 to 700 °C. The structural, morphological, and optical properties of materials were investigated using X-ray powder diffraction (XRD), scanning electron microscopy (SEM), UV-Vis diffuse reflectance (UV-Vis DRS), Raman spectroscopies, and Fourier transform infrared (FTIR) spectroscopies. The study demonstrates that calcination temperature significantly influences the photocatalytic activity of ZnO nanoparticles by altering their size, surface properties, shape, and optical behavior. Optimal decomposition efficiencies of Rhodamine B were achieved at 400 °C, with yields of 24%, 92%, and 91% under visible, UV, and sunlight irradiation, respectively. Additionally, the surface area decreased from 12.556 to 8.445 m^2/g, the band gap narrowed slightly from 3.153 to 3.125 eV, and crystal growth increased from 0.223 to 0.506 μm as the calcination temperature rose. The photocatalytic properties of ZnO nanoparticles were assessed to determine their efficiency in decomposing Rhodamine B dye under operational parameters, including pollutant concentration (C0), sample amount, pH level, and reaction time. The sample exhibited the best breakdown rates with C0 = 5 mg/L, solid-to-liquid ratio (S/L) = 50 mg/L, pH = 7, and reaction time = 1 h. Additionally, we combined two oxidation processes, namely H_2O_2 and photocatalytic oxidation processes, which significantly improved the Rhodamine B removal efficiency, where 100% of RhB was degraded after 60 min and 100 μL of H_2O_2.

Keywords: nanoparticles; photocatalytic activity; rhodamine B; sunlight; UV light; visible light; ZnO

1. Introduction

Solar energy is one of the most important renewable energy sources, if not the most important, in terms of availability, abundance, and affordability [1–3]. Researchers are exploring photocatalysis as a transformative technology to address environmental challenges. Photocatalysis harnesses sunlight, a renewable energy source in water treatment, by effectively breaking down harmful pollutants like dyes into less toxic substances, reducing their environmental and health risks and minimizing their impact on ecosystems. This eco-friendly solution contributes to cleaner air, water, and soil. It aligns with global efforts toward greener, more energy-efficient technologies [4–7]. Moreover, photocatalysis enables the direct transformation of solar energy into usable heat or other valuable energy resources, sparking significant interest and applications across various sectors. Concretely, photocatalytic technologies have found utility in sensitized solar cells [8,9], chemical pollutant oxidation, and reduction [10], as well as the inactivation of microbial pathogens. Experts claim that photocatalysis is a complex process that fundamentally depends on the presence of a catalyst, usually a semiconductor. This process influences the kinetic rate of the photochemical reaction involving reactive species and light, while maintaining its fundamental thermodynamic limitations. Photons serve as the energy source, aligning with the band gap energy of the semiconductor, to excite electrons from the valence band to the conduction band, resulting in the generation of electron–hole pairs. Charge carriers migrate to the photocatalyst surface, where they participate in redox reactions that generate reactive oxygen species (ROS), including hydroxyl radicals ($^{\bullet}OH$) and superoxide anions (O_2^-), essential for pollutant degradation [11–13]. Zinc oxide (ZnO) is a widely recognized, and highly attractive semiconductor, valued for its remarkable properties such as high crystallinity, tunable optical and electronic properties, exceptional quantum efficiency, low electrical conductivity, and excellent thermal stability [14,15]. These properties can be tailored by synthesizing ZnO in various nanostructural forms, such as nanoparticles, nanorods, nanosheets, and hierarchical architectures [16], each offering distinct advantages for diverse applications like anti-corrosion [17], electronics and optoelectronics [18], industrial applications [19], nanomedicine applications like cancer therapy [20], antioxidant [21] and antibacterial uses [22], and environmental protection like photocatalysis for pollutant degradation [23,24]. Zinc oxide ZnO is often considered superior to titanium dioxide TiO_2 for sunlight photocatalysis [9,25,26].

Zinc oxide is classified as an n-type semiconductor owing to the presence of extra zinc atoms and oxygen vacancies in the interatomic sites. Its crystalline structure adopts the hexagonal wurtzite form [27,28]. It is recognized that its activity in the ultraviolet region is attributed to its broad band gap of 3.3 eV. However, as the band gap energy decreases, the energy threshold for electronic transitions is lowered, allowing the absorption of lower energy photons and shifting the material's radiation absorption towards the visible spectrum [29,30]. This transition is essential for improving the performance of ZnO in sunlight [31], an objective actively targeted by researchers in the field by exploring several strategies, including doping with metal or non-metal elements, creating heterojunctions with other materials, and modifying defect density to optimize light absorption and charge carrier dynamics [32]. Another crucial consideration is the assessment of the suitability of a material as a catalyst by subjecting it to specific conditions for gaining a higher specific surface area and a larger number of active sites to achieve more efficiency with less quantity because we should mention that the big quantity of using ZnO pose ecological risks to aquatic and terrestrial life, affecting their health, disrupting ecological processes, and altering ecosystem balance. They accumulate in aquatic organisms, disrupt soil microbiota, and pose risks to non-target organisms. Green synthesis of biocompatible ZnO NPs offers potential for environmental remediation, but comprehensive ecotoxicological assessments are needed for safety, as well as improved thermal and mechanical stability.

It is important to recognize that the performance of the material can be tuned by adjusting its particle size and morphology [33]. The surface characteristics of metal oxides, including oxygen vacancies and surface defects, are essential in improving photocatalytic activity by offering active sites for reactions. As the calcination temperature rises, particle aggregation diminishes the surface area and defect density, resulting in a reduction of active sites. This process enhances crystallinity while marginally increasing the band gap, resulting in a shift of light absorption towards the UV region and restricting the utilization of visible light. Optimizing photocatalytic performance necessitates a balance among surface area, defect density, and band gap energy [34–37]. It is important to find a compromise between band gap energy and surface area where a catalyst exhibits its optimal photocatalytic performance. Therefore, achieving an efficient photocatalyst requires implementing two contradictory conditions while controlling the thermal treatment. Extensive research has been conducted on the effects of thermal treatments on ZnO on properties such as optical properties [34,36], structural morphology [38,39], and photonic efficiency [33,40]. However, there is a void in the understanding of the inverse relationship between band gap, area, and photonic energy. This study provides a unique perspective. By evaluating how this ratio affects photonic efficiency across different light sources, the work bridges the gap between material characterization and performance optimization. The focus of our study is a comprehensive comparison of the effect of calcination temperature on the photonic performance of ZnO nanoparticles under different light sources UV, visible, and solar radiation. This approach allows us to determine the optimal calcination conditions for specific light sources. This research is based on the fabrication of ZnO nanostructures using the deposition method with the calcination temperature varying from 400 °C to 700 °C. Zinc oxide (ZnO) nanoparticles can be synthesized using various methods, including sol–gel [41], hydrothermal [42], chemical vapor deposition [43], microwave-assisted [44], solution combustion [45], and co-precipitation [38]. Each method offers unique advantages and applications, with the choice depending on the specific application requirements and desired properties. Precipitation is considered one of the most efficient processes for producing ZnO nanoparticles. The preference for this method stems from its straightforward synthesis, low formation temperature, high level of sample homogeneity, and the ability to precisely control the chemical composition [46,47]. The primary goal is to achieve the optimal calcination temperature, ensuring the highest possible photocurrent yield under UV, visible light, and sunlight. The secondary goal is to determine a favorable ratio between small band gap energy and large specific surface area, resulting in high efficiency in the degradation of the organic pollutant [44]. After determining the optimal light source and calcination temperature, the catalytic process is optimized by adjusting both the ZnO mass and the pollutant concentration, along with pH adjustments. A complementary investigation is also conducted on the effect of adding H_2O_2 to the photocatalytic system to achieve maximum efficiency. Rhodamine B was chosen as a model pollutant of choice to evaluate the photocatalytic performance of the synthesized samples. This dye is often used in various industrial applications and has been observed to pose potential environmental hazards. Trace experiments have indicated its potential ecotoxicity towards freshwater aquatic organisms [48,49]. Lethal doses of this substance have the potential to cause toxicity, and its introduction into the natural ecosystem may pose risks to aquatic organisms, as well as carcinogenic effects on both humans and animals [50].

2. Materials and Methods

The materials used in this study included zinc chloride (anhydrous) ($ZnCl_2$, >97%, MW = 81.89 g/mol, Sigma-Aldrich, Germany), sodium hydroxide (NaOH, 99%, Merck, Germany), and Rhodamine B dye (RhB) ($C_{28}H_{31}ClN_2O_3$, purity 95%, Biochem, France). All materials were used without any additional purification steps. Deionized water (DI) was obtained from local sources and used in the experimental procedures.

2.1. Synthesis of ZnO Nanopowders

ZnO was prepared by precipitation with a key modification in the addition of a saturated NaOH solution all at once rapidly instead of gradual addition [51] involving the following steps. At room temperature (26 °C), 30 g of zinc chloride ($ZnCl_2$) was dissolved in 100 mL of deionized water (DI) to form a solution. The solution underwent continuous agitation using a stirrer plate at room temperature, and upon achieving homogeneity, 15 mL of a sodium hydroxide-saturated solution (NaOH) was added at once. The mixture was vigorously stirred for 15 min at room temperature, resulting in the formation of a white precipitate. This precipitate was filtered by filter paper with a pore size equal to 0.45 μm and meticulously rinsed with DI to eliminate all salt and chloride residues. The obtained zinc hydroxide was then dried at 100 °C and divided into five parts. Each part was then treated in air at a specific temperature in the range of 400 °C to 700 °C for 2 h starting from room temperature with a heating rate of 5 °C/min. The synthesized samples were designated as follows: ZnO-400, ZnO-500, ZnO-600, and ZnO-700, corresponding to the calcination temperature.

2.2. Characterization

Powder X-ray diffraction (XRD) patterns were recorded in the range of 20–80° (2θ) in the scan mode (0.028 step size, 2 s counting time) using an Empyrean Panalytical (Germany) powder diffractometer employing CuKα radiation operating at 40 kV and 40 mA. The unit cell parameters were refined within the hexagonal system through the measured peak positions of the XRD pattern, with a least squares method and employing the CELREF program. For the determination of peak characteristics, the peak profiles were fitted with the WINPLOTR program. The crystallite sizes and microstrain contributions were estimated from analyses of Williamson–Hall plots [52]. For morphological studies, a scanning electron microscope (SEM) of the Thermo Scientific Quattro S (USA) field emission type was utilized. The functional groups in samples were identified using a Fourier-transform infrared spectrometer (FT-IR), specifically the Thermo Scientific Nicolet IS50 FT-IR spectrometer (USA), covering the wavenumber range of 400–4000 cm^{-1}. Raman spectroscopy was performed using an Avantes (AvaSpec-HSC-TEC spectrometer, Apeldoorn, The Netherlands), ensuring precise detection of Raman signals over the spectral range of 200–4000 cm^{-1}. Additionally, optical absorbance in the range of 200–800 nm was measured using an Evolution 220 (Thermo Scientifi, Waltham, MA, USA) UV-visible spectrophotometer. Surface area measurements were conducted at 77 K using an Autosorb iQ Station Analyzer (Quantachrome, Boynton Beach, FL, USA), providing detailed insights into the surface properties of the samples.

2.3. Photocatalytic Experiments

Rhodamine B (RhB) served as the model dye to evaluate the photocatalytic performance of the various samples of ZnO. The photodegradation tests were conducted under distinct irradiation sources, including solar (1800 Lux), visible (using 150 W LED lamps), and UV light (365 nm, rated at 15 W). The tests were performed in beakers containing 50 mg of photocatalyst and 100 mL of an aqueous solution of RhB dye (5 mg/L). Prior to the photocatalytic process, the solutions underwent magnetic stirring for 120 min in the dark to assess the absorbance yield of sample even the adsorption equilibrium was reached within 30 min. Subsequently, the mixture was exposed to different light irradiation sources to evaluate the impact of heat treatment conditions on the photocatalytic efficiency.

The influence of pH on photocatalytic degradation was examined by modifying the pH of a 5 mg/L Rhodamine B solution with 0.1 M NaOH and 0.1 M HCl, measuring pH values (2, 7, 10) using a calibrated pH meter. For H_2O_2, 100, 300, and 500 microliters were added to the reaction mixture while keeping the catalyst dosage and reaction volume constant. Kinetic studies were conducted by collecting samples at 30-min intervals for those irradiated by visible and UV light, and at 15-min intervals for samples exposed to sunlight. For the effects of H_2O_2 and pH, the interval was 5 min. The samples were centrifuged to remove the catalyst, and the supernatant was analyzed using UV-Vis spectrophotometry.

The photodegradation yield (D (%)) was estimated using:

$$D(\%) = \frac{(C - C_t)}{C} \times 100 \tag{1}$$

where C and C_t represent the RhB concentration after adsorption equilibrium on the semiconductor before irradiation and at a given time, respectively. The photodegradation kinetics of RhB were evaluated using the Langmuir–Hinshelwood equation [53] (Equation (2))

$$r = -\frac{dC_t}{dt} = \frac{bKC_t}{1 + KC_t} \tag{2}$$

In photocatalysis experiments, the concentrations used are generally low, so the integrated Langmuir–Hinshelwood equation can be simplified as follows:

$$C_t = C \times e^{-kt} \tag{3}$$

where b ($mg \cdot min^{-1} \cdot L^{-1}$), K ($L \cdot mg^{-1}$), and k (= b× K, min^{-1}) are the photocatalytic constant, the adsorption constant and the first-order reaction rate constant, respectively.

3. Results

3.1. Material Characterization

3.1.1. X-Ray Powder Diffraction

The X-ray diffraction patterns are shown in Figure 1. ZnO is shown as a single pure phase. Refinement of the lattice parameters for this phase as a function of the calcination temperature shows no important differences (Table 1); the observed values are in accordance with those of JCPDS reference. Their very slight decrease upon rising calcination temperature is associated with the increase of crystal sizes. In this respect, analysis by means of Williamson–Hall method of the evolution of crystallite size and microstrain confirms this tendency. As expected, the crystallite size D and microstrain ε (Δd/d%) are found to vary in opposite directions vs. calcination temperature (Table 2). Both results reveal the decrease in the number of defects and the enhancement of crystallinity while calcination temperature increases, as corroborated by other research. This observation will have a profound impact on the optical properties. It is important to note that the grain sizes in a polycrystalline sample is independent of the crystallite sizes [54], although in some cases these two quantities are correlated. The former is deduced from electron microscope observations while the latter are estimated from X-ray diffraction peak profiles.

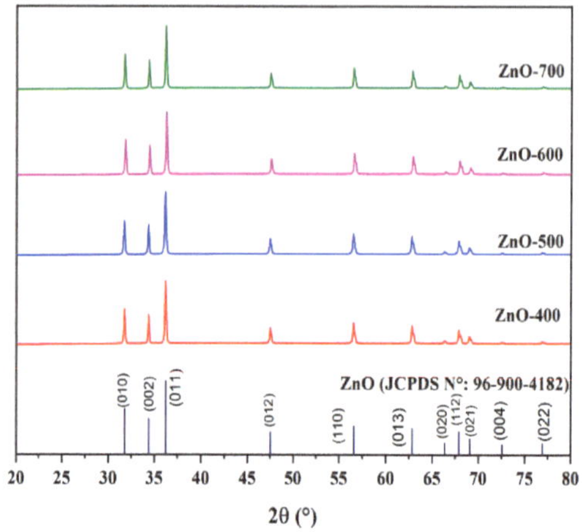

Figure 1. XRD spectra of ZnO calcined at 400 °C, 500 °C, 600 °C, and 700 °C.

Table 1. Lattice parameters, crystallite size D, and microstrain ε of ZnO phase in the four samples.

Calcination Temperature (°C)	Crystallite Size D (nm)	Crystal Parameters (Å)		Microstrain ε (%)
		A	C	
ZnO-400	63	3.2502(1)	5.2068(2)	0.0335
ZnO-500	66	3.2498(1)	5.2062(3)	0.0177
ZnO-600	67	3.2497(1)	5.2056(2)	0.0139
ZnO-700	72	3.2498(1)	5.2058(2)	0.0061

Table 2. BET analysis results of synthesized ZnO samples.

Samples	BET Surface Area (m^2/g)	Langmuir Surface Area (m^2/g)	BJH Adsorption Average Pore Diameter (nm)	BJH Desorption Average Pore Diameter (nm)	Pore Volume (cm^3/g)
ZnO-400	2.793	12.556	27.238	26.453	0.0048
ZnO-500	2.640	10.733	32.039	25.904	0.0054
ZnO-600	2.518	9.126	32.268	22.965	0.0036
ZnO-700	2.201	8.445	15.467	9.518	0.0026

3.1.2. Scanning Electron Microscopy (SEM)

In this investigation, we consequently explored the impact of diverse calcination temperatures, ranging from 400 °C to 700 °C, on the morphology and size of zinc oxide (ZnO) nanoparticles (Figure 2). At 400 °C, the nanoparticles homogeneously exhibited roughly spherical shapes. At intermediate calcination temperatures, an evolution towards larger agglomerations and well-faced crystals is evident, reflecting enhanced sintering and crystal growth from 0.112 to 0.506 μm. Parallelly, the spread of the particle size distribution is more extended. At 700 °C, these tendencies are more pronounced. These results correspond to those presented in a similar study [55]. This comprehensive approach allowed us to discern nuanced changes in morphology and particle size at different temperature regimes, providing valuable insights for tailoring the synthesis of ZnO nanoparticles to specific applications, whether requiring finer particles at lower temperatures or larger, faceted structures at higher temperatures. Additionally, our findings shed light on how these temperature variations impact the optical properties and crystal structure of ZnO, enriching our understanding of the multifaceted implications for various applications in catalysis and photocatalysis. The reduced sizes and higher defect levels at lower temperatures suggest higher specific surface area and density of active sites. The correlation between the nuanced morphological changes induced by varying calcination temperatures and the subsequent effects on catalytic and photocatalytic activities provides a comprehensive framework for optimizing the synthesis of ZnO nanoparticles for specific functional applications.

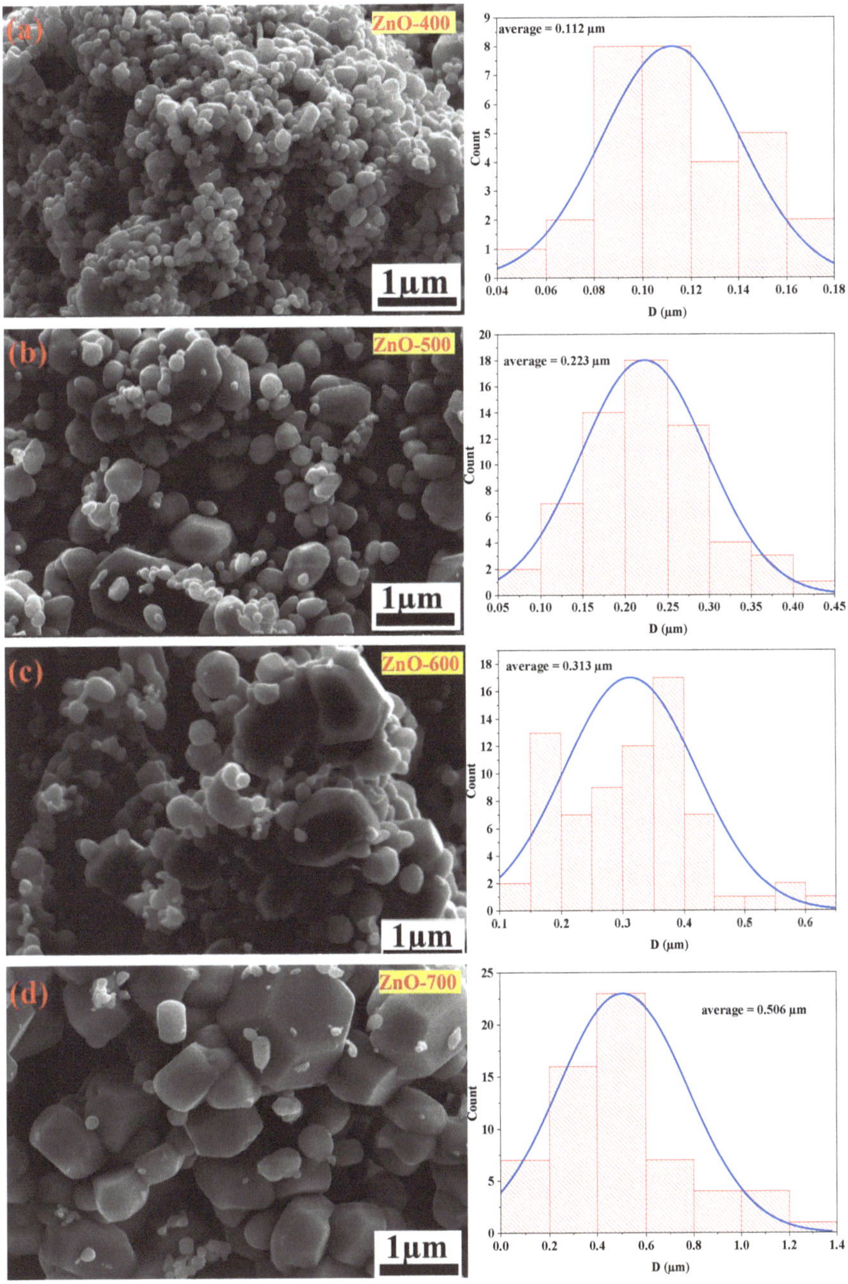

Figure 2. SEM image for different samples of ZnO nanoparticles calcined at 400 °C, 500 °C, 600 °C, and 700 °C.

3.1.3. Fourier Transform Infrared (FTIR)

Figure 3 shows the FTIR spectra of the samples synthesized under different heat treatment conditions for ZnO nanoparticles. In the spectral range between 1600 and 1700 cm^{-1}, the zinc oxide peak became sharper and smoother, indicating increased crystallinity as the calcination temperature increased. A slight peak shift was observed, specifically between 435.06 and 459.16 cm^{-1}. Furthermore, in the calcined samples above 400 °C, the disappearance of the OH band at 3492.70 cm^{-1} suggests the achievement of the transformation from $Zn(OH)_2$ to ZnO [56]. During the first step of preparation, the presence of excess OH^- ions led to the formation of $Zn(OH)_2$, which further transformed into ZnO during the calcination process. These findings support the results obtained from XRD analysis confirming the transformation of $Zn(OH)_2$ into ZnO and enhancing the overall reliability of the experimental observations. The examination of the FTIR and XRD results shows that chemical changes in the ZnO nanoparticles occur during calcination.

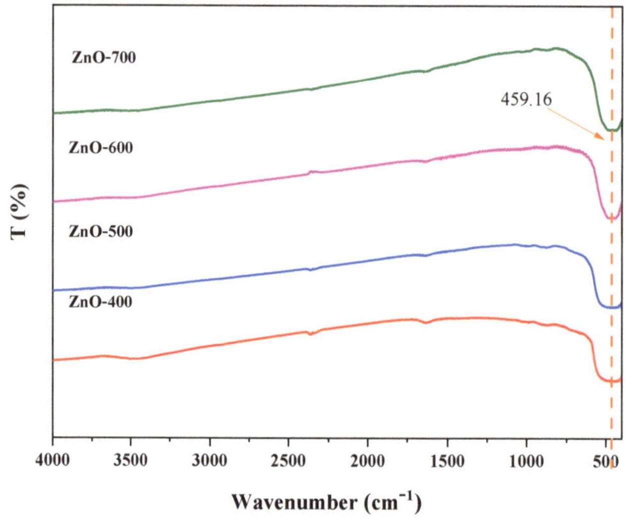

Figure 3. FTIR spectra of ZnO nanoparticles calcined at 400 °C, 500 °C, 600 °C, and 700 °C.

3.1.4. BET Surface Area

The BET analysis of the four ZnO samples, subjected to different calcination temperatures (400 °C, 500 °C, 600 °C, 700 °C), revealed distinctive patterns in their surface properties. Figure 4 shows the N_2 adsorption–desorption isotherms for both the ZnO precursor and the samples treated at different temperatures. The observations indicated a continuous decrease in the specific surface area from 12.556 to 8.445 m^2/g with the increase in calcination temperature from 400 °C to 700 °C, as detailed in Table 2. This decrease indicates a reduction in the available surface area, resulting from the fusion and agglomeration of particles at high temperatures, as clearly shown by scanning electron microscope images that highlight the increase in the size of crystalline particles.

When analyzing the pore size distribution, it was observed that the samples exposed to high calcination temperatures showed a significant shift towards larger pores, reflecting a relative increase in larger pores at the expense of smaller ones. However, the analysis showed a decrease in the total pore volume with increasing calcination temperature, due to particle coalescence and the shrinkage of small pores.

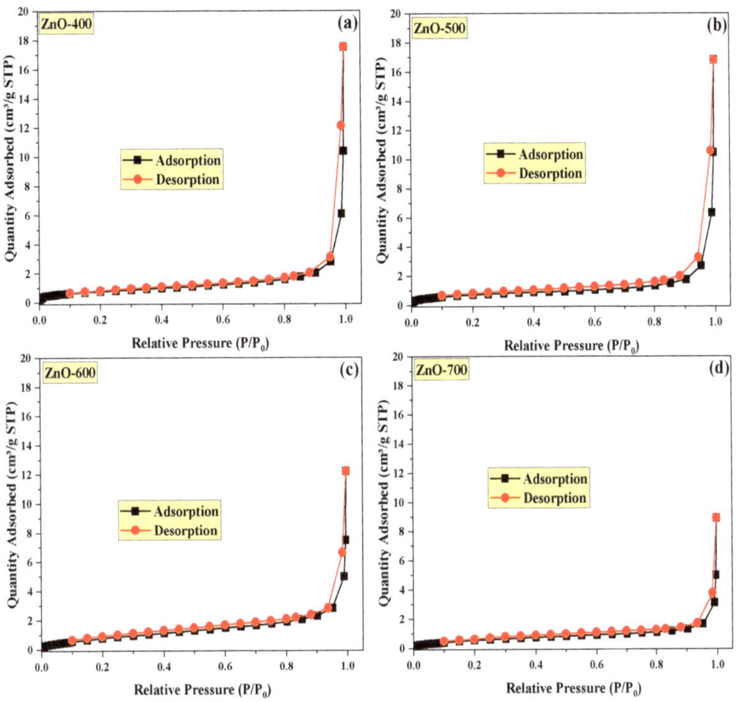

Figure 4. Nitrogen adsorption–desorption isotherms for synthesized ZnO samples calcined at (**a**) 400 °C, (**b**) 500 °C, (**c**) 600 °C, and (**d**) at 700 °C.

Similar to the current results, another study indicated similar effects on different materials. For example, a study showed a significant decrease in surface area from 25.4 to 8.8 m^2/g and pore volume from 0.35 to 0.03 cm^3/g with an increase in temperature up to 600 degrees Celsius [57]. Furthermore, a study on zirconia gel showed a significant decrease in surface area from 142 to 93 m^2/g when calcined up to 500 degrees Celsius, with the complete elimination of micropores at 700 degrees Celsius [58]. This indicates a reduction in the smaller pores, along with a significant increase in crystallization.

3.1.5. Raman Spectroscopy

The Raman spectrum of ZnO nanoparticles calcined at 400 °C, 500 °C, 600 °C, and 700 °C demonstrates the progression of crystal quality, defect concentration, and structural characteristics as the temperature rises. The notable E_2 (high) mode (~437 cm^{-1}), illustrated in Figure 5, is indicative of the wurtzite ZnO lattice structure [59], which sharpens and intensifies at 600 °C, indicating improved crystallinity and well-ordered lattice vibrations. At lower temperatures (400 °C, 500 °C), broader peaks, and higher defect-related signals [60], the A_1(LO) mode, approximately at 600 cm^{-1}, is detected, indicating reduced crystallite dimensions and the existence of oxygen vacancies, along with lattice imperfections. At 700 °C, the high level of crystallinity is maintained; however, the minor reduction in peak intensity indicates that grain growth is occurring alongside a decrease in surface scattering effects, potentially leading to a weaker Raman signal. These trends underscore the interaction among defect annealing, grain size, and lattice ordering [61].

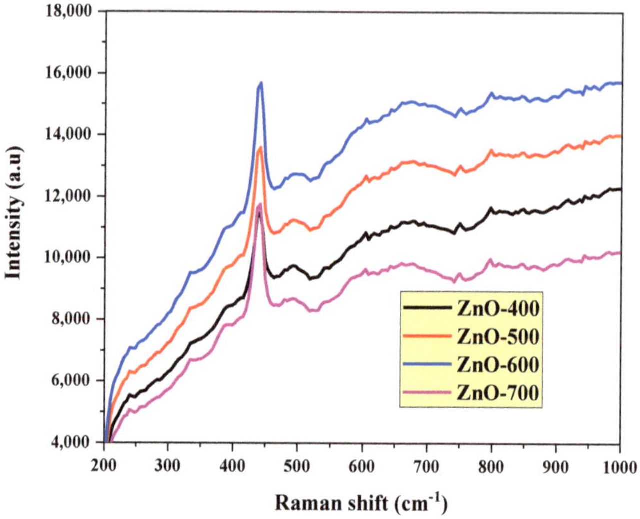

Figure 5. Raman spectrum of ZnO nanoparticles calcined at 400 °C, 500 °C, 600 °C, and 700 °C.

3.1.6. Optical Properties

The optical absorption characteristics of ZnO nanoparticles were explored using UV-Vis spectroscopy within the wavelength range of 200 to 800 nm (Figure 6). Evidently, ZnO, being a high band gap semiconductor, showcases distinctive properties in the visible and near-ultraviolet spectra. The determination of the optical band gap of the nanoparticles was calculated using the Tauc formula which analyzes the linear portion of the graph between the square of the absorption coefficient (αhν) and photon energy (hν) [62]:

$$(\alpha h\nu)^n = B(h\nu - E_g) \tag{4}$$

where α is the absorption coefficient, hν is the photon energy, E_g is the optical band gap energy in eV, B is a proportionality constant, n is the exponent depending on the electronic transition type, n = 1/2 n for directly allowed transitions, and n = 2 for indirectly allowed transitions [63]. The optical band gap of the ZnO nanoparticles was found to be temperature-dependent, exhibiting a decrease in band gap energy as the calcination temperature increased. The absorption maxima of the nanoparticles, calcined at various temperatures (400 °C, 500 °C, 600 °C, and 700 °C), shifted towards lower wavelengths, specifically at 371, 373, 375, and 376 nm Figure 7, respectively. Correspondingly, the band gaps were measured as 3.153, 3.139, 3.128, and 3.125 eV, respectively, as presented in Table 3 and Figure 6. These findings are consistent with various similar studies [33,63]. The observed blue shift in the band edge absorption wavelength of ZnO nanopowders produced at high temperatures is generally attributed to the presence of shallow levels within the band gap, which can be linked to impurity atoms or disruptions in the ZnO lattice [37,64,65]. However, it is suggested that this red shift is mainly due to the decrease in band gap resulting from an increase in particle size at higher calcination temperatures. This increase in particle size leads to a reduction in specific surface area, causing an increase in the coordination between surface atoms and a decrease in the coordination between surface atoms and underlying atoms. This, in turn, contributes to the observed blue shift phenomenon, attributed to the reduction in interatomic distance in the surface.

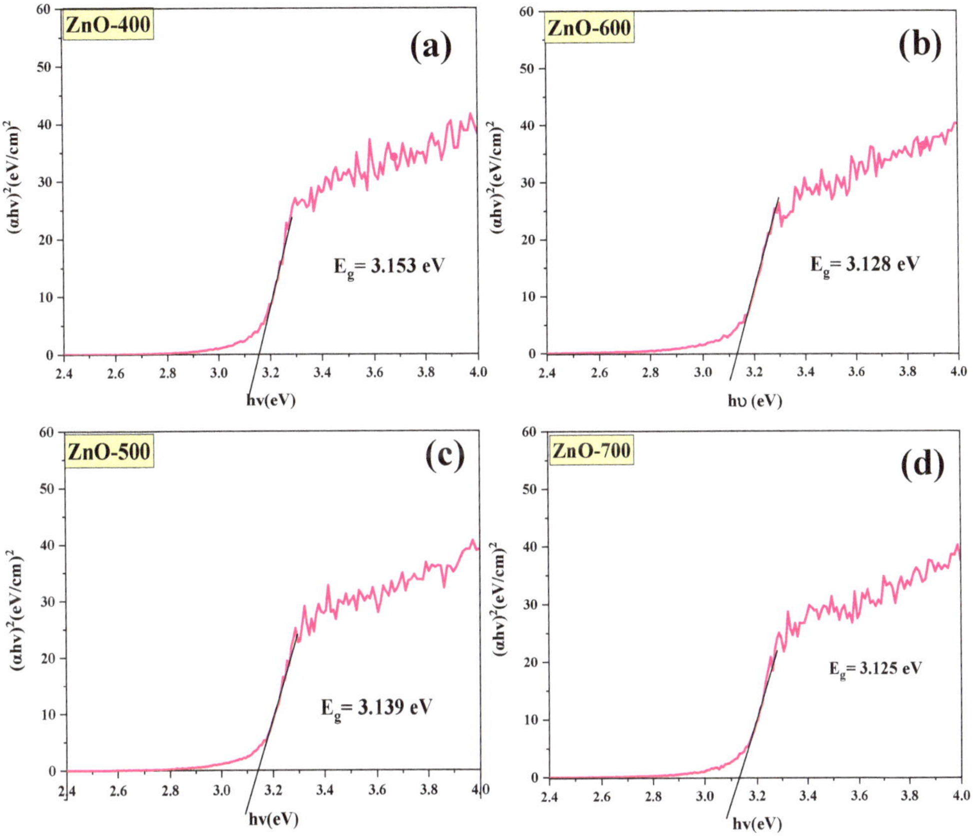

Figure 6. The optical properties and Tauc plots for determining the band gap energy of the photocatalyst calcined at 400 °C, 500 °C, 600 °C, and 700 °C.

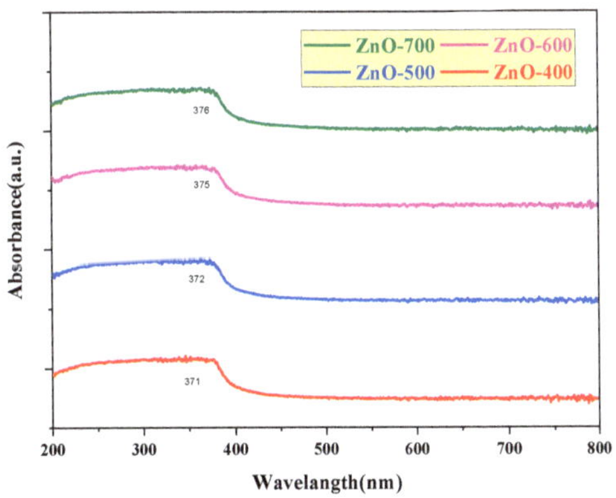

Figure 7. UV-visible absorption spectra of ZnO nanoparticles calcined at 400 °C, 500 °C, 600 °C, and 700 °C.

The surface-to-band gap energy ratio (S/E) was calculated to determine the optimal ratio for achieving the maximum rate of RhB breakdown, as outlined in Table 3. This strategic analysis recognizes the critical interplay between surface properties and band gap energy in influencing the effectiveness of the degradation process.

Table 3. Optical properties of the synthetized materials.

Calcination Temperature	Wavelength of Light Required for Photoactivation (nm)	Band Gap (eV)	S/E Ratio
ZnO-400	371	3.153	3.982
ZnO-500	372	3.139	3.419
ZnO-600	375	3.128	2.917
ZnO-700	376	3.125	2.699

3.2. Adsorption Test

The interaction between ZnO nanoparticle samples and Rhodamine B was examined in the absence of light, and the results are illustrated in Figure 8. The adsorption capacity of ZnO nanoparticles decreases with an increase in calcination temperature. This decrease can be attributed to various factors, including heightened agglomeration (as evident in SEM data), reduced specific surface area (confirmed in Figure 2), and diminished dispersion. Moreover, the structural transformation of ZnO nanoparticles during calcination led to a reduction in surface defects [8,37], oxygen vacancies, and surface hydroxyl groups. These defects and functional groups play a fundamental role in the adsorption process, emphasizing that the control of calcination temperature enables the tailoring of ZnO materials with specific and customized adsorption capabilities.

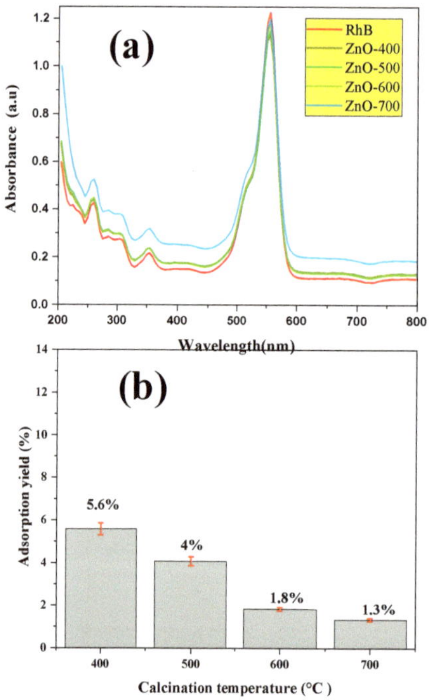

Figure 8. Adsorption capacity of ZnO calcined at different temperature for Rhodamine B at different calcination temperatures (pH = 6, S/L = 50 mg/100 mL, C_0 = 5 mg/L, time = 60 min).

3.3. Photocatalysis Tests

To evaluate the effect of calcination temperature and light source on the optical performance of ZnO, the photoactivity includes the study of the photolysis of different sources of light of Rhodamine beta. As shown in (Figure 9), the study revealed that under different light sources (sunlight, UV, and visible light), the photoactivity of ZnO changed differently with the increase of calcination temperature from 400 °C to 700 °C from one type of light source to another. Under UV light, the photoactivity of ZnO samples showed a significant decrease with the increase of calcination temperature from (92%) at 400 °C to (27%) at 700 °C (Figure 9b). The significant decrease is related to the decrease in the area effect ratio compared to the gap effect (S/E) with the increase in calcination temperature. At 400 °C, the high S/E ratio results in a high surface area with a large availability of active sites, making ZnO very efficient under UV light due to both the ideal band gap and area. As the calcination temperature increases to 700 °C, the particle size increases as confirmed in XDR and SEM, which greatly reduces the surface area ratio and results in fewer active sites [36] for UV-driven photocatalysis, while also narrowing the band gap. The narrowing of the band gap at high temperatures is not beneficial for UV-driven photocatalysis, in contrast to visible light [66], where we observe that the low S/E ratios of ZnO-600 and ZnO-700 have efficiencies similar to those observed at higher ratios at 400 °C for visible light activity of (24%) at 700 °C to (16%) at 400 °C (Figure 9a), which is explained by the narrowing of the band gap allowing ZnO nanoparticles to partially utilize visible light for photocatalytic reactions, facilitated by localized defect states [67], where even minor absorption in the visible spectrum results in measurable photocatalytic activity [68,69]. However, the overall efficiency remains low due to the limited area. In contrast, sunlight includes all wavelengths of UV and visible light [70], allowing both S and E to influence performance simultaneously, thus mitigating the large decrease in photocatalytic efficiency observed with UV alone. The moderate decrease in the slope from 91% at 400 °C to 61% at 700 °C (Figure 9c, Table 4) indicates a compensatory effect between S and E. The slight narrowing of the band gap at higher temperatures enhances visible light absorption, resulting in stable sunlight efficiency despite reduced active sites. In addition to utilizing a wide range of wavelengths provided by natural sunlight [28], the photocatalytic charging efficiency is significantly improved, several critical factors also contribute to this phenomenon. One such factor is the presence of an aerobic environment, where oxygen acts as a redox mediator and an efficient electron acceptor [71], which resulted in the double performance in sunlight compared to other sources, where samples were exposed to UV radiation for two hours to achieve similar results like under sunlight, which was within one hour (the case of the sample calcined at 400 °C). The S-to-E ratio (S/E) is a key criterion for evaluating photocatalytic performance, integrating the effects of surface area and band gap energy.

Table 4. Kinetics parameters for the degradation of RhB dye under UV, visible, and sunlight irradiation.

Samples	k (min^{-1})	R^2	D (%)
	Under visible light		
ZnO-400	0.0045	0.909	24
ZnO-500	0.0018	0.933	20
ZnO-600	0.0017	0.965	18
ZnO-700	0.0017	0.844	16
	Under sunlight		
ZnO-400	0.01492	0.984	91
ZnO-500	0.00964	0.975	78
ZnO-600	0.00781	0.980	73
ZnO-700	0.00218	0.862	61
	Under UV light		
ZnO-400	0.0231	0.885	92
ZnO-500	0.0331	0.978	70
ZnO-600	0.0245	0.990	63
ZnO-700	0.0153	0.870	27

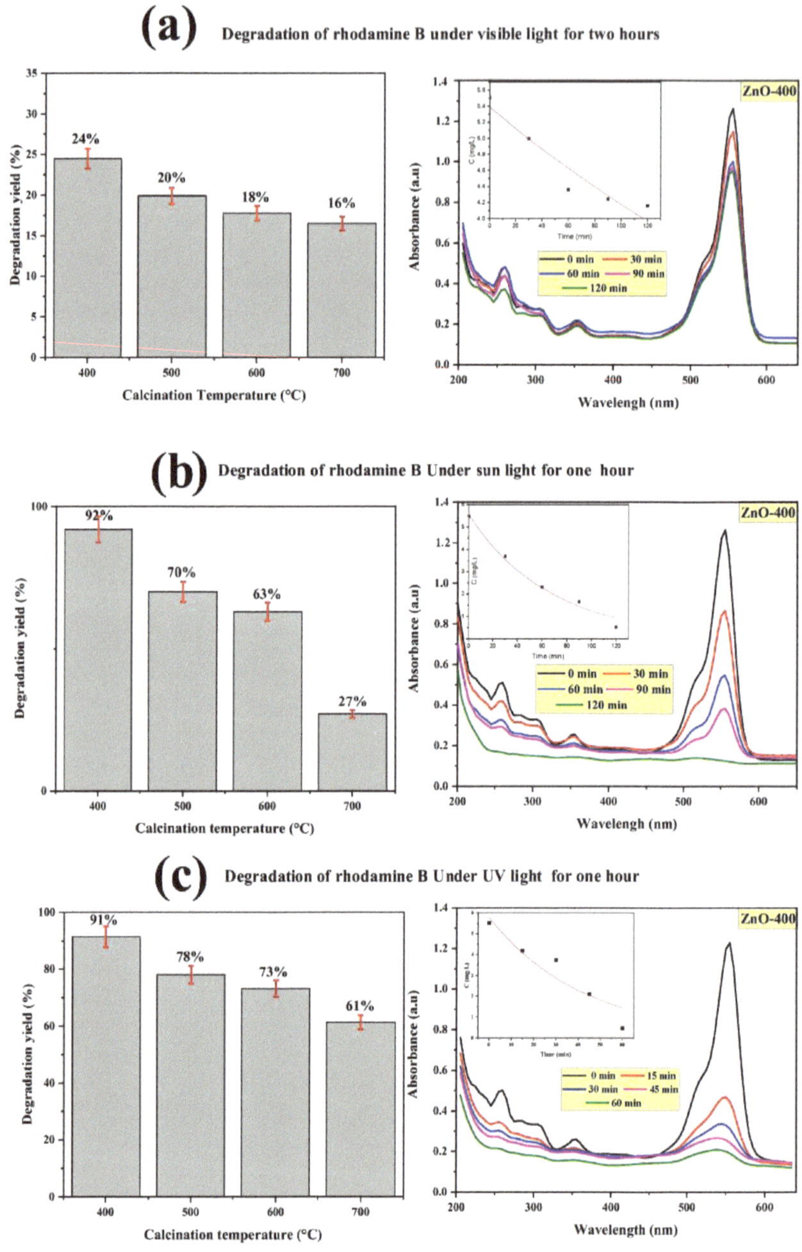

Figure 9. Degradation of RhB under visible (**a**), UV (**b**), and sunlight (**c**) irradiation (pH = 6, S/L = 50 mg/100 mL, C_0 = 5 mg/L, time = 60 min).

The following reactions (Equations (5)–(11)) and Figure 10 are suggested for the photodegradation of RhB by ZnO:

$$ZnO + h\upsilon \rightarrow ZnO\ (h^{+}_{VB} + e^{-}_{CB}) \quad (5)$$

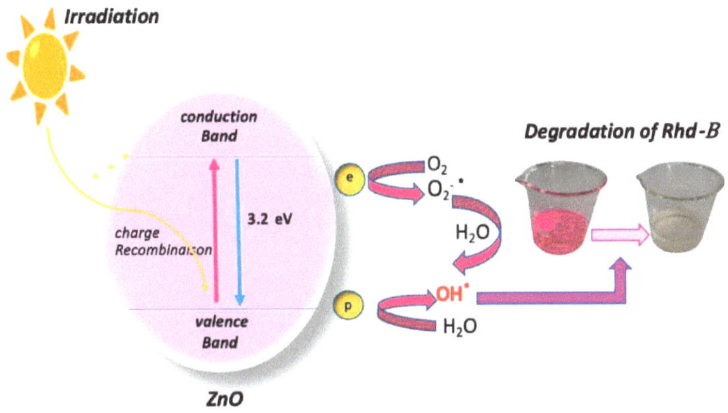

Figure 10. Degradation mechanism of RhB by ZnO.

The electrons and holes formed during the reaction in Equation (5) migrate to the surface and can react with H_2O and O_2 to selectively produce OH• radicals:

$$h^{+}{}_{VB} + H_2O \rightarrow OH^{-} + H^{+} \quad (6)$$

$$e^{-}{}_{CB} + O_2 \rightarrow \bullet O_2{}^{-} \quad (7)$$

$$\bullet O_2{}^{-} + H^{+} \rightarrow HO_2\bullet \quad (8)$$

$$HO_2\bullet + HO_2\bullet \rightarrow H_2O_2 + O_2 \quad (9)$$

$$e^{-}{}_{CB} + H_2O_2 \rightarrow OH\bullet + OH^{-} \quad (10)$$

The OH• radicals further react with Rhodamine B:

$$OH\bullet + RhB \rightarrow \text{Degradation products} \rightarrow CO_2 + H_2O \quad (11)$$

3.4. *Optimization*

The ZnO sample calcined at 400 °C was selected for optimization experiments as it exhibited the best combination of surface area and band gap energy, making it highly effective for photocatalytic activity. Sunlight was chosen as the light source for these experiments due to its superior performance compared to other sources. Its broad wavelength spectrum, encompassing both UV and visible light, enhances photocatalytic efficiency, and its availability as a natural, sustainable energy source makes it an ideal choice.

To further optimize the performance of this sample, additional experiments were conducted by varying parameters such as catalyst mass, pH, and the concentration of H_2O_2. These optimization steps aim to maximize the degradation efficiency of Rhodamine B under sunlight, ensuring the ZnO photocatalyst performs at its full potential.

3.4.1. Effect of ZnO Amount

The data show a correlation between ZnO amount and RhB degradation efficiency. Degradation rates increase from 56% (solid-to-liquid ratio (S/L) = 10 mg/100 mL) to 84% (S/L = 50 mg/100 mL), stabilizing at around 85% when the weight reaches 100 mg in 100 mL of Rhodamine B solution, as shown in (Figure 11a). This observed pattern suggests that increasing the mass of ZnO results in a more pronounced enhancement in photocatalytic degradation, likely attributed to the availability of more active sites for photocatalytic activity. However, the incremental rise from 50 mg to 100 mg indicates that the degradation rate has reached a limit value, light penetration limitations, particle agglomeration, competing reactions, and diffusion limitations [72], indicating that 50 mg of ZnO is the most effective dosage for achieving the highest level of degrading efficiency. This result carries practical implications, underscoring the importance of optimizing the ZnO loading for both cost-effectiveness and efficient removal of pollutants.

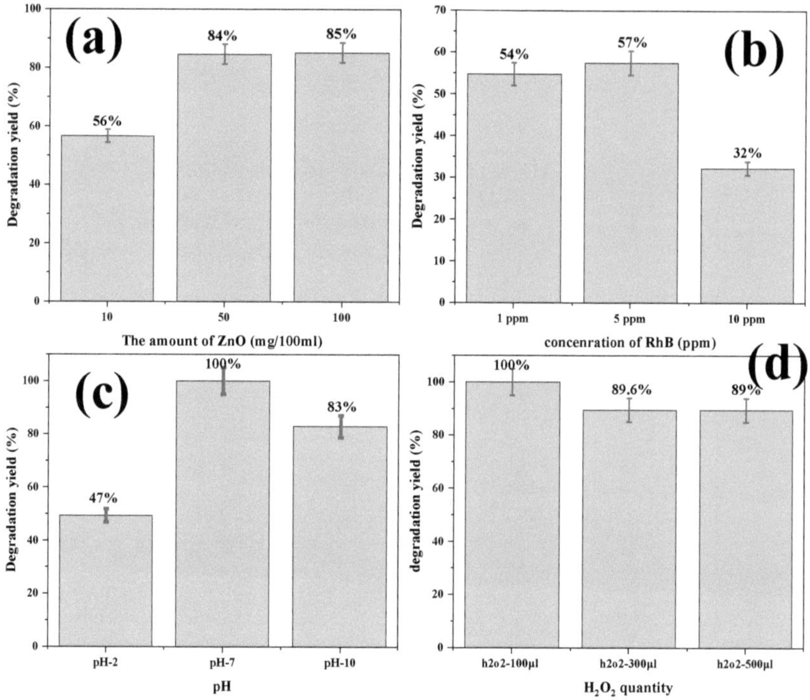

Figure 11. (**a**) Mass effect of ZnO degradation of RhB (pH = 6, C_0 = 5 mg/L), (**b**) effect of initial concentration of RhB on degradation rate (pH = 6, S/L = 50 mg/100 mL), (**c**) pH Effect for the degradation of RhB by ZnO under sunlight (S/L= 50 mg/100 mL, C_0 = 6 mg/L), and (**d**) H_2O_2 effect on the degradation of RhB (pH = 7.4, S/L = 50 mg/100 mL, C_0 = 6 mg/L).

3.4.2. Effect of the Rhodamine B Concentration

Figure 11b show the effect of varying the initial concentration of RhB on the photodegradation yield. The data reveal a relationship between the concentration of RhB and its degradation efficiency by ZnO. As the concentration increases from 1 ppm to 5 ppm and further to 10 ppm, a significant variation in the degradation yield is observed, decreasing from 54% to 32% at the maximum concentration (Figure 11b). This would suggest that higher initial concentrations of RhB caused a reduction in the efficiency of ZnO-mediated degradation. Upon closer examination of the yield at 1 ppm and 5 ppm, a slight difference becomes apparent, with yields of 54% and 57%, respectively. Although the yield is similar, the slightly elevated degradation yield at 5 ppm implies that this concentration would be somewhat more efficient than 1 ppm under the provided circumstances because, at 5 ppm, RhB molecules provide an optimal balance of reactants, allowing more species for interaction with ZnO catalysts. Efficiency in photocatalysis depends on reactant availability and active sites on the catalyst surface. Lower concentrations may lead to underutilization of the catalyst capacity; in contrast, higher concentrations of pollutant molecules at higher densities might result in a light-blocking phenomenon, reducing the entry of light into the solution and affecting the photocatalytic activity. The marginal disparity in degradation rates among these concentrations suggests that ZnO may effectively decompose RhB even at lower concentrations, though a slight increase to 5 ppm seems to enhance the degradation process. This finding emphasizes the importance of precisely adjusting the initial concentration to achieve the possible results in studies using ZnO-mediated degradation.

3.4.3. pH Effect

The pH in the reaction environment plays a crucial role in determining the degradation of Rhodamine B through ZnO photocatalysis, because the electrostatic interactions between ZnO nanoparticles and Rhodamine B molecules are closely linked to their charges, where the pH of the solution can play a double role in the electrostatic interactions between the solid and the dye: on the one hand, the charges on the surface of the solid change at the point of zero charge (pH_{zpc}) and, on the other hand, the dye itself takes on different forms depending on the pH relative to its pK_a [73]. We found $pH_{zpc} = 7$ for ZnO (Figure S1), in accordance with the literature [74]. Additionally, we have a positively charged species RhB^+ with, for pH < 3.22 (with $=NEt_2^+$) and a zwitterion $RhB^{\pm}$ (with $=NEt_2^+$ and COO^-) for pH > 3.22 [72]. At pH values higher than pH_{zpc}, the surface of ZnO tends to acquire a negative charge, enhancing its ability to attract and bond with Rhodamine B molecules, which are positively charged. This phenomenon occurs during the adsorption process, similar to what happens at a lower pH of 7 (Figure 11c). However, it is important to note that excessively elevated pH levels, reaching pH 10, reduce the photocatalytic effectiveness of the ZnO surface. Excessive pH conditions might lead to an abundance of hydroxide ions (OH^-) in the system, resulting in an environment with a large negative charge that can interact with and potentially capture the positively charged holes (h^+) [6], causing the electron–hole couples to become unstable. The literature emphasizes that pH 7 is the ideal acidity level for achieving an efficient breakdown of RhB [75,76]. Even in this experimental research, the optimal efficiency of ZnO photocatalysis for degrading RhB is achieved by the careful balance of surface charges, photocatalytic activity, and reactive oxygen species (ROS) dynamics at pH = 7. At this specific pH level, the interactions between ZnO and Rhodamine B are optimized, resulting in a robust adsorption process and concurrently facilitating the production of OH• radicals.

4. Effect of H_2O_2 Dose

The interaction of H_2O_2 with ZnO leads to complex formation, increasing the surface area of ZnO nanoparticles and offering additional active sites for photocatalytic reactions. The enhanced surface area, resulting from the complex structure [77], facilitates a greater likelihood of interactions, leading to effective photocatalysis. The process requires an optimal concentration of H_2O_2 (Figure 11), as evidenced by studies indicating that 100 µL of H_2O_2 markedly improves photocatalytic activity in the degradation of Rhodamine B (RhB). Higher concentrations, specifically 300 µL and 500 µL, do not lead to a proportional increase in reactive oxygen species (ROS) generation, suggesting the existence of a saturation point where competition for active sites restricts further advantages [6]. This equilibrium indicates an optimal balance that enhances the efficiency of ROS generation while preventing excess H_2O_2 from hindering the reaction.

The generation of hydroxyl radicals (·OH) from H_2O_2 is fundamental to this enhanced degradation process and can be expressed by the reaction:

$$H_2O_2 + h\nu \rightarrow 2\,{\cdot}OH$$

In this context, hv denotes photons possessing adequate energy to activate the ZnO photocatalyst, thereby commencing the reaction. When exposed to UV or sunlight, ZnO absorbs photons with energy at or above its band gap, resulting in the excitation of electrons (e^-) from the valence band to the conduction band, thereby generating holes (h^+):

$$ZnO + h\nu \rightarrow e^-CB + h^+VB$$

The conduction band electrons interact with H_2O_2 to produce hydroxyl radicals:

$$H_2O_2 + e^-CB \rightarrow \cdot OH + OH^-$$

Simultaneously, the valence band holes oxidize water (H_2O) or hydroxide ions (OH^-) to generate additional hydroxyl radicals [78]:

$$H_2O + h^+VB \rightarrow \cdot OH + H^+$$

$$OH^- + h^+VB \rightarrow \cdot OH$$

These hydroxyl radicals are highly reactive and attack the chromophore structure of RhB, initiating its breakdown into intermediate products, which are further degraded into CO_2 and H_2O:

$$\cdot OH + RhB \rightarrow \text{Intermediates} \rightarrow CO_2 + H_2O$$

The combined mechanism elucidates the high efficiency noted in the H_2O_2 + ZnO system, resulting in complete degradation within 25 min of sunlight exposure.

5. Scavenger Effect

This part of our work examines the impact of scavenger species, namely methanol and benzoquinone, on the effectiveness of ZnO photocatalysis in breaking down RhB (Figure 9). This study seeks to clarify the precise processes involved in pollutant degradation and provide insights for enhancing photocatalytic systems for environmental remediation applications by understanding the influence of these scavengers on the photocatalytic process.

The exceptional efficacy of ZnO photocatalysis in breaking down RhB under the stated circumstances is highlighted by the total destruction of the compound in the absence of scavengers. Nevertheless, the recorded decrease of 30% in degradation when methanol is present indicates that methanol behaves as an electron scavenger, hence exerting a detrimental influence on the photocatalytic process by constraining the accessibility of electrons. Furthermore, it is well recognized that methanol effectively scavenges •OH radicals, whereas benzoquinone effectively scavenges $O_2^{\bullet -}$ [79]. The histogram presented in Figure 12 indicates that the •OH radicals demonstrate the highest level of significance. The incorporation of benzoquinone results in a slightly increased degradation rate of 55%, indicating that, although it acts as a scavenger, its effect is relatively less significant than that of methanol. Benzoquinone selectively eliminates or accumulates holes, thereby promoting segregation between electrons and holes, which ultimately leads to the generation of reactive oxygen species (ROS) [80].

In order to evaluate the degradation efficiency of the ZnO synthesized in this study, we compared it to that of ZnO reported in the literature. Table 5 provides a comparison of the key findings from our study with those reported in the literature.

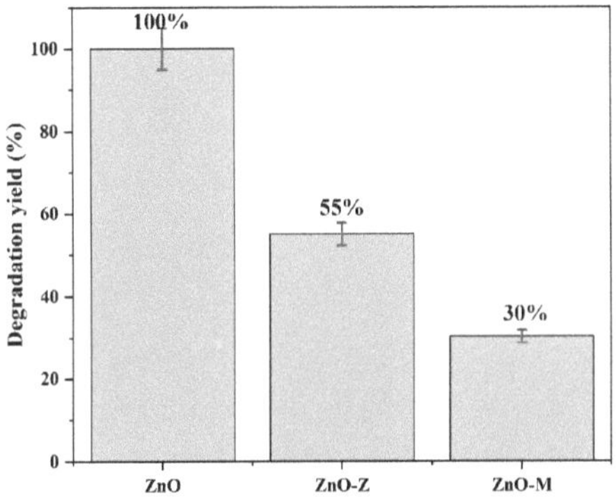

Figure 12. Scavenger effect on photodegradation of RhB by ZnO under sunlight, pH = 6, S/L = 50 mg/100 mL, C_0 = 3 mg/L and [Scavenger] = 1×10^{-5} M.

Table 5. Comparison of different studies on the degradation of Rhodamine B by ZnO prepared with different methods.

Preparation Method	Focus	Light Sourse	Condition of Photocatalysis	D (%)	Reference
green synthesis method using Argyreia nervosa leaf calcined at 500 °C, 4 h	The focus is on the biosynthesis of ZnO nanoparticles using Argyreia nervosa leaf extract and their photocatalytic efficiency in removing Rhodamine B dye	visible (100 W)	C_0: 5 ppm pH = 11 S/L = 1.4 g/L	48% in 180 min	[78]
sol–gel synthesis technique calcined from 190 t o 550 °C	The study focuses on the effect of calcination temperature on the UV light-driven photocatalytic degradation of Rhodamine B by ZnO nanoparticles	UV	C_0: 25 ppm pH: 7 S/L = 2 g/L	95% in 120 min	[79]
Sol–gel calcined at different temperatures 400 °C	The focus of the study is on the effect of calcination temperature in photocatalytic properties of ZnO nanoparticles under UV light	UV (330 w)	C_0: 10 ppm S/L: 2 g/L	95% in 160 min	[80]

Table 5. *Cont.*

Preparation Method	Focus	Light Sourse	Condition of Photocatalysis	D (%)	Reference
Green synthesis using the leaf extract of Cyanometraramiflora	ZnO nanoparticles synthesized using the leaf extract of Cyanometra ramiflora demonstrated effective photocatalytic degradation of Rhodamine B dye under sunlight, showcasing their potential for eco-friendly and sustainable dye remediation	sun light	C_0: 10 ppm S/L = 20 mg in 100 mL of solution	98% in 200 min	[81]
hydrothermal method	This study highlights a simple hydrothermal method to synthesize ZnO microcrystals with high photocatalytic efficiency and easy recovery, optimized for sustainable industrial applications under uv	UV (15 w)	C_0:10 ppm S/L: 1 g/L	97% in 40 min	[82]
Precipitation Calcined 400 °C to 700 °C	Our study demonstrates the optimized photocatalytic performance of ZnO under UV, visible, and solar light by introducing the S/E metric and leveraging controlled precipitation and calcination; optimization of the samples calcined at 400 °C under sun light	sun light uv light visible light	C_0: 5 ppm sun light S/L: 0.5 g/L	92% in 60 min	This study

6. Conclusions

Renewable natural sunlight is a more effective and ecologically sustainable energy source for photocharging applications compared to standard artificial light sources such as visible and UV light. Our results are based on a thorough investigation highlighting the higher photocatalytic performance of zinc oxide (ZnO) nanoparticles under natural sunlight conditions; this stresses the benefits of using solar energy and showcases the impressive capability of ZnO to use various sunshine wavelengths. The work improves the capability for removing pollutants and creates a more realistic testing setting for measuring photocatalytic effectiveness by using natural sunshine conditions. The study provides interesting information on the ideal circumstances for ZnO nanoparticle efficiency. Zinc oxide samples heated to 400 °C provide the best results. This temperature enhances crystal formation, improves surface area, and promotes the effective creation and usage of electron–hole pairs under light, while it is important to recognize that higher temperatures result in decreasing reactivity, emphasizing the intricate relationship among calcination temperature, surface properties, and photocatalytic effectiveness in ZnO nanoparticles. Where the study found peak decomposition efficiency correlates with the highest surface area to gap energy ratio, emphasizing surface optimization's importance

For applications that rely on UV or sunlight, maintaining a high S/E ratio is critical. Low calcination temperatures can preserve surface area and active sites, thereby maximizing photocatalytic efficiency. Conversely, visible light applications are enhanced by methods that lower the band gap energy, including doping or the formation of hybrid structures. These methods improve performance despite a reduced S/E effect, facilitating the creation of specialized ZnO materials designed for photonic applications.

Comprehending and enhancing these elements are crucial for creating very effective photocatalytic devices for practical use in environmental cleanup and sustainable energy generation. In addition, in this work, the effect of hydrogen peroxide (H_2O_2) on the degradation of RhB was studied. It was shown that H_2O_2 was a very good oxidant for improving the degradation of RhB under sunlight. ZnO has great promise for tackling environmental issues and promoting green technology by using the plentiful and clean energy from sunshine. Continuing study in this area might result in the creation of more effective and expandable photocatalytic devices, which would help create a more sustainable and environmentally friendly future.

Supplementary Materials: The following supporting information can be downloaded at: https://www.mdpi.com/article/10.3390/w17010032/s1. Table S1: Effect of H_2O_2 addition on the kinetic parameters for the degradation of RhB dye under sunlight. Table S2: Effect of solution pH on the kinetic parameters for the degradation of RhB dye under sun light. Figure S1: PZC of ZnO Nanoparticles Determined by pH Titration.

Author Contributions: Conceptualization, A.S., R.B., L.B. and K.F.; methodology, A.S., R.B., K.F., L.M., A.B. and A.D.; software, R.B. and A.B., validation, L.M., J.-C.B. and S.S.; formal analysis, S.S., L.M. and R.B.; investigation, A.A.A. and A.S., resources, A.A.A., L.B. and K.F.; data curation A.S. and R.B.; writing—original draft preparation, A.S., R.B., K.F. and L.B.; writing—review and editing, L.M., J.-C.B., A.A.A. and S.S.; visualization, L.M., J.-C.B., S.S. and A.B.; supervision, A.B., L.M. and J.-C.B.; project administration, K.F., A.B. and K.F. All authors have read and agreed to the published version of the manuscript.

Funding: This research was funded by the Researchers Supporting Project number (RSPD2024R527), King Saud University, Riyadh, Saudi Arabia.

Data Availability Statement: Data is contained within the article.

Acknowledgments: The authors acknowledge the financial support through the Researchers Supporting Project, number (RSPD2024R527), King Saud University, Riyadh, Saudi Arabia.

Conflicts of Interest: The authors declare no conflict of interest.

References

1. Kabir, E.; Kumar, P.; Kumar, S.; Adelodun, A.A.; Kim, K.-H. Solar Energy: Potential and Future Prospects. *Renew. Sustain. Energy Rev.* **2018**, *82*, 894–900. [CrossRef]
2. Schmalensee, R. The Future of Solar Energy: A Personal Assessment. *Energy Econ.* **2015**, *52*, S142–S148. [CrossRef]
3. Jaiswal, K.K.; Chowdhury, C.R.; Yadav, D.; Verma, R.; Dutta, S.; Jaiswal, K.S.; Karuppasamy, K.S.K. Renewable and Sustainable Clean Energy Development and Impact on Social, Economic, and Environmental Health. *Energy Nexus* **2022**, *7*, 100118. [CrossRef]
4. Crabtree, G.W.; Lewis, N.S. Solar Energy Conversion. *Phys. Today* **2007**, *60*, 37–42. [CrossRef]
5. Rekhila, G.; Saidani, A.; Hocine, F.; Hariz, S.H.B.; Trari, M. Characterization of the Hetero-System $ZnCo_2O_4$/ZnO Prepared by Sol Gel: Application to the Degradation of Ponceau 4R under Solar Light. *Appl. Phys. A* **2020**, *126*, 620. [CrossRef]
6. Nezamzadeh-Ejhieh, A.; Salimi, Z. Solar Photocatalytic Degradation of O-Phenylenediamine by Heterogeneous CuO/X Zeolite Catalyst. *Desalination* **2011**, *280*, 281–287. [CrossRef]
7. Nezamzadeh-Ejhieh, A.; Moazzeni, N. Sunlight Photodecolorization of a Mixture of Methyl Orange and Bromocresol Green by CuS Incorporated in a Clinoptilolite Zeolite as a Heterogeneous Catalyst. *J. Ind. Eng. Chem.* **2013**, *19*, 1433–1442. [CrossRef]
8. Golsheikh, A.M.; Kamali, K.Z.; Huang, N.M.; Zak, A.K. Effect of Calcination Temperature on Performance of ZnO Nanoparticles for Dye-Sensitized Solar Cells. *Powder Technol.* **2018**, *329*, 282–287. [CrossRef]
9. Liu, R.; Yang, W.-D.; Qiang, L.-S.; Liu, H.-Y. Conveniently Fabricated Heterojunction ZnO/TiO_2 Electrodes Using TiO_2 Nanotube Arrays for Dye-Sensitized Solar Cells. *J. Power Sources* **2012**, *220*, 153–159. [CrossRef]

10. Boudraa, R.; Talantikite-Touati, D.; Djermoune, A.; Souici, A.; Kebir, M.; Merzeg, F.A.; Amrane, A.; Bollinger, J.-C.; Mouni, L. Comprehensive Characterization and Unprecedented Photocatalytic Efficacy of TiO_2-CuO-La_2O_3 and TiO_2-CuO-Bi_2O_3 Nanocomposites: A Novel Approach to Environmental Remediation. *Mater. Sci. Eng. B* **2025**, *312*, 117863. [CrossRef]
11. Cheikh, S.; Imessaoudene, A.; Bollinger, J.C.; Hadadi, A.; Manseri, A.; Bouzaza, A.; Assadi, A.; Amrane, A.; Zamouche, M.; El Jery, A.; et al. Complete Elimination of the Ciprofloxacin Antibiotic from Water by the Combination of Adsorption–Photocatalysis Process Using Natural Hydroxyapatite and TiO_2. *Catalysts* **2023**, *13*, 336. [CrossRef]
12. Merdoud, R.; Aoudjit, F.; Mouni, L.; Ranade, V.V. Degradation of methyl orange using hydrodynamic Cavitation, H_2O_2, and photo-catalysis with TiO_2-Coated glass Fibers: Key operating parameters and synergistic effects. *Ultrason. Sonochemistry* **2024**, *103*, 106772. [CrossRef]
13. Imessaoudene, A.; Mechraoui, O.; Aberkane, B.; Benabbas, A.; Manseri, A.; Moussaoui, Y.; Bollinger, J.-C.; Amrane, A.; Zoukel, A.; Mouni, L. Synthesis of a TiO_2/zeolite composite: Evaluation of adsorption-photodegradation synergy for the removal of Malachite Green. *Nano-Struct. Nano-Objects* **2024**, *38*, 101191. [CrossRef]
14. Sharma, D.K.; Shukla, S.; Sharma, K.K.; Kumar, V. A Review on ZnO: Fundamental Properties and Applications. *Mater. Today Proc.* **2022**, *49*, 3028–3035. [CrossRef]
15. Nezamzadeh-Ejhieh, A.; Khorsandi, S. Photocatalytic Degradation of 4-Nitrophenol with ZnO Supported Nano-Clinoptilolite Zeolite. *J. Ind. Eng. Chem.* **2014**, *20*, 937–946. [CrossRef]
16. Mandal, A.K.; Katuwal, S.; Tettey, F.; Gupta, A.; Bhattarai, S.; Jaisi, S.; Bhandari, D.P.; Shah, A.K.; Bhattarai, N.; Parajuli, N. Current Research on Zinc Oxide Nanoparticles: Synthesis, Characterization, and Biomedical Applications. *Nanomaterials* **2022**, *12*, 3066. [CrossRef] [PubMed]
17. Hasnidawani, J.; Hassan, N.A.; Norita, H.; Samat, N.; Bonnia, N.N.; Surip, S.N. ZnO Nanoparticles for Anti-Corrosion Nanocoating of Carbon Steel. *Trans. Tech. Publ.* **2017**, *894*, 76–80. [CrossRef]
18. Saravade, V.; Feng, Z.C.; Nafisa, M.T.; Zhou, C.; Lu, N.; Klein, B.; Ferguson, I. Advances in Growth, Doping, and Devices and Applications of Zinc Oxide. *J. Vac. Sci. Technol. A* **2024**, *42*, 020802. [CrossRef]
19. Ravindranadh, M.R.K.; Mary, T.R. Development of ZnO Nanoparticles for Clinical Applications. *J. Chem. Biol. Phys. Sci. (JCBPS)* **2013**, *4*, 469.
20. Sharma, P.; Jang, N.-Y.; Lee, J.-W.; Park, B.C.; Kim, Y.K.; Cho, N.-H. Application of ZnO-Based Nanocomposites for Vaccines and Cancer Immunotherapy. *Pharmaceutics* **2019**, *11*, 493. [CrossRef] [PubMed]
21. Sari, L.M.; Rilda, Y.; Armaini. Biosynthesis of ZnO Nanoparticles Using Spirulina Platensis Based on Calcination Temperature Changes and Its Antioxidant Activity. *Chem. Sci. Int. J.* **2023**, *32*, 1–8. [CrossRef]
22. Hendili, R.; Alatrache, A.; Ben-Attia, M.; Pons, M.-N. Antibacterial Inactivation of Spiramycin after Titanium Dioxide Photocatalytic Treatment. *Comptes Rendus. Chim.* **2017**, *20*, 710–716. [CrossRef]
23. Raja, A.; Ashokkumar, S.; Marthandam, R.P.; Jayachandiran, J.; Khatiwada, C.P.; Kaviyarasu, K.; Raman, R.G.; Swaminathan, M. Eco-friendly preparation of zinc oxide nanoparticles using Tabernaemontana divaricata and its photocatalytic and antimicrobial activity. *J. Photobiol.* **2018**, *181*, 53–58. [CrossRef]
24. Shirzadi, A.; Nezamzadeh-Ejhieh, A. Enhanced Photocatalytic Activity of Supported CuO–ZnO Semiconductors towards the Photodegradation of Mefenamic Acid Aqueous Solution as a Semi Real Sample. *J. Mol. Catal. A Chem.* **2016**, *411*, 222–229. [CrossRef]
25. A Barzinjy, A. Zinc Oxide NPs as Solar Photo-Catalysis for Water Disinfection. *Eurasian J. Sci. Eng.* **2022**, *8*. [CrossRef]
26. AlMohamadi, H.; Awad, S.A.; Sharma, A.K.; Fayzullaev, N.; Távara-Aponte, A.; Chiguala-Contreras, L.; Amari, A.; Rodriguez-Benites, C.; Tahoon, M.A.; Esmaeili, H. Comparative Study of Photocatalytic Activity of Metals-and Non-Metals Doped ZnO and TiO_2 Nanocatalysts for Advanced Photocatalysis. *Chem. Eng.* **2024**. [CrossRef]
27. Di Mauro, A.; Cantarella, M.; Nicotra, G.; Privitera, V.; Impellizzeri, G. Low Temperature Atomic Layer Deposition of ZnO: Applications in Photocatalysis. *Appl. Catal. B Environ.* **2016**, *196*, 68–76. [CrossRef]
28. Chang, X.; Li, Z.; Zhai, X.; Sun, S.; Gu, D.; Dong, L.; Yin, Y.; Zhu, Y. Efficient Synthesis of Sunlight-Driven ZnO-Based Heterogeneous Photocatalysts. *Mater. Des.* **2016**, *98*, 324–332. [CrossRef]
29. Sarkar, N.; Ghosh, S. Photoluminescence Spectroscopy of Band Gap Shrinkage in GaN. *arXiv* **2006**, arXiv:cond-mat/0604221.
30. Hamilton, C.; Kowalski, O.; McIlvaney, K.; Bryce, A.; Marsh, J.; Button, C. Bandgap Tuning of Visible Laser Material. *Electron. Lett.* **1998**, *34*, 665–666. [CrossRef]
31. Mayer, M.A.; Yu, K.M.; Speaks, D.T.; Denlinger, J.D.; Reichertz, L.A.; Beeman, J.W.; Haller, E.E.; Walukiewicz, W. Band Gap Engineering of Oxide Photoelectrodes: Characterization of $ZnO_{1-x}Se_x$. *J. Phys. Chem. C* **2012**, *116*, 15281–15289. [CrossRef]
32. Qiao, F.; Sun, K.; Chu, H.; Wang, J.; Xie, Y.; Chen, L.; Yan, T. Design Strategies of ZnO Heterojunction Arrays towards Effective Photovoltaic Applications. *Battery Energy* **2022**, *1*, 20210008. [CrossRef]
33. He, L.; Tong, Z.; Wang, Z.; Chen, M.; Huang, N.; Zhang, W. Effects of Calcination Temperature and Heating Rate on the Photocatalytic Properties of ZnO Prepared by Pyrolysis. *J. Colloid Interface Sci.* **2018**, *509*, 448–456. [CrossRef] [PubMed]

34. Sugihartono, I.; Retnoningtyas, A.; Rustana, C.; Umiatin, U.; Yudasari, N.; Isnaeni, I.; Imawan, C.; Kurniadewi, F. *The Influence of Calcination Temperature on Optical Properties of ZnO Nanoparticles*; AIP Publishing: Melville, NY, USA, 2019; Volume 2169.
35. Kayani, Z.N.; Saleemi, F.; Batool, I. Effect of Calcination Temperature on the Properties of ZnO Nanoparticles. *Appl. Phys. A* **2015**, *119*, 713–720. [CrossRef]
36. Mornani, E.; Mosayebian, P.; Dorranian, D.; Behzad, K. Effect of Calcination Temperature on the Size and Optical Properties of Synthesized ZnO Nanoparticles. *J. Ovonic Res.* **2016**, *12*, 75–80.
37. Khayatian, A.; Kashi, M.A.; Azimirad, R.; Safa, S.; Akhtarian, S.A. Effect of Annealing Process in Tuning of Defects in ZnO Nanorods and Their Application in UV Photodetectors. *Optik* **2016**, *127*, 4675–4681. [CrossRef]
38. Baharudin, K.B.; Abdullah, N.; Derawi, D. Effect of Calcination Temperature on the Physicochemical Properties of Zinc Oxide Nanoparticles Synthesized by Coprecipitation. *Mater. Res. Express* **2018**, *5*, 125018. [CrossRef]
39. Gandhi, S.; Kaur, R.; Sharma, V.; Mandal, S.K. Effect of Calcination Temperature on the Morphology and Catalytic Properties of ZnO Nanostructures Fabricated from a Chiral Precursor for Photodegradation of Both Cationic and Anionic Dyes. *New J. Chem.* **2022**, *46*, 3645–3657. [CrossRef]
40. Suguna, A.; Prabhu, S.; Selvaraj, M.; Geerthana, M.; Silambarasan, A.; Navaneethan, M.; Ramesh, R.; Sridevi, C. Annealing Effect on Photocatalytic Activity of ZnO Nanostructures for Organic Dye Degradation. *J. Mater. Sci. Mater. Electron.* **2021**, *33*, 8868–8879. [CrossRef]
41. Hasnidawani, J.; Azlina, H.; Norita, H.; Bonnia, N.; Ratim, S.; Ali, E. Synthesis of ZnO Nanostructures Using Sol-Gel Method. *Procedia Chem.* **2016**, *19*, 211–216. [CrossRef]
42. Baruah, S.; Dutta, J. Hydrothermal Growth of ZnO Nanostructures. *Sci. Technol. Adv. Mater.* **2009**, *10*, 013001. [CrossRef]
43. Purica, M.; Budianu, E.; Rusu, E.; Danila, M.; Gavrila, R. Optical and Structural Investigation of ZnO Thin Films Prepared by Chemical Vapor Deposition (CVD). *Thin Solid Film.* **2002**, *403*, 485–488. [CrossRef]
44. Rana, A.U.H.S.; Kang, M.; Kim, H.-S. Microwave-Assisted Facile and Ultrafast Growth of ZnO Nanostructures and Proposition of Alternative Microwave-Assisted Methods to Address Growth Stoppage. *Sci. Rep.* **2016**, *6*, 24870. [CrossRef] [PubMed]
45. Nagaraju, G.; Shivaraju, G.; Banuprakash, G.; Rangappa, D. Photocatalytic Activity of ZnO Nanoparticles: Synthesis via Solution Combustion Method. *Mater. Today Proc.* **2017**, *4*, 11700–11705. [CrossRef]
46. Raoufi, D. Synthesis and Microstructural Properties of ZnO Nanoparticles Prepared by Precipitation Method. *Renew. Energy* **2013**, *50*, 932–937. [CrossRef]
47. Kołodziejczak-Radzimska, A.; Markiewicz, E.; Jesionowski, T. Structural Characterisation of ZnO Particles Obtained by the Emulsion Precipitation Method. *J. Nanomater.* **2012**, *2012*, 656353. [CrossRef]
48. Wang, L.; Wang, J.; Zhao, M. Kinetic Studies on Electrochemical Degradation of Rhodamine B. *J. Water Chem. Technol.* **2021**, *43*, 123–130. [CrossRef]
49. Skjolding, L.M.; Dyhr, K.; Köppl, C.; McKnight, U.; Bauer-Gottwein, P.; Mayer, P.; Bjerg, P.; Baun, A. Assessing the Aquatic Toxicity and Environmental Safety of Tracer Compounds Rhodamine B and Rhodamine WT. *Water Res.* **2021**, *197*, 117109. [CrossRef]
50. Khan, S.; Noor, T.; Iqbal, N.; Yaqoob, L. Photocatalytic Dye Degradation from Textile Wastewater: A Review. *ACS Omega* **2024**, *9*, 21751–21767. [CrossRef]
51. Williamson, G.K.; Hall, W.H. X-Ray Line Broadening from Filed Aluminium and Wolfram. *Acta Metall.* **1953**, *1*, 22–31. [CrossRef]
52. Boudraa, R.; Talantikite-Touati, D.; Souici, A.; Djermoune, A.; Saidani, A.; Fendi, K.; Amrane, A.; Bollinger, J.-C.; Tran, H.N.; Hadadi, A. Optical and Photocatalytic Properties of TiO_2–Bi_2O_3–CuO Supported on Natural Zeolite for Removing Safranin-O Dye from Water and Wastewater. *J. Photochem. Photobiol. A Chem.* **2023**, *443*, 114845. [CrossRef]
53. Rida, K.; Benabbas, A.; Bouremmad, F.; Pena, M.A.; Sastre, E.; Martínez-Arias, A. Effect of Calcination Temperature on the Structural Characteristics and Catalytic Activity for Propene Combustion of Sol–Gel Derived Lanthanum Chromite Perovskite. *Appl. Catal. A Gen.* **2007**, *327*, 173–179. [CrossRef]
54. Wang, M.; Zhou, Y.; Zhang, Y.; Hahn, S.H.; Kim, E.J. From Zn $(OH)_2$ to ZnO: A Study on the Mechanism of Phase Transformation. *CrystEngComm* **2011**, *13*, 6024–6026. [CrossRef]
55. Ismail, M.; Taha, K.; Modwi, A.; Khezami, L. ZnO Nanoparticles: Surface and X-Ray Profile Analysis. *J. Ovonic Res.* **2018**, *14*, 381–393.
56. Arean, C.O.; Garcia, M.V.; Colinas, J.F. High-Temperature Study of Zirconia Gels: Evolution of Surface Area and Porosity on Calcination in a Vacuum. *Mater. Chem. Phys.* **1985**, *13*, 163–170. [CrossRef]
57. Muniraja, P.; Sunil Kumar, K.; Ramanadha, M.; Sudharani, A.; Ravi, M.; Vijayalakshmi, R. Effect of Synthesis Temperature on Structural, Optical, and Magnetic Properties of ZnO Nanoparticles Synthesized by Combustion Method. *J. Supercond. Nov. Magn.* **2019**, *32*, 2175–2183. [CrossRef]

58. Khachadorian, S.; Gillen, R.; Choi, S.; Ton-That, C.; Kliem, A.; Maultzsch, J.; Phillips, M.R.; Hoffmann, A. Effects of Annealing on Optical and Structural Properties of Zinc Oxide Nanocrystals. *Phys. Status Solidi (B)* **2015**, *252*, 2620–2625. [CrossRef]
59. Mizoguchi, K.; Harima, H.; Nakashima, S.; Hara, T. Raman Image Study of Flash-lamp Annealing of Ion-implanted Silicon. *J. Appl. Phys.* **1995**, *77*, 3388–3392. [CrossRef]
60. Boudraa, R.; Talantikite-Touati, D.; Souici, A.; Djermoune, K.; Amrane, A.; Bollinger, J.-C.; Tran, H.N.; Mouni, L. Breaking new grounds: Solid-state synthesis of TiO_2–La_2O_3–CuO nanocomposites for degrading brilliant green dye under visible light. *J. Clean. Prod.* **2024**, *481*, 144126. [CrossRef]
61. Farsi, M.; Nezamzadeh-Ejhieh, A. A Z-Scheme Cobalt (II) Oxide-Silver Tungstate Nano Photocatalyst: Experimental Design and Mechanism Study for the Degradation of Methylene Blue. *Surf. Interfaces* **2022**, *32*, 102148. [CrossRef]
62. Sharma, V.; Sharma, J.; Kansay, V.; Sharma, V.D.; Sharma, A.; Kumar, S.; Sharma, A.; Bera, M. The Effect of Calcination Temperatures on the Structural and Optical Properties of Zinc Oxide Nanoparticles and Their Influence on the Photocatalytic Degradation of Leather Dye. *Chem. Phys. Impact* **2023**, *6*, 100196. [CrossRef]
63. Makama, A.B.; Salmiaton, A.; Choong, T.S.Y.; Hamid, M.R.A.; Abdullah, N.; Saion, E. Influence of Parameters and Radical Scavengers on the Visible-Light-Induced Degradation of Ciprofloxacin in ZnO/SnS_2 Nanocomposite Suspension: Identification of Transformation Products. *Chemosphere* **2020**, *253*, 126689. [CrossRef]
64. Zhang, H.; Li, C.; Lang, F.; Li, M.; Liu, H.; Zhong, D.-C.; Qin, J.-S.; Di, Z.; Wang, D.-H.; Zeng, L. Precisely Tuning Band Gaps of Hexabenzocoronene-Based MOFs Toward Enhanced Photocatalysis. *Angew. Chem.* **2024**, e202418017. [CrossRef]
65. Sena, H.; Fuji, M. Band Gap Engineering of Semiconductors and Ceramics by Severe Plastic Deformation for Solar Energy Harvesting. *Mater. Trans.* **2023**, *64*, 1497–1503. [CrossRef]
66. Ogawa, K.; Abe, R.; Walsh, A. Band Gap Narrowing by Suppressed Lone-Pair Activity of Bi^{3+}. *J. Am. Chem. Soc.* **2024**, *146*, 5806–5810. [CrossRef] [PubMed]
67. Derikvandi, H.; Nezamzadeh-Ejhieh, A. An Effective Wastewater Treatment Based on Sunlight Photodegradation by SnS_2–ZnS/Clinoptilolite Composite. *Solid State Sci.* **2020**, *101*, 106127. [CrossRef]
68. Kuvykin, I.; Vershubskii, A.; Ptushenko, V.; Tikhonov, A. Oxygen as an Alternative Electron Acceptor in the Photosynthetic Electron Transport Chain of C_3 Plants. *Biochemistry* **2008**, *73*, 1063–1075. [CrossRef] [PubMed]
69. Mohammadyari, P.; Nezamzadeh-Ejhieh, A. Supporting of Mixed ZnS–NiS Semiconductors onto Clinoptilolite Nano-Particles to Improve Its Activity in Photodegradation of 2-Nitrotoluene. *RSC Adv.* **2015**, *5*, 75300–75310. [CrossRef]
70. Mazi, W.; Adhikari, R.; Zhang, Y.; Xia, S.; Fang, M.; Luck, R.L.; Tajiri, M.; Tiwari, A.; Tanasova, M.; Liu, H. Fluorescent Probes with High pKa Values Based on Traditional, near-Infrared Rhodamine, and Hemicyanine Fluorophores for Sensitive Detection of Lysosomal pH Variations. *Methods* **2019**, *168*, 40–50. [CrossRef] [PubMed]
71. Hasan, I.M.; Tawfik, A.R.; Assaf, F.H. GC/MS Screening of Buckthorn Phytochemicals and Their Use to Synthesize ZnO Nanoparticles for Photocatalytic Degradation of Malachite Green Dye in Water. *Water Sci. Technol.* **2022**, *85*, 664–684. [CrossRef] [PubMed]
72. Saravanan, S.; Kumar, P.S.; Chitra, B.; Rangasamy, G. Biodegradation of Textile Dye Rhodamine-B by Brevundimonas Diminuta and Screening of Their Breakdown Metabolites. *Chemosphere* **2022**, *308*, 136266. [CrossRef] [PubMed]
73. Hou, M.-F.; Ma, C.-X.; Zhang, W.-D.; Tang, X.-Y.; Fan, Y.-N.; Wan, H.-F. Removal of Rhodamine B Using Iron-Pillared Bentonite. *J. Hazard. Mater.* **2011**, *186*, 1118–1123. [CrossRef] [PubMed]
74. Wallen, C.M.; Bacsa, J.; Scarborough, C.C. Hydrogen Peroxide Complex of Zinc. *J. Am. Chem. Soc.* **2015**, *137*, 14606–14609. [CrossRef] [PubMed]
75. Nezamzadeh-Ejhieh, A.; Karimi-Shamsabadi, M. Decolorization of a Binary Azo Dyes Mixture Using CuO Incorporated Nanozeolite-X as a Heterogeneous Catalyst and Solar Irradiation. *Chem. Eng. J.* **2013**, *228*, 631–641. [CrossRef]
76. Fonagy, O.; Szabo-Bardos, E.; Horvath, O. 1, 4-Benzoquinone and 1, 4-Hydroquinone Based Determination of Electron and Superoxide Radical Formed in Heterogeneous Photocatalytic Systems. *J. Photochem. Photobiol. A Chem.* **2021**, *407*, 113057. [CrossRef]
77. Omrani, N.; Nezamzadeh-Ejhieh, A. A Ternary $Cu_2O/BiVO_4/WO_3$ Nano-Composite: Scavenging Agents and the Mechanism Pathways in the Photodegradation of Sulfasalazine. *J. Mol. Liq.* **2020**, *315*, 113701. [CrossRef]
78. Abd, S.F.; Alshamsi, H.A. Enhanced Photocatalytic Degradationof Rhodamine-B Dye Using Photocatalyst ZNO Sensitized by Visible Light. *Pak. J. Med. Health Sci.* **2022**, *16*, 716. [CrossRef]
79. Lal, M.; Sharma, P.; Singh, L.; Ram, C. Photocatalytic Degradation of Hazardous Rhodamine B Dye Using Sol-Gel Mediated Ultrasonic Hydrothermal Synthesized of ZnO Nanoparticles. *Results Eng.* **2023**, *17*, 100890. [CrossRef]
80. Dodoo-Arhin, D.; Asiedu, T.; Agyei-Tuffour, B.; Nyankson, E.; Obada, D.; Mwabora, J. Photocatalytic Degradation of Rhodamine Dyes Using Zinc Oxide Nanoparticles. *Mater. Today Proc.* **2021**, *38*, 809–815. [CrossRef]

81. Varadavenkatesan, T.; Lyubchik, E.; Pai, S.; Pugazhendhi, A.; Vinayagam, R.; Selvaraj, R. Photocatalytic Degradation of Rhodamine B by Zinc Oxide Nanoparticles Synthesized Using the Leaf Extract of *Cyanometra ramiflora*. *J. Photochem. Photobiol. B Biol.* **2019**, *199*, 111621. [CrossRef] [PubMed]
82. Júnior, E.A.A.; Nobre, F.X.; da Silva Sousa, G.; Cavalcante, L.S.; de Morais Chaves Santos, M.R.; Souza, F.L.; de Matos, J.M.E. Synthesis, Growth Mechanism, Optical Properties and Catalytic Activity of ZnO Microcrystals Obtained via Hydrothermal Processing. *RSC Adv.* **2017**, *7*, 24263–24281. [CrossRef]

water

MDPI

Article

The Pyrolysis of Biosolids in a Novel Closed Coupled Pyrolysis and Gasification Technology: Pilot Plant Trials, Aspen Plus Modelling, and a Techno-Economic Analysis

Nimesha Rathnayake [1,2], Savankumar Patel [1,2], Ibrahim Gbolahan Hakeem [1,2], Ganesh Veluswamy [1,2], Ibrahim Al-Waili [1,2], Shivani Agnihotri [1,2], Arun Krishna Vuppaladadiyam [2,3], Aravind Surapaneni [2,4], David Bergmann [4] and Kalpit Shah [1,2,*]

1 Chemical & Environmental Engineering, School of Engineering, RMIT University, Melbourne, VIC 3000, Australia; nimesha.rathnayake@rmit.edu.au (N.R.); savankumar.patel@rmit.edu.au (S.P.); ibrahim.hakeem@rmit.edu.au (I.G.H.)
2 ARC Training Centre for the Transformation of Australia's Biosolids Resource, RMIT University, Bundoora, VIC 3083, Australia
3 School of Civil and Mechanical Engineering, Curtin University, Perth, WA 6102, Australia
4 South East Water, Frankston, VIC 3199, Australia
* Correspondence: kalpit.shah@rmit.edu.au

Citation: Rathnayake, N.; Patel, S.; Hakeem, I.G.; Veluswamy, G.; Al-Waili, I.; Agnihotri, S.; Vuppaladadiyam, A.K.; Surapaneni, A.; Bergmann, D.; Shah, K. The Pyrolysis of Biosolids in a Novel Closed Coupled Pyrolysis and Gasification Technology: Pilot Plant Trials, Aspen Plus Modelling, and a Techno-Economic Analysis. *Water* **2024**, *16*, 3399. https://doi.org/10.3390/w16233399

Academic Editors: Issam A. Al-Khatib, Rehab O. Abdel Rahman, Tsuyoshi Imai and Yung-Tse Hung

Received: 30 September 2024
Revised: 28 October 2024
Accepted: 28 October 2024
Published: 26 November 2024

Abstract: Pyrolysis is gaining recognition as a sustainable solution for biosolid management, though scaling it commercially presents challenges. To address this, RMIT developed a novel integrated pyrolysis and gasification technology called PYROCO™, which was successfully tested in pilot-scale trials. This study introduces PYROCO™ and its application to produce biochar, highlighting the biochar properties of the results of the initial trials. In addition, an energy analysis using semi-empirical Aspen Plus modelling, paired with a preliminary techno-economic assessment, was carried out to evaluate the feasibility of this technology. The results show that the PYROCO™ pilot plant produced biochar with a ~30 wt% yield, featuring beneficial agronomic properties such as high organic carbon (210–220 g/kg) and nutrient contents (total P: 36–42 g/kg and total N: 16–18 g/kg). The system also effectively removed contaminants such as PFASs, PAHs, pharmaceuticals, and microplastics from the biochar and scrubber water and stack gas emissions. An energy analysis and Aspen Plus modelling showed that a commercial-scale PYROCO™ plant could operate energy self-sufficiently with biosolids containing >30% solids and with a minimum calorific value of 11 MJ/kg. The process generates excess energy for drying biosolids and for electricity generation. Profitability is sensitive to biochar price; prices rise from AUD 300 to AUD 1000 per tonne, the NPV improves from AUD 0.24 million to AUD 4.31 million, and the payback period shortens from 26 to 12 years. The low NPV and high payback period reflect the use of a relatively high discount rate of 8%, chosen to be on the conservative side given the novel nature of the technology.

Keywords: biochar; pilot scale; sewage sludge; techno-economic feasibility; energy analysis

1. Introduction

Biosolids are stabilised sewage sludge derived from wastewater treatment and contain valuable macro- and micro-nutrients and organic carbon [1]. Globally, about 40–70% of biosolids are applied to agricultural land [2]; in particular, ~80% of biosolids produced in 2023 (372 kT of dry solids) in Australia were beneficiated to agricultural soils [1]. However, the presence of contaminants, such as per- and polyfluoroalkyl substances (PFAS), microplastics, heavy metals, pharmaceuticals, and pesticides, are causing significant challenges for the land application of biosolids [2,3]. Thermal techniques, such as pyrolysis and gasification, have gained significant interest in transforming biosolids to biochar and bioenergy in the last few years while safely removing PFAS and other contaminants [4–10].

The pyrolysis of biosolids has gathered significant attention over the past decade; however, most studies that focus on the pyrolysis of biosolids are lab-scale experiments primarily aimed at identifying the effects of operating conditions, such as temperature, heating rate, and residence times, on the pyrolysis of biosolids [4,11–13]. Existing works have also investigated the product yields and product compositions of the pyrolysis of biosolids and the properties of biochar through ultimate and proximate analyses, surface areas, and heavy metal concentrations [4,7,11–14]. Additionally, the suitability of biosolid-derived biochar for various agricultural and non-agricultural applications has been extensively studied [15–17]. For example, Figueiredo et al. demonstrated that the biochar of biosolids enhances organic, inorganic, and available phosphorus fractions in soil applications [18]. Similarly, Nicomel et al. showed that the biochar of biosolids effectively removes copper from citric acid-rich aqueous media, highlighting its potential as an adsorbent in non-agricultural contexts [19].

The findings from lab-scale studies are crucial for the determination of optimum process conditions, intrinsic kinetics, product distribution, and properties. Lab-scale studies provide limited information about the techno-economic and environmental feasibility of the pyrolysis of biosolids. Large-scale experiments such as pilot- or semi-pilot-scale studies are essential to evaluate the practical feasibility of the pyrolysis of biosolids on a commercial scale, including the performance of pyrolysis reactors, energy requirements, maintenance needs, the control of emissions, and heat transfer dynamics [20,21].

However, only a few studies have reported the semi-pilot- and pilot plant-scale pyrolysis of biosolids [22–25]. Some pilot and semi-pilot studies have focused on the contaminant reduction potential of their technologies during the pyrolysis of biosolids [22,23]. For example, Thoma et al. reported that their auger-based pyrolysis system achieved removal efficiencies for 41 targeted PFAS compounds ranging from >81.3% to >99.9%, with an average efficiency of over 97.4% [22]. Most pilot-scale pyrolysis studies have utilised auger-type reactors [24–26]. Auger reactors, which use a rotating screw to transport feedstock, are effective in radial mixing but lack any axial mixing [27]. This may lead to uneven heat exposure and inconsistent pyrolysis reactions, resulting in variability in product quality [27]. Furthermore, heat transfer in auger reactors primarily occurs through the reactor walls, which can be inefficient, particularly in larger reactors [27]. The outer layers of feed materials may heat up, while the material closer to the centre of the screw remains cooler due to a non-uniform temperature distribution in the reactor. Additionally, the mechanical parts of auger reactors, especially the screw, are subject to wear and tear and require additional mechanical energy. Liu et al. [24] and Sarvi et al. [24,25] have also noted that further modifications are necessary to improve mixing and heat transfer in auger pyrolysis reactors [24,25]. Given these challenges, assessing the feasibility of other types of reactors that offer improved heat transfer and the ability to produce homogeneous products in large-scale pyrolysis units is crucial. This exploration is essential to advance the practical application of the pyrolysis of biosolids at a commercial scale.

The current paper focuses on a fluidised-bed pyrolysis reactor. Fluid-bed heat exchangers mix well due to bed fluidisation, which can produce a more homogenous product. Fluidisation also increases heat transfer and avoids hot and cold spots [28]. Fluid-bed reactors are easy to scale up and can be operated in batch and continuous mode [28]. Another advantage of fluid reactors is that they do not have any moving parts, reducing maintenance and mechanical energy requirements [28]. Conventional fluid-bed reactors also have several drawbacks, such as the requirement of feedstock sizing, the generation of fines, and the requirement of a high flow rate of inert fluidisation gas [29]. Inert fluidisation gas, such as nitrogen, is expensive and increase operating costs. In addition, fluidisation gas increases the overall volume of flue gas to be treated and dilutes the energy content of the outlet gas stream from the reactor [29].

The novel fluidised-bed pyrolysis technology PYROCO™, developed and patented by RMIT University, Australia, modifies the conventional fluid-bed reactor to overcome or minimise these limitations. Therefore, compared to existing large-scale pyrolysis reactors,

PYROCO™ offers several benefits such as (1) uniform temperature inside the reactor, hence ensuring consistent product quality; (2) no moving parts for mixing, hence reducing maintenance costs; (3) improved heat and mass transfer, hence reducing the reactor volume and, therefore, the capital and operating costs; and (4) flexibility to produce more bioenergy versus biochar, hence helping to achieve thermal energy neutrality without the need for any external thermal energy for the process. The pilot plant processed 1 ton of biosolids per day (50% of total solids).

The current study presents the results from the PYROCO™ pilot plant trials, which characterised the produced biochar, air emissions, and scrubber water. A semi-empirical process model was developed using the Aspen Plus (v12.1) software to simulate the PYROCO™ process for processing 10 tonnes/day of wet biosolids, providing detailed mass and energy balance data. The developed process model was validated using data from the pilot trial. Then, the data from the model was used in techno-economic and sensitivity analyses to explore the economic feasibility of the PYROCO™ process under different operational conditions.

2. Materials and Methods

2.1. PYROCO™ Pilot Plant Trials

2.1.1. Sample and Preparation of Biosolids

The biosolids used in this study were collected from Mount Martha Water Recycling Plant (38°16′06″ S, 145°03′31″ E), South East Water Corporation, Victoria, Australia. This plant primarily processes domestic and trade sewage using an activated sludge process followed by anaerobic digestion. After digestion, solids are dewatered with a centrifuge and are dried in a solar drying facility before stockpiling. The biosolid samples were collected from the solar dryer shed, ground with a pin mill, and then sieved to <1 mm particle sizes using a vibrating screen. Images of the pin mill and vibrating screen are shown in Figure S1. The proximate and ultimate properties of the biosolid samples are provided in Table 1.

Table 1. Properties of biosolid samples.

Proximate Analysis (wt%) [c]				Ultimate Analysis (wt%) [c]				
Moisture [a]	Volatiles	Ash	Fixed Carbon	C	H	N	S	O [b]
11.0	60.6	29.0	10.4	38.3	4.7	6.02	0.96	21.02

Notes: [a] value on as-received basis, [b] obtained by difference, [c] value on dry weight basis.

2.1.2. Description of the Pilot Plant and Experimental Matrix

A schematic of the PYROCO™ pilot plant is shown in Figure 1. It comprises a gas producer (fluidised-bed gasifier), a PYROCO™ reactor (fluidised-bed pyrolysis integrated with a combustor and heat exchanger), a venturi scrubber, a water tank, an activated carbon bed, and a stack. At the current scale, the drying and sieving unit operations are not yet integrated with the plant. Therefore, feed preparation (drying and sizing) operations were performed separately in batch mode, while the rest of the thermal plant operates in continuous mode. Dried and sieved biosolids can be fed into both the gas producer (GP) and the PYROCO™ pyrolysis zone. During the pilot trial, the feed rate of the biosolids, with an 85% total solid content, ranged from 12 to 16 kg/h. The GP generates oxygen-free producer gas, which is then used as the fluidising gas for pyrolysis in the PYROCO™ reactor. The oil and gas vapours from the pyrolysis unit and the producer gas are channelled from the pyrolysis zone to the adjacent combustor, where combustion occurs with the help of tangentially introduced air. The combustion of these vapours generates thermal energy, which is transferred to the pyrolysis zone through heat-exchanger tubes. The flue gas leaving the combustor is scrubbed in a venturi scrubber.

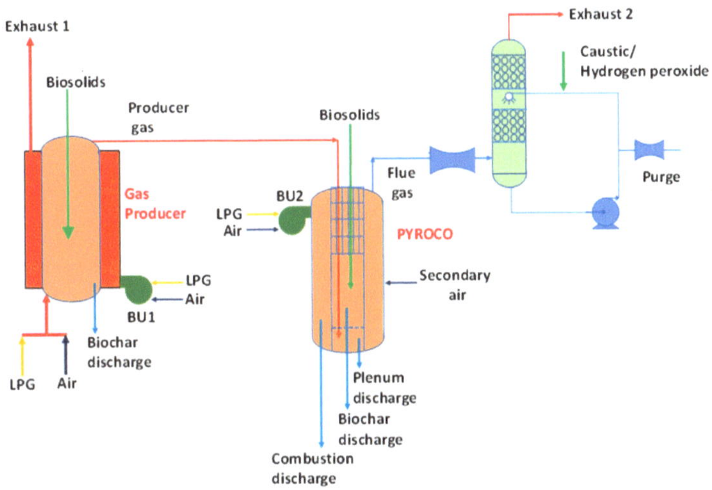

Figure 1. Process flow diagram of the PYROCO™ pilot plant.

The PYROCO™ reactor, as a fluidised-bed system, offers a more uniform temperature distribution and better mixing, leading to a more homogenous biochar product. Gases generated during gasification in the GP are used for fluidisation in the PYROCO™ reactor, enhancing process efficiency. The PYROCO™ reactor features two concentric chambers—pyrolysis occurs in the inner chamber, while combustion of pyrolysis gases in the outer chamber provides the necessary heat for pyrolysis. The inner chamber also contains internal heat-exchanger tubes, where combusted flue gases circulate, further improving heat transfer. The design is also advantageous because it contains no moving parts, resulting in lower capital and maintenance costs and higher plant availability.

The venturi scrubber serves multiple purposes, such as removing particulate matter (fines carried over from the pyrolyser), quenching the flue gas to lower temperatures (flue gas enters at >700 °C), and generating low-temperature water vapour. It also captures emissions, such as NOx, SOx, Hg, and other impurities. In a full-scale PYROCO™ plant, the venturi scrubber would mainly focus on capturing gas emissions, capturing fines, and heat recovery, aided by a cyclone and heat exchanger before the scrubber. The treated flue gas then passes through an activated carbon bed filter before being released into the exhaust. Water is continuously circulated to the venturi scrubber via a pump from a water tank. Caustic soda or hydrogen peroxide can be added to the water tank to capture SOx and NOx effectively.

During current pilot plant trials, GP operated at 800 °C, the PYROCO™ pyrolysis reactor operated between 600 and 700 °C, and the PYROCO™ combustion reactor operated at 900 °C. The experimental matrix for the pilot plant trials is included in Table 2. The pilot plant operated for 118 h (including startup) in two daily shifts. It was kept under a nitrogen/inert environment overnight without heating during the shutdown period. On one occasion, it was operated continuously for 36 h.

Table 2. Experimental matrix for the pilot plant trials.

Equipment	Temperature (°C)	Reaction Environment	Biosolids (kg/h)	LPG (kg/h)
PYROCO®	600 and 700	Producer gas	12–16	-
Gas producer	800	Partial air	-	0.9

Composite scrubber water sampling was performed, and biochar samples were stored in sealed containers for further analysis, while the scrubber water samples were sent to

ALS Laboratories, Australia. The scrubber water was stored in two IBC tanks for future disposal at a wastewater treatment plant. Neutralisation will be performed if required.

The biochar produced during the trial was characterised by surface imaging using a scanning electron microscope (SEM) of the FEI Quanta 200 model (FEI, Hillsboro, OR, USA) and a Brunauer–Emmett–Teller (BET) analyser (Micromeritics 2000/2400, Micrometrics, Norcross, GA, USA). Analyses related to electrical conductivity (EC), pH, total solids, total organic carbon, nutrients, heavy metals, and targeted PFAS were performed externally (by ALS Environmental Science, Melbourne, Australia), according to the methods APHA 2510B, APHA 4500 H+ -B, APHA 2540B, VIC EPA 1981, 1.139, USEPA 6010 ICP/AES, and EP231X, respectively. Microplastics analyses were carried out by Eurofins Environment Testing Australia Pty Ltd., Melbourne, Australia using LTM-MPS-9050 MPs in Soils Testing method. Pharmaceuticals, PAHs, endocrine-disrupting chemicals, and siloxanes analyses were performed at Leeder Analytical, Melbourne Australia. Scrubber water analyses for PFAS, polycyclic aromatic hydrocarbons (PAHs), heavy metals, and halides were analysed by ALS Environmental Sciences, Melbourne, Australia according to methods EP231X, EG 093F, and USEPA 8270, respectively. All laboratories are accredited by NATA (National Association of Testing Authorities, Australia).

The real-time temperature, pressure, and stack gas emissions (CO, CO_2, O_2, NOx, and SOx) were recorded using a data logger and RMIT's NDIR-type online gas analyser. Ektimo, a NATA-accredited lab, was engaged to conduct comprehensive stack gas-emission testing, covering pollutants such as COx, SOx, NOx, hydrocarbons, VOCs, HCl, HF, dioxins, and furans.

2.2. *Aspen Plus Modelling*

This study developed a comprehensive semi-empirical process model using Aspen Plus to simulate a full-scale PYROCO™ plant and to assess technical feasibility regarding operational adaptability and energy self-sufficiency. The chemical compositions of biosolids and biochar, derived from the pilot plant trial data and literature data, were used as input for the simulation.

2.2.1. Process Flow Diagram of the Proposed Fully Integrated PYROCO™ Plant

For process modelling, a fully integrated PYROCO™ process for the pyrolysis of biosolids, including front-end and back-end unit operations, was considered, as shown in Figure 2. Alongside the gas producer and PYROCO™ reactor, the full-scale integrated unit includes a feed pretreatment unit for drying and sizing biosolids. As in the pilot plant trials, dried and sieved biosolids are divided into two fractions, one fed into the PYROCO™ reactor and the other into the GP. The ratio between the GP and PYROCO™ reactors (GP:PYROCO) can be adjusted depending on the feedstock characteristics (total solids and calorific value) or by prioritising a higher biochar yield against greater surplus energy.

The GP operates in auto-thermal mode at a 0.3 ER (equivalence ratio) for the gasification of biosolids. In the full-scale PYROCO™ process, volatile compounds generated during pyrolysis are only partially combusted in the PYROCO™ reactor's combustion chamber, providing the necessary energy for pyrolysis. The resulting flue gas is fully combusted in a thermal oxidiser to eliminate contaminants per regulatory mandates. This is achieved by maintaining 12 vol% of O_2 in the flue gas exiting the thermal oxidiser, which operates at 900 °C with a minimum flue gas residence time of 2 s.

The flue gas then passes through a thermal energy recovery unit, which includes heat exchangers, boilers, or hot water generators. The recovered energy is primarily used to dry the incoming wet biosolids. Any surplus energy after drying the biosolids is assumed to be used in electricity generation. After the thermal energy recovery unit, the flue gas enters a flue gas cleaning system consisting of a wet venturi scrubber for quenching, SOx and NOx removal, and a unit for particulate-matter removal.

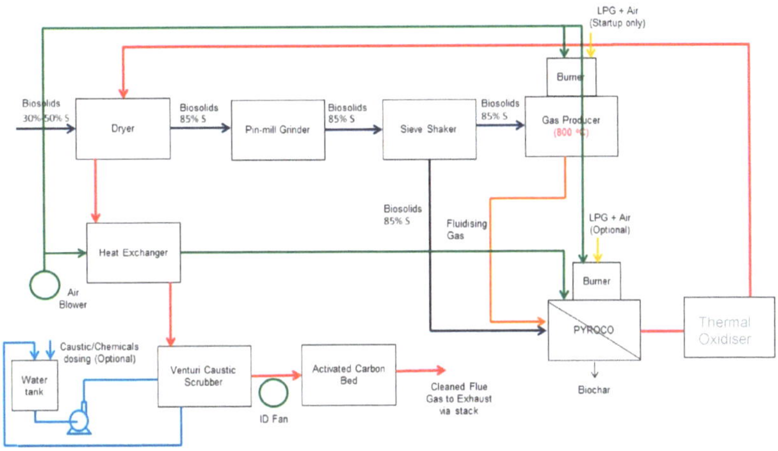

Figure 2. Process flow diagram of the proposed full-scale PYROCO™ process.

2.2.2. Aspen Flowsheet Development

The process modelling was carried out using the Aspen Plus (V.12) software. The developed process flowsheet from the Aspen Plus modelling is shown in Figure 3. The simulation used FSplit, RGibbs, Heater, HeatX, Compressor, Flash2, and Tank blocks to model the proposed process. The process simulator does not have a built-in model for pyrolysis or gasification. However, it does have several built-in block models that can be used collectively to model pyrolysis and gasification processes. As pyrolysis involves complicated multiphase reactions, biosolids and biochar are not standard components and do not have fixed molecular structures. A non-stoichiometric approach is used to model pyrolysis. Therefore, RYield and RGibbs reactors were used to model pyrolysis. The first RYield reactor transforms the feed material into its constituting elements, including carbon, hydrogen, oxygen, nitrogen, sulphur, and biochar, based on the ultimate analysis of biosolids and biochar and the mass yield of pyrolysis of biosolids. The second RYield reactor transforms elements from the first RYield reactor into oil and gas components based on lab-scale experimental data from our previous studies. The RGibbs reactor operates at 700 °C and calculates the composition of non-condensable and condensable vapours in the resulting stream based on minimising the Gibbs energy for the system. A gasification reaction occurring in the GP operating at 800 °C was also modelled with an RYield and an RGibbs reactor.

In this work, three classes of components were used, namely, conventional components, non-conventional components, and solids. In this simulation, HCOALGEN and DCOALIGT require the proximate, ultimate, and sulphur compositions to calculate the enthalpy and density of the non-conventional components used. Based on the literature, UNIQUAC was chosen as the physical property method to calculate the thermodynamic properties [30]. This property method applies the UNIQUAC activity coefficient method in the liquid phase, and for the vapour phase, it applies the ideal-gas fugacity coefficient method. The basic assumptions applied in the Aspen simulation are listed below.

1. The biosolids were homogeneous, and the chemical composition was uniform throughout the material.
2. The input data, including heat capacity, enthalpy, and entropy, represented the thermodynamic properties of the biosolids.
3. The pyrolysis reactions followed thermodynamic (equilibrium) rate conversions based on experimental data, and the reaction rate was not defined in the pyrolysis reactor. However, it is assumed that close-to-equilibrium conversion values would be obtained due to high-temperature reactions.

4. The pyrolysis process operated under steady-state conditions; the operating parameters, such as temperature, pressure, and feed rate, were constant throughout the reactor.

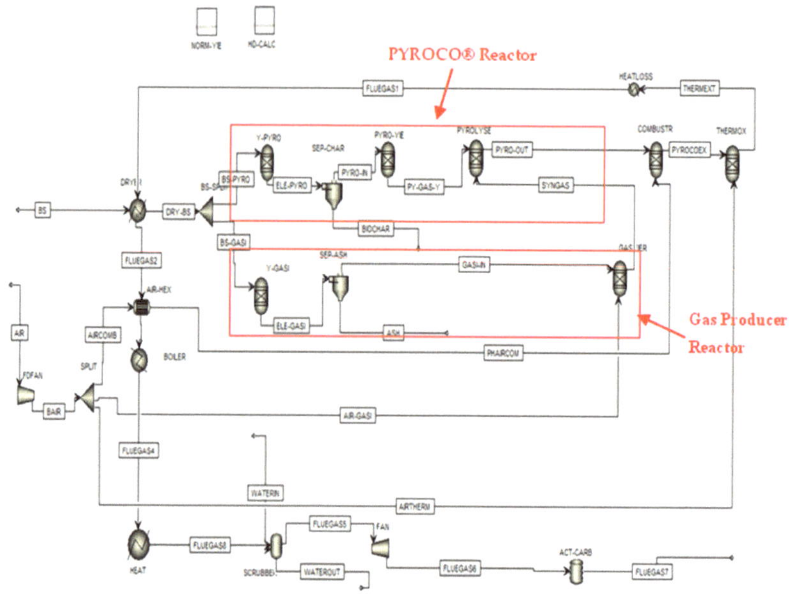

Figure 3. Aspen Plus flowsheet of the proposed full-scale PYROCO™ plant for the pyrolysis of biosolids.

2.2.3. Model Validation

The Aspen Plus model developed in the current study contained two main reactor units: the PYROCO™ reactor and gas GP. To validate this model, a comparison was made between the simulated data and the gas profiles collected from the pilot plant trials at the GP and PYROCO™ reactor exits. This comparison aimed to ensure that the simulated gas profiles accurately reflect the results of the pilot plant trial, thereby confirming the reliability and predictive capabilities of the Aspen Plus model.

2.2.4. Energy Analysis

An energy analysis of the process was performed in this research to study the influence of the feedstock flowrate ratio between the GP and PYROCO™ reactors (GP:PYROCO) and the biosolids' calorific value and solid content on the amount of energy generated/required (kW) by the pyrolysis process. This analysis used three different biosolid feedstocks with different heating values. The proximate and ultimate analysis, heating value, and biochar yield of these biosolids are included in the supplementary data in Table S1. The biochar yield in different types of biosolids was calculated from correlations developed by Li et al. [31]. GP:PYROCO ratios of 0.2, 0.4, 0.6, and 0.8 and biosolids with a 30%, 40%, and 50% solid content were employed in the energy analysis.

2.3. Economic Analysis

A high-level techno-economic analysis was conducted for a fully integrated PYROCO™ plant designed to process 10 tonnes of wet feedstock per day. The objective was to evaluate the techno-commercial feasibility using economic indicators such as the Net Present Value (NPV) and payback period for biosolid processing. In the methodology for this analysis, the pilot plant trial data were used to develop an Aspen Plus simulation for the PYROCO™ plant. This simulation facilitated accurate material and energy balance calculations, which were critical for estimating the capital and operating costs of the plant. Subsequently, a discounted cash flow (DCF) analysis was performed to provide a preliminary economic assessment based on financial assumptions. The NPV was calculated to determine the economic viability of the plant.

2.3.1. Capital Cost Estimation

In order to estimate the capital cost, two main factors were taken into account: (i) the base cost year and (ii) the capacity of the equipment. The capital cost estimates were derived using the CAPCOST software version 2017, vendor quotes, and the NREL report published in 1996 [32]. Therefore, these costs were several years old and were updated to the targeted year. The magazine *Chemical Engineering* regularly publishes the Chemical Engineering Plant Cost Index (CEPCI) to help estimate capital costs [33]. The equation below was used for predicting the current price of the equipment:

$$\text{Cost in 2023 \$} = \text{Cost in base year} \times \left(\frac{\text{2023 index : 790.8}}{\text{base year index}}\right) \tag{1}$$

The CEPCI values were either published in the literature or were calculated using previous years' data [32].

As mentioned above, the other factor is equipment capacity, meaning that the cost of the instrument should be scaled based on its capacity using the following equations.

$$\text{Scaled cos t} = \text{Cost of original scale} \times (\text{Scaled capacity}/\text{Original capacity})^{n} \tag{2}$$

The scaling exponent, n, is normally between 0.6 and 0.8 and depends on the type of equipment [34]. The above equation can calculate the scaled cost in the targeted year. Accordingly, the scaled installed equipment costs were calculated, as shown in Table S2. These costs include installation, covering piping, instrumentation, and control. The sum of the installed costs for all pieces of equipment and the cost for the balance of the plant (known as the additional components required to deliver energy) are defined as the Total Installed Cost (TIC). When established, capital costs are of two types—direct and indirect. The Total Direct Cost (TDC) includes the TIC and costs for the land and building, instrument and control (10% of the TIC), as well as additional piping (15% of the TIC), electrical systems (10% of the TIC), and site preparation (3.5% of the TIC) [35]. The land is expected to be provided by the plant at no cost. Indirect costs are also imposed on the plant in explicit ways, including engineering and supervision expenses (5% of TDC), legal expenses (1% of the TDC), contractors' fees (10% of the TDC), and project contingency (5% of the TDC). Fixed Capital Investment (FCI) is the sum of the total direct and indirect costs [34]. The working capital is assumed to be 15% of the FCI, as mentioned in the Nth-plant assumptions [34]. Ultimately, if we sum the FCI and working capital (plus the land), this gives the total capital investment (TCI), as detailed in Table S2.

Additionally, this study calculated the required processing cost of biosolids per kilogram in the PYROCO process using the following equation.

$$\text{Cost per tonne of biosolids processing in PYROCO}\left(\frac{\$}{\text{tonne}}\right) = \frac{\text{CAPEX} + \text{OPEX per annum}}{\text{Biosolids processed per annum}}$$

2.3.2. Estimation of Operating Costs

Apart from the capital investment paid to establish the plant, operating costs also play a role in the analysis of process economics. The operating costs considered in this analysis include labour, material procurement, utilities (electricity and LPG), and maintenance. The costs for traditional biosolids management (transport and landfilling) were considered cost-saving. It is important to note that these estimates are based on specific assumptions (detailed in Table S3) and are intended for benchmarking. A more granular financial assessment with detailed cost and process information from technology providers will be possible. The analysis also does not include other site costs, including complete process automation, material-storage costs, substation capital, grid connection costs, and project management and consultancy costs. Material-handling and drying costs might be underestimated, as biosolids may require more appropriate dust- and odour-management equipment and controls [36]. The main revenue from the plant is from biochar and electricity sales.

2.3.3. Calculations of Net Present Value (NPV), Payback Period, and Benefit-to-Cost Ratio (BCR)

NPV represents the difference between the present value of cash inflows and outflows, helping assess whether a project generates more value than its cost. The payback period is the time it takes for an investment to recover its initial cost through generated cash flows, indicating how quickly a project becomes profitable. The NPV and payback period were calculated for the above assumptions in Table S3 using the formulas below [34].

$$NPV = \sum_{n=1}^{T} \frac{CF_n}{(1+i)^n} - I_0$$

$$\text{Payback period} = (\text{Total CAPEX})/(\text{Average annual cash flow})$$

CF_n: cash flow generated in year n
I_0: total capital investment in year 0
T: project lifetime
i: discount rate
The cash flow generated in year n, CF_n, can be estimated using the following formula:

$$CF_n = (1-t) \times (R-E) + t \times D$$

where R is the revenue, E is the expenditure, t is the tax rate, and D is depreciation. D is calculated using a straight-line method.

BCR compares the present value of benefits to the present value of costs, and a BCR greater than 1 indicates that the project's benefits exceed the costs, making it economically viable. It was calculated using the formula below.

$$BCR = \frac{\text{Present Value of Capital and Operating Costs}}{\sum_{n=1}^{T} \frac{CF_n}{(1+i)^n}}$$

3. Results and Discussion

3.1. PYROCO™ Pilot Plant Trials

3.1.1. Mass and Energy Balance

The mass and energy balance data from the pilot trials are depicted in Table 3. It was observed that the PYROCO™ reactor predominantly utilised the heat generated by the combustion of volatile gases from the gas producer and the combustion chamber. The external energy input was minimal, with only 8% of PYROCO®'s total energy requirement being supplied by LPG. This LPG input was primarily necessary to meet the pilot burner's minimum operational requirements. It is anticipated that in a commercial-scale PYROCO™ plant, the percentage of energy supplied by LPG would decrease even further. Additionally, the pilot trials estimated about 15% heat losses during stable operations, which is relatively high and likely attributable to the smaller scale of the operation.

Table 3. Mass and energy balance of PYROCO™ trial.

Details	Mass Flow Rate (kg/h)		Energy (MJ/h) on a Dry Basis		
	In	Out	In	Loss/Consumed	Out
Biosolids	12		175		
LPG in gas producer	0.9		42		
LPG in PYROCO™ pilot burner	0.1		5		
Biochar from PYROCO®		3.2			32
Secondary air in PYROCO®	5.7				
Air in gas producer	14.3				
Air in PYROCO®	104.6				
PYROCO®				62	
Heat loss				33	
Exhaust		134.4			95
Total	137.6	137.6	222	95	127
Difference	0			0	

3.1.2. Estimation of Biochar Yield

During the pilot plant trials, a total of 830 kg of biosolids was processed, leading to the production of approximately 250 kg of biochar. This corresponds to a biochar yield of around 30 wt%. Consequently, the PYROCO™ process achieved a significant volume reduction of ~70% (on a dry basis) in biosolids. Compared with the yield data from our lab-scale experiments reported by Patel et al. (2019), a 38 wt% biochar yield was obtained at 700 °C; thus, about a 25% difference in yield was noticeable [37]. This discrepancy can be attributed to an increased generation of fines due to the top-feeding arrangement and the varying operating parameters between lab-scale and pilot trials. Lab-scale experiments usually offer more controlled conditions, while temperature and fluidising gas flow rates fluctuate more during pilot trials, impacting the overall yield. Despite these differences, the pilot-scale results demonstrate that a reasonable biochar product can be produced using the PYROCO™ process. Furthermore, optimising the pretreatment of biosolids by introducing a granulating or pelletising step to reduce the generation of fines and fine-tuning the operating parameters can lead to even better biochar yields in future trials [38,39].

3.1.3. Properties of Biochar

Physicochemical and Surface Properties

The pH, electrical conductivity (EC), volatile solids, and organic carbon content of biochar are important chemical properties routinely measured to evaluate biochar for soil applications [40,41]. The biochar produced through the PYROCO™ process exhibited favourable agronomic characteristics, including a higher pH value, lower electrical conductivity (EC), lower volatile solids, and a high organic carbon content, as detailed in Table 4.

Table 4. Properties of biosolids and biochar.

Parameters	Units	Biosolids	Biochar (600 °C)	Biochar (700 °C)
Electrical conductivity (EC)	uS/cm	6700	1100	1000
pH	Units	7.3	11.7	11.6
Total solids	% w/wet	81	100	100
Volatile total solids	%	64	20	20
Total organic carbon	mg/kg	240,000	220,000	210,000
C:N ratio		5.4	12.2	13.1
BET-specific surface area	m^2/g	2.3	11.7	25.1

The biochar's pH was higher than that of the biosolids, as acidic functional groups were removed during pyrolysis, and alkali and alkaline earth-element salts became concentrated. This elevated pH can be advantageous for improving acidic soils due to its acid-neutralising capacity when biochar is applied to agricultural land [42]. Electrical conductivity is a standard soil salinity test that measures the concentration and nature of salts in a solution [40]. Understanding the amount of soluble salts in biochar is crucial, because excessive biochar applications may negatively affect salt-sensitive plants [40]. Compared to biosolids, the EC of the biochar reduced by approximately six-fold, indicating that the pyrolysis process effectively lowered the concentration of soluble inorganic salt. Furthermore, the biochar samples demonstrated a reduction of approximately 65–70% in volatile content compared to the biosolids. This reduced volatile content indicates greater stability, eliminates off-gassing and unpleasant odours, and offers potential nutritional benefits due to stable volatile matter that can enhance soil fertility [36,40].

While the total organic carbon content in the biochar was slightly reduced compared to that of the biosolids (8.3% and 12.5% reduction at 600 °C and 700 °C, respectively), the carbon-to-nitrogen (C/N) ratio was about twice as high (5.4 in biosolids vs 12.2-13.1 in biochar). This increase resulted from nitrogen reduction during the PYROCO™ reactor's pyrolysis process. This higher C/N ratio improves the biochar stability and carbon sequestration potential, making it suitable for long-term soil amendment [41]. However, the lower nitrogen content means that the biochar may need to be supplemented with nitrogen-rich fertilisers to supply adequate nutrients for plants.

The surface characteristics of the biochar produced through the PYROCO™ process at 600 °C and 700 °C were analysed by measuring the Brunauer–Emmett–Teller (BET) surface area, and the results are presented in Table 4. The surface area of the biochar increased significantly when compared to raw biosolids, from 2.3 m^2/g to 11.7 m^2/g at 600 °C and 25.1 m^2/g at 700 °C. This increase is attributed to the devolatilisation intensity at high temperatures, which creates a more porous structure within the biochar [42,43]. A higher surface area is crucial for biochar's efficacy in adsorption applications, as it allows for a greater interaction with and the retention of various substances [44]. Additionally, a larger surface area enhances biochar's use in agriculture, improving water- and nutrient-retention capacities and fostering beneficial microbial activity in the soil [13,45]. The surface morphology was assessed via SEM images, shown in Figure 4. There was a clear difference in the pore structure between the biochar produced at 600 °C and 700 °C. The images reveal a more developed pore structure in the biochar processed at 700 °C compared to that processed at 600 °C, improving the biochar's quality.

Compositions of Nutrients

Table 5 shows the nutrient content of the biosolids and biochar. Biochar contains a significant amount of macro-nutrients (such as nitrogen (N), phosphorus (P), and potassium (K)) and micro-nutrients (including calcium (Ca), magnesium (Mg), and sodium (Na)). The phosphorus and potassium levels were notably higher in the biochar than in the biosolids, with phosphorus increasing from 25,000 mg/kg in the biosolids to 36,000 mg/kg and 42,000 mg/kg in the biochar produced at 600 and 700 °C, respectively. Similarly, in the biochar samples, the potassium content rose from 4400 mg/kg in the biosolids to 6600 mg/kg and 8100 mg/kg. The nitrogen content in the biochar reduced by around 50% compared to that of the biosolids. The total nitrogen content in the biochar was 18,000 mg/kg at 600 °C and decreased slightly to 16,000 mg/kg at 700 °C compared to the 44,000 mg/kg in the biosolid feed. The micro-nutrient levels also increased in the biochar relative to the biosolids. Calcium increased from 51,000 mg/kg in the biosolids to 72,000 mg/kg (600 °C) and 87,000 mg/kg (700 °C) in the biochar. A similar trend was observed for magnesium and sodium. An increase in the concentration of inorganic nutrient elements, such as K, Mg, Ca, P, and Na, in the biochar compared to the biosolids is due to the thermal stability of these elements under pyrolysis conditions and the substantial volume reduction caused by the decomposition of organic matter.

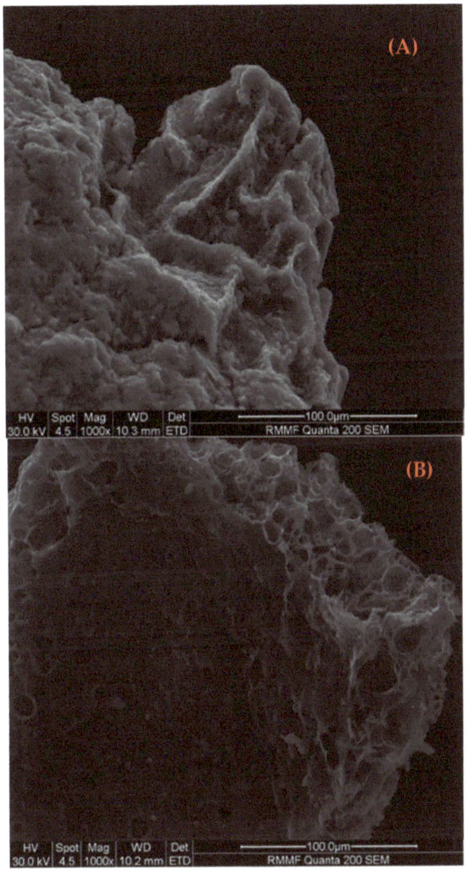

Figure 4. SEM images of PYROCO™ biochar produced at (**A**) 600 °C and (**B**) 700 °C.

Table 5. Nutrients in biosolids and biochar.

Nutrients	Units	Biosolids	Biochar (600 °C)	Biochar (700 °C)
Nitrogen (N)				
Total Kjeldahl nitrogen (TKN)	mg/kg	43,000	18,000	16,000
Ammonia (NH_3)	mg N/kg	7400	170	30
Nitrite (NO_2)	mg/kg	1.3	<0.2	<0.2
Nitrate (NO_3)	mg/kg	740	<0.2	<0.2
NOx	mg N/kg	740	<0.1	0.1
Total nitrogen	mg N/kg	44,000	18,000	16,000
Phosphorus (P)				
Total P	mg/kg	25,000	36,000	42,000
Olsen P	mg/kg	1400	83	65
Total potassium (K)	mg/kg	4400	6600	8100
Sulphur (S)				
OESEXTRA/S	mg/kg	10,000	6700	7800
Total sulfur	%	0.4	0.95	1.05
Total calcium (Ca)	mg/kg	51,000	72,000	87,000
Total magnesium (Mg)	mg/kg	5800	6900	7400
Total sodium (Na)	mg/kg	3300	4400	5100

Heavy Metal Contents

The heavy metal concentration in the biosolids and biochar are given in Table 6. The heavy metal concentrations, except for that of mercury, were higher in the biochar samples than in the biosolid ones. This increase is due to the removal of volatile components and the resultant mass loss during pyrolysis, concentrating the remaining non-volatile elements [42]. The mercury levels, however, significantly reduced in the biochar, as the element evaporated (at a boiling point of ~355 °C) during the pyrolysis process [42]. The mercury concentration in the biochar was found to be below detectable limits. Table 6 also outlines the Victorian EPA biosolid guidelines' upper limits for heavy metals [46]. These guidelines classify contaminants into two grades: C1, the more stringent classification allowing unrestricted land applications, and C2, which is suitable for restricted land applications. The arsenic, chromium, nickel, and lead levels were within the limits for the C1 grade under these guidelines [46]. However, the cadmium, copper, selenium, and zinc concentrations in the biosolids and biochar exceeded the C1 limits but remained within the C2 thresholds. Thus, the biosolids and biochar produced via the PYROCO™ process fall under the C2 contaminant grade. It should be noted that these guidelines are specifically for biosolids in land applications and not for biochar, and the stability of heavy metals in biochar might lead to different permissible limits. However, when compared with the International Biochar Initiative (IBI) Guidelines, both biosolids and biochar remained within the required limits [17].

Table 6. Heavy metal concentration (mg/kg) in biosolids and biochar.

Heavy Metals	Biosolids	Biochar (600 °C)	Biochar (700 °C)	VIC EPA [46]		International Biochar Initiative Guidelines [17]
				C1 Grade	C2 Grade	
Arsenic	<5	<5	5	20	60	13–100
Cadmium	1.4	1.2	1.1	1	10	1.4–39
Chromium	28	49	45	400	3000	93–1200
Copper	720	900	1000	100	2000	143–6000
Mercury	0.94	<0.05	<0.05	1	5	1–17
Nickel	23	28	28	60	270	47–420
Lead	20	39	40	300	500	121–300
Selenium	6	6	7	3	50	2–200
Zinc	1100	1500	1700	200	2500	416–7400

Contaminants of Emerging Concerns

The concentration of micropollutants such as PFAS, microplastics, pharmaceuticals, oestrogen-disrupting chemicals, siloxanes, and PAHs was measured in the biosolids and their derived biochar. These contaminants are typically present in biosolids in minute concentrations; however, their presence is of increasing concern due to their environmental toxicity and potential mobility to soil, air, and water media through biosolid land applications. The concentration of 28 targeted PFAS compounds in the biosolids and biochar is presented in Table 7. The total PFAS concentration significantly decreased from 134 ng/g in the biosolids to below detectable levels in the biochar. This remarkable reduction indicates that the PYROCO™ process is highly effective in reducing PFAS levels, thereby enhancing the environmental safety of the biochar produced. In addition to the PFAS concentrations, we also analysed the levels of pharmaceuticals, oestrogens, microplastics, siloxanes, and PAHs in the biosolids and biochar samples. The results are presented in Tables S4–S7 in the Supplementary Material. These comprehensive analyses provide further insights into the safety and environmental impact of the biochar produced through the PYROCO™ process. Microplastics were detected in the PYROCO™ biochar, as shown in Table S5, yet the results are considered inconclusive due to potential errors in sampling and the analytical methods employed by the laboratory.

Table 7. PFAS concentration (ng/g) in biosolids and biochar.

PFAS Compound Name	Limit of Reporting (LOR)	Biosolids	Biochar (600 °C)	Biochar (700 °C)
10:2 Fluorotelomer Sulfonic Acid	<2.5	<2.5	<2.5	<2.5
4:2 Fluorotelomer Sulfonic Acid	<2.5	<2.5	<2.5	<2.5
6:2 Fluorotelomer Sulfonic Acid	<2.5	<2.5	<2.5	<2.5
8:2 Fluorotelomer Sulfonic Acid	<2.5	<2.5	<2.5	<2.5
N-Ethyl Perfluorooctane Sulfonamido Acetic Acid (EtFOSAA)	<5	<5	<5	<5
N-Ethyl Perfluorooctane Sulfonamidoethanol (EtFOSE)	<5	<5	<5	<5
N-Methyl Perfluorooctane Sulfonamido Acetic Acid (MeFOSAA)	<10	12	<10	<10
N-Methyl Perfluorooctane Sulfonamidoethanol (MeFOSE)	<5	<5	<5	<5
N-Ethyl Perfluorooctane Sulfonamide (EtFOSA)	<5	<5	<5	<5
N-Methyl Perfluorooctane Sulfonamide (MeFOSA)	<5	<5	<5	<5
Perfluorobutane Sulfonic Acid (PFBS)	<2.5	9.6	<2.5	<2.5
Perfluorobutanoic Acid (PFBA)	<10	<10	<10	<10
Perfluorodecane Sulfonic Acid (PFDS)	<2.5	<2.5	<2.5	<2.5
Perfluorodecanoic Acid (PFDA)	<2.5	42	<2.5	<2.5
Perfluorododecanoic Acid (PFDoDA)	<2.5	5	<2.5	<2.5
Perfluoroheptane Sulfonic Acid (PFHpS)	<2.5	<2.5	<2.5	<2.5
Perfluoroheptanoic Acid (PFHpA)	<2.5	<2.5	<2.5	<2.5
Perfluorohexane Sulfonic Acid (PFHxS)	<2.5	<2.5	<2.5	<2.5
Perfluorohexanoic Acid (PFHxA)	<2.5	11	<2.5	<2.5
Perfluorononanoic Acid (PFNA)	<2.5	2.8	<2.5	<2.5
Perfluorooctane Sulfonamide (FOSA)	<2.5	<2.5	<2.5	<2.5
Perfluorooctane Sulfonic Acid (PFOS)	<2.5	26	<2.5	<2.5
Perfluorooctanoic Acid (PFOA)	<2.5	21	<2.5	<2.5
Perfluoropentane Sulfonic Acid (PFPeS)	<2.5	<2.5	<2.5	<2.5
Perfluoropentanoic Acid (PFPeA)	<2.5	10	<2.5	<2.5
Perfluorotetradecanoic Acid (PFTeDA)	<5	<5	<5	<5
Perfluorotridecanoic Acid (PFTrDA)	<2.5	<2.5	<2.5	<2.5
Perfluoroundecanoic Acid (PFUnDA)	<2.5	<2.5	<2.5	<2.5
Sum of PFAS	**<2.5**	**134**	**<2.5**	**<2.5**

Similarly, siloxanes were found to not be removed entirely from the biochar, as indicated in Table S6. This persistence could be attributed to siloxanes' thermal stability at pyrolysis temperatures. Given these uncertainties, it is essential to conduct more thorough investigations with controlled sampling and precise analyses in future studies to ascertain whether microplastics and siloxanes can be effectively destroyed during pyrolysis.

3.1.4. Stack Gas Emission Analysis

The stack gas emission composition comprising NOx, CO, CO_2, and O_2 was measured in real time, and the gas profile between 8 and 10 h of plant operations under stable conditions is shown in Figure 5. The NOx levels generally remained below 320 ppm under stable conditions but occasionally rose as high as 500 ppm. This increase was due to the unstable operation of the pilot plant and manual handling of the scrubber, which led to inadequate dosing of caustic soda or hydrogen peroxide. Additionally, high NOx emissions may result from fuel or prompt NOx (formed under fuel-rich conditions). This suggests that NOx emissions can be further reduced by a more controlled and automated caustic soda/hydrogen peroxide injection system with a modified scrubber (with an improved structured packing) in the next stage to ensure that NOx emissions are fully captured in the scrubber. The CO emission levels were generally below 150 ppm. However, there were instances where the CO emissions spiked up to 6000 ppm, primarily due to the manual control of the feed and process. The over- or underfeeding of biosolids led to fluctuations in CO levels. Improved process control and automation around feed and operations in the next stage of PYRO pilot trials will help regulate CO emissions more efficiently. The CO_2 levels generally ranged between 5 and 10 vol.%, while the O_2 fluctuated between 11 and 17 vol.%. These fluctuations were mainly caused by manual handling.

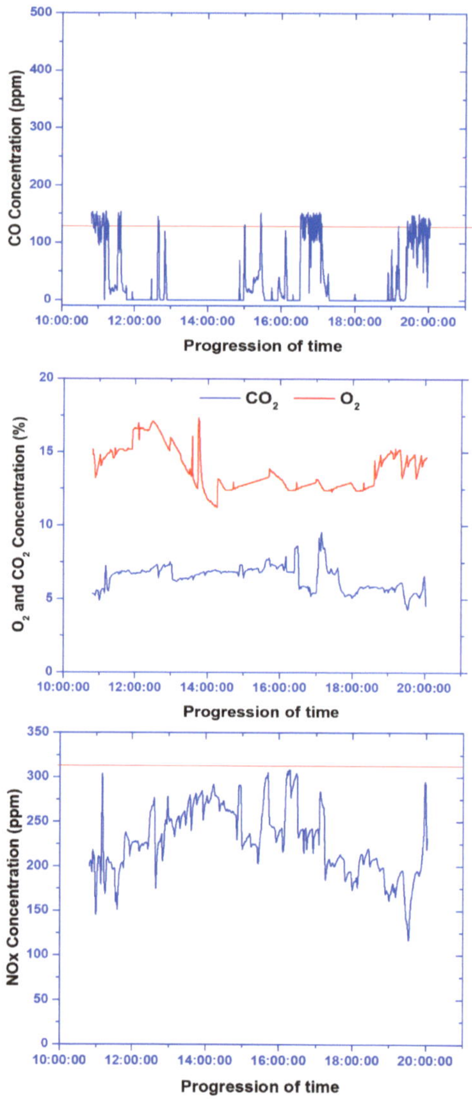

Figure 5. Gas emission profile under stable conditions.

3.1.5. Scrubber Water Analysis

The scrubber water produced in the PYROCO™ reactor system contains particulate matter, emissions such as NOx and SOx, and other impurities captured during flue gas scrubbing. To ensure environmental safety and regulatory compliance, scrubber water samples were collected and analysed for contaminants such as heavy metals, PFASs, polycyclic aromatic hydrocarbons (PAHs), and halides. This analysis, performed by accredited laboratories, ensures the water is properly characterised before disposal. Additionally, the scrubber water is stored in IBC tanks and will be treated or neutralised as required prior to disposal at a wastewater treatment plant. The potential environmental impacts of releasing untreated scrubber water are significant, as it may contain harmful contaminants that could affect the groundwater or ecosystems. Therefore, proper treatment and disposal are critical to meet regulatory standards and to minimise environmental risks.

We first evaluated the PFAS concentrations in the scrubber water against the NEMP 3.0 draft guidelines to ensure compliance, to protect human health, and to prevent ecological harm when the water is sent to a recycling plant. This assessment ensures that the water meets safety thresholds, preventing contamination and safeguarding the integrity of the recycling process. Figures 6 and 7 illustrate the PFAS concentrations in the scrubber water, evaluated against the NEMP 3.0 draft guidelines [47]. Specifically, Figure 6 compares the PFAS levels in the scrubber water with the thresholds for the drinking-water quality human health guidelines outlined in the NEMP 3.0 draft, while Figure 7 compares these concentrations to the Ecological Water Quality Guidelines specified in the same document [47]. The accepted threshold for the sum of the PFOS and PFHxS concentrations in drinking water is below 0.07 μg/L, while for recreational water quality, it is below 2 μg/L, according to these guidelines. Similarly, the accepted threshold for the sum of the PFOA concentrations in drinking water is below 0.56 μg/L, while for recreational water quality, it is below 10 μg/L (according to the NEMP 3.0 draft guidelines) [47]. Despite the presence of some PFASs in the scrubber water, their concentrations remained below the threshold limits of the NEMP 3.0 draft guidelines in both assessments [47].

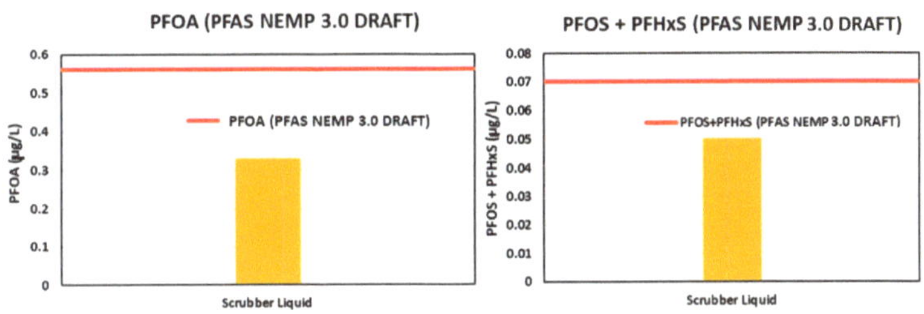

Figure 6. PFAS in scrubber water and comparison with Human Health Guidelines.

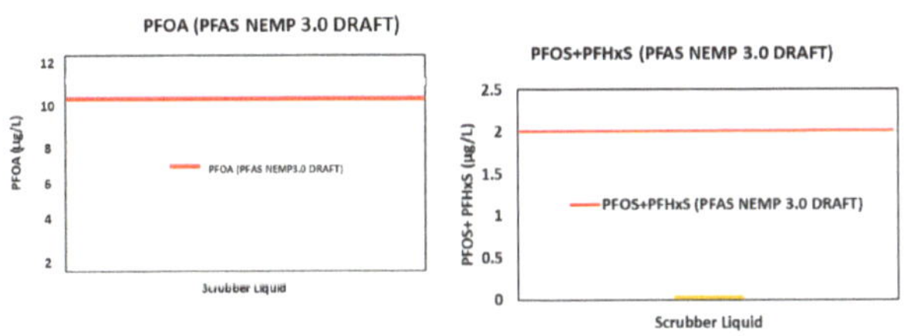

Figure 7. PFAS in scrubber water and comparison with Ecological Water Quality Guidelines (95% species protection).

The results of an analysis of the PAHs in the scrubber water obtained from the PYROCO™ pilot plant are detailed in Table S8 of the supplementary data. The PAHs detected in the scrubber water likely originate from the biosolids or the thermal processes (i.e., PYROCO®). The presence of PAHs may be attributed to the incomplete combustion of pyrolysis oil and gas vapours during the pilot plant trials. This incomplete combustion was evidenced by CO concentrations reaching as high as 6000 ppm, primarily due to process disturbances caused by manual operations. To mitigate PAH levels, achieving complete combustion in subsequent stages is crucial. This can be accomplished by improving process automation and incorporating an afterburner or thermal oxidiser post-PYROCO®.

Several PAH levels in the scrubber water exceeded the recommended limits set by the National Institute for Public Health and the Environment (NIPHE), rendering the water potentially unsafe for direct discharge into natural water bodies due to its aquatic, benthic, and terrestrial toxicity.

The US EPA guidelines do not specify minimum reporting requirements for PAHs, but given their toxicity, the scrubber water should be treated as trade wastewater. Moreover, the heavy metal concentration in the scrubber water was also analysed, and the results are presented in Table S9 in the Supplementary Materials. Furthermore, the halides and organic compounds in the scrubber water were also analysed, and the results are shown in Table S10 in the Supplementary Materials.

3.2. Validation and Results of ASPEN Process Model

Table 8 provides the mole compositions of CO_2, H_2, CO, and O_2 at the GP and PYROCO™ exit, as obtained from the Aspen Plus simulation and the pilot plant trials. The results show that the Aspen Plus simulation data are within the range of the pilot plant trial data, demonstrating consistency and validating the accuracy of the simulation model.

Table 8. Comparison between pilot plant trial data and simulation data of GP and PYROCO™ exit gas composition.

	GP Gas Profiles (%)		PYROCO™ Exit Gas Profile (%)	
	Pilot Plant Data	Simulation Data	Pilot Plant Data	Simulation Data
CO_2	9–11.5	11.2	11–17	12
H_2	4–5	5		
O_2	0–1	5.5×10^{-12}	5–10	7.0
CO	10–13	12.9	0–0.015	9.80×10^{-7}
NOx			0.014–0.03	0.015

3.2.1. Mass and Energy Balance

The mass and energy balance data for the pyrolysis of the biosolids are shown in Table 9. Here, the solid content in the biosolids was assumed to be 50 wt.%, and the GP:PYROCO™ ratio was assumed to be 0.4. As shown in Table 9, this plant can process 10 tons of biosolids per day and can produce 1.7 tons of biochar from PYROCO™ and 0.7 tons of biochar from GP.

As shown in Table 9, the GP operates in autothermal mode without requiring an external energy input. The PYROCO™ reactor also functions without an external energy supply, as the partial combustion of volatile products in its outer shell provides the heat required for the pyrolysis reactor. The remaining volatiles are then combusted in the thermal oxidiser using excess air. The heat from the flue gas exiting the thermal oxidiser is used to dry the incoming wet biosolids. Any residual heat remaining in the flue gas after indirect drying is recovered, leaving an exhaust temperature of 400 °C. This low-grade heat can be used for various operations in wastewater treatment plants or office buildings. According to Table 9, the total exergy from this plant is 294 kW, providing an energy efficiency of 84% when operating at a 0.4 GP:PYROCO ratio with the biosolids containing 50% solids and a calorific value of 16 MJ/kg.

Table 9. Mass and energy balance for pyrolysis of biosolids at a 50 wt.% solid content and 0.4 GP:PYROCO™ feed fraction.

Details	Mass Flow Rate (kg/h)		Energy (kW)		
	In	Out	In	Out: Loss	Out: Gen.
Biosolids	417	-	943.6		
Dryer	-	172		107.8	
Gas producer (autothermal)	147	29			
Pyrolyzer+thermal oxidizer	2145	69			
Heat loss in GP and PYROCO®				19.6	
Thermal energy in biochar				192.4	
Thermal energy in ash				15.9	
Low-grade heat recovery (LGHR)					313.7
Exergy losses				294.1	
Exhaust	-	2439			
Total	2708	2708	943.6	629.9	313.7
Difference (mass in–out)	0				

3.2.2. Extended Energy Analysis Using Validated ASPEN Process Model

The energy analysis provided insights into the PYROCO™ process's energy requirements and consumption patterns. In the proposed PYROCO™ process, the energy required for pyrolysis is supplied by the partial combustion of volatiles generated in the pyrolysis unit. The energy required for drying is supplied by the heat available in the flue gas, leaving the thermal oxidiser. The energy generated or required in this process varies with the biosolids' calorific value and the feed ratio between the GP and PYROCO™ reactors (GP:PYROCO). Figure 8 shows how these parameters affect the amount of energy generated/required in the process.

Energy production in the PYROCO™ process depends on the calorific value of biosolids. The source of biosolids, their compositions, and stabilisation methods affect biosolids' heating value. Hence, a biosolid's calorific value typically ranges from 10–20 MJ/kg. The results of the analysis, presented in Figure 8, considers four biosolids having 11, 14, 16, and 20 MJ/kg of HHV. Figure 8 shows that the PYROCO™ process can usually operate without external energy for most biosolids with varying calorific values, and the thermal energy production increases with the biosolids' calorific value. According to Figure 8, at 0.2 and 0.4 GP:PYROCO ratios, the process cannot operate without an external energy source when the heating value of the biosolids is as low as 11 MJ/kg and when the solid content in the biosolids is lower than 30%. However, biosolids with such low-calorific values can still be processed without external energy in the PYROCO™ process by adjusting the GP:PYROCO ratio to a higher value.

As shown in Figure 8, the energy generated by the process increases with a higher GP:PYROCO ratio. At ratios greater than 0.5, all four feedstocks can be considered self-sufficient. As mentioned above, even when biosolids have a lower solid content or heating value, increasing the GP:PYROCO ratio allows the system to operate efficiently. However, higher GP:PYROCO ratios mean more biosolids are fed to the GP instead of the PYROCO™ reactor. This reduces the overall biochar yield, because gasification produces less biochar than pyrolysis.

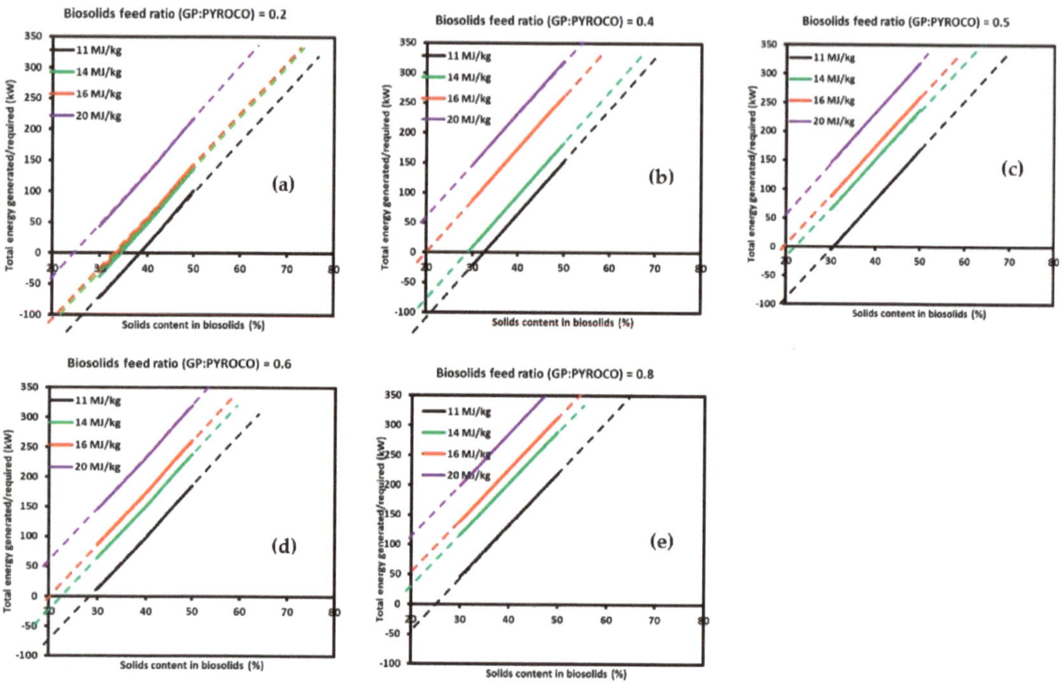

Figure 8. Energy analysis in PYROCO™ process at a GP:PYROCO ratio of (**a**) 0.2, (**b**) 0.4, (**c**) 0.5, (**d**) 0.6, and (**e**) 0.8 for biosolids with heating values of 11 MJ/kg, 14 MJ/kg, 16 MJ/kg, and 20 MJ/kg.

Furthermore, biochar produced in GP has a high ash content, lower carbon content, and lower calorific value than those of biochar from pyrolysis, leading to a potentially lower market value. Thus, adjusting the GP:PYROCO ratio depends on the following specific requirements: either a higher biochar yield or surplus energy. As a higher GP:PYROCO ratio directs more biosolids to the GP, it increases the energy surplus, because gasification in the GP generates more volatiles than in pyrolysis. However, this also leads to less energy generated in the PYROCO™ reactor from the partial combustion of volatiles, which has a negative effect on the energy surplus. However, the net effect of these two factors was still positive, increasing the surplus energy in the process.

3.3. Economic Analysis

This section examines the influence of key economic parameters, such as the biochar sale price, the calorific value of biosolids, and the GP:PYROCO ratio on the NPV of the pyrolysis of biosolids. A positive NPV was observed with the current economic assumptions and parameters considered in all the scenarios addressed. NPV = 0 when the rate of return generated by the project matches the cost of the capital or the required rate of return. At this point, the project neither gains nor loses value from a financial perspective, as the expected returns are exactly offset by the costs, leaving no net benefit or loss in present value terms.

3.3.1. Effect of Biochar Sale Price on NPV

Figure 9 presents the NPV results for varying biochar sale prices from AUD 300 to AUD 1000 per tonne under the operational settings of a 0.5 GP:PYROCO ratio and 50% solid content of biosolids having a calorific value of 20 MJ/kg. The financial analysis maintained the discount rate at 8% and the tax rate at 25%. When the biochar price increases from AUD 300 per tonne to AUD 1000 per tonne, the NPV improves from AUD 0.24 million to

AUD 4.31 million, and the payback period decreases from 26 years to 12 years. However, the current market price for biochar is not as high as AUD 1000 per tonne. To enhance the economic viability of the investment, exploring higher-value applications of biochar, such as in adsorption processes, catalysis, or as additives in rubber compounding, could open up new market opportunities for biosolid-derived biochar.

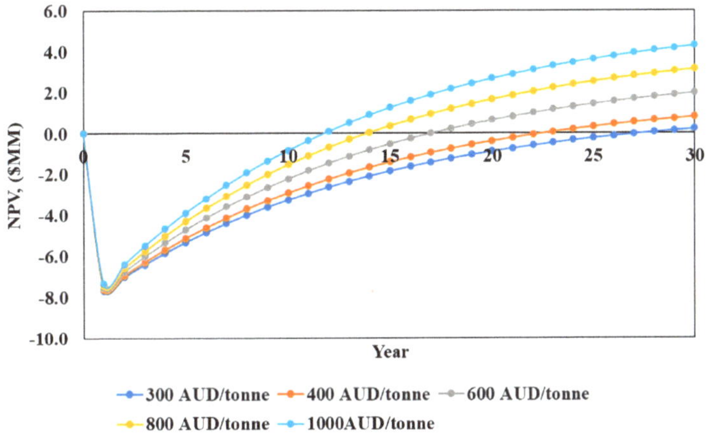

Figure 9. Effect of the biochar sale price on NPV at a 0.5 GP:PYROCO ratio and 50% solid content of biosolids and 20 MJ/kg of HHV.

Furthermore, the BCR increased from 0.96 to 1.8 when the biochar sale price rose from AUD 300/tonne to AUD 1000/tonne, reflecting the strong impact of biochar pricing on project viability. For the project to achieve a BCR greater than 1 (indicating economic viability), the biochar sale price must be at least AUD 400/tonne or higher. This shows the importance of biochar market development and pricing strategies in ensuring the financial sustainability of the PYROCO process, as the price point of biochar directly influences the project's profitability.

3.3.2. Effect of Biosolids' Calorific Value on NPV

Figure 10 shows the NPV values for biosolids with varying calorific values, from 11 MJ/kg to 20 MJ/kg, under the same operational settings used in the biochar pricing analysis but with a fixed biochar sale price of AUD 300 per tonne. The effect of the biosolids' calorific value on the NPV is more subdued than the impact of the biochar sale price. As the calorific value of the biosolids increases from 11 MJ/kg to 20 MJ/kg, the NPV rises from AUD 0.31 million to AUD 0.82 million, and the payback period shortens from 26 to 23 years. Although higher calorific values lead to more energy generation, thereby increasing the revenue from electricity production, they also result in a lower biochar yield. This decrease in biochar production subsequently reduces the revenue generated from biochar sales, balancing out the benefits from increased energy generation. This nuanced impact suggests that while enhancing the calorific value does contribute to economic performance, it does not have a dramatic effect compared to changes in the biochar sale price.

The BCR also improved from 0.95 to 1 when the calorific value of the biosolids increased from 11 MJ/kg to 20 MJ/kg, demonstrating that biosolids with a high calorific value improve the profitability of the investment.

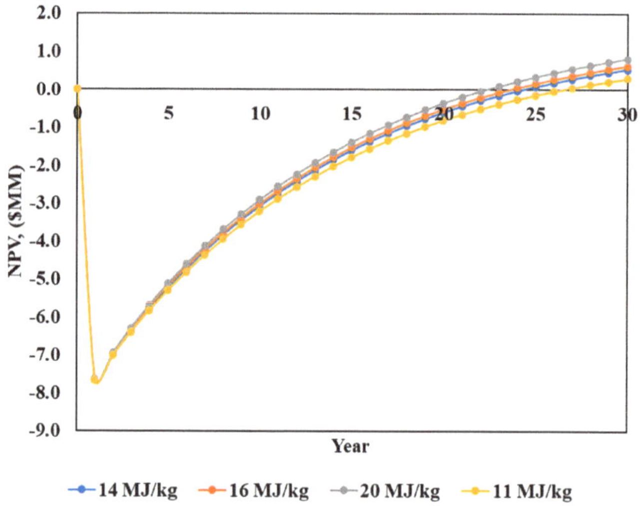

Figure 10. Effect of the biosolids' calorific value on NPV at a 0.5 GP:PYROCO ratio and 50% solid content of biosolids with AUD 400/tonne biochar sale price.

3.3.3. Effect of GP:PYROCO Ratio on NPV

The GP:PYROCO ratio is another important parameter in the PYROCO™ process. Therefore, the influence of this ratio on the NPV and the payback period is depicted in Figure 11. The data reveal that a decrease in the GP:PYROCO ratio leads to an increase in the NPV, suggesting that an economically optimal operation involves feeding more biosolids to PYROCO™ and less to GP. This operational adjustment results in a higher biochar yield and a reduction in electricity generation, underscoring that biochar sales yield a more substantial impact on the economic feasibility of the PYROCO™ plant. Therefore, directing more biosolids to PYROCO®, when feasible, enhances the plant's financial performance by maximising revenue from biochar sales over energy generation.

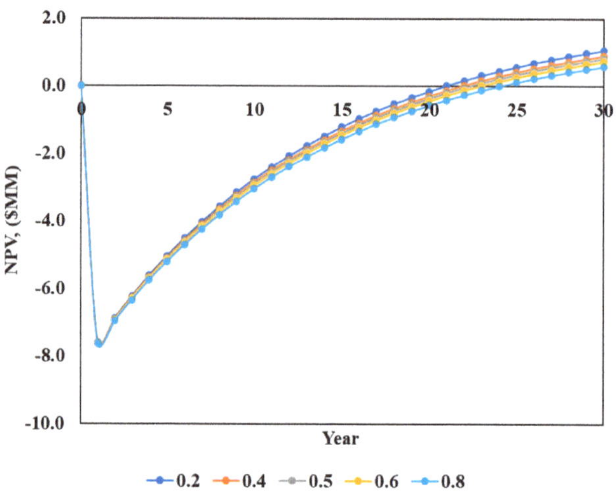

Figure 11. Effect of GP:PYROCO ratio on NPV at a 50% solid content of biosolids with a 20 MJ/kg calorific value and AUD 400/tonne biochar sale price.

The BCR increased from 0.97 to 1.02 when the GP:PYROCO ratio decreased from 0.8 to 0.2, indicating that a lower GP:PYROCO ratio improves the financial viability of the process. For the BCR to consistently exceed 1, the GP:PYROCO ratio must be kept below 0.5.

In addition, it was calculated that the cost to process biosolids through the PYROCO process was only AUD 205/ton. This is comparable to the gasification study by Alves et al., which reported a cost range of AUD 158–363/ton [48]. Both processes involve the thermal treatment of biosolids, making the cost range in Alves et al.'s study relevant. This suggests that the PYROCO process falls within an established cost range for similar gasification technologies, supporting its economic feasibility.

3.3.4. Sensitivity Analysis of Different Economic Factors

A sensitivity analysis was performed on various economic factors, as shown in Figure 12. OPEX is the most significant factor influencing the techno-economic estimate of the PYROCO™ process, as indicated by the steepest slope among the evaluated parameters. An increase in OPEX of around 35% would render the process non-feasible, with a negative NPV. CAPEX is the second most influential factor, and a 40% increase in CAPEX generates a negative NPV. A by-product credit from electricity sales has the least influence on the plant's NPV. A ±50% variation in the revenue from electricity sales caused a <20% change in the NPV. The income tax rate and biochar selling price have a similar effect on the process; however, even with a 50% variation, these parameters would not result in a negative NPV. Therefore, optimising the economic performance of PYROCO™ primarily depends on improving both OPEX and CAPEX.

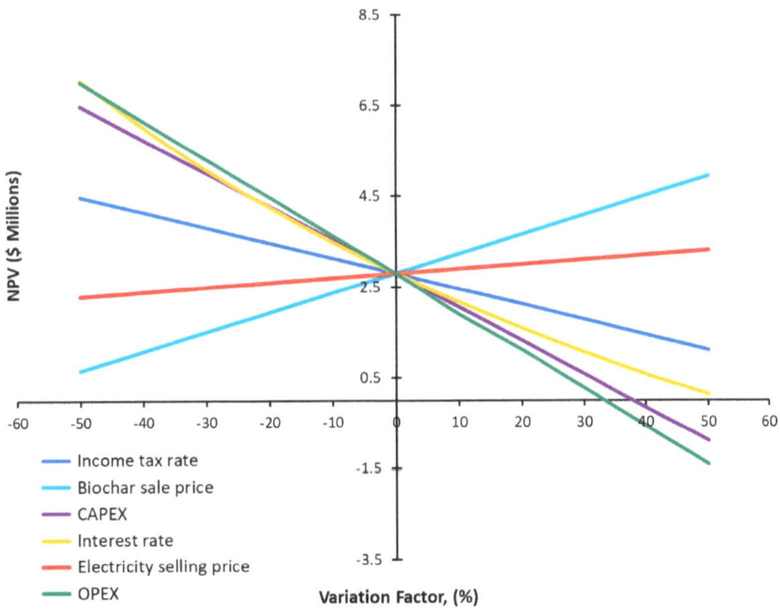

Figure 12. Results of sensitivity analysis conducted to identify the effects of a varying income tax rate, biochar sale price, CAPEX, interest rate, electricity selling price, and OPEX on the NPV of the PYROCO™ process. The base value of the parameters is as follows: income tax rate, 30%; biochar sale price, 600 AUD/tonne; CAPEX, AUD 8.38 million; interest rate, 5.3%; electricity selling price, 15 c/kWh; OPEX, AUD 0.81 million; and NPV, AUD 2.8 million.

4. Conclusions

- The PYROCO™ pilot trials successfully produced biochar with a 30% mass yield and desirable agronomic and physicochemical properties. The process produced biochar with high organic carbon and nutrient contents, the complete elimination of PFASs, and undetectable levels of PAHs, pathogens, and oestrogens. However, challenges in accurately analysing microplastics and siloxanes in biochar were identified, highlighting the need for enhanced analytical methods in subsequent trials.
- Mass and energy balances confirmed the PYROCO™ reactor's capability for energy self-sufficiency with minimal reliance on external energy sources.
- The semi-empirical process modelling of a commercial-scale PYROCO™ plant further demonstrated that the process could operate in a thermal energy neutral mode with biosolids containing a solid content above 30%. For biosolids with a lower solid content, the system could still achieve energy self-sufficiency by adjusting the GP:PYROCO feed ratio, albeit at the cost of reduced biochar production.
- Higher calorific values of biosolids generated greater surplus energy, enhancing the process's overall energy surplus. This finding underscores the importance of the biosolid feed quality in optimising this process's performance.
- We conducted a preliminary economic analysis that focused on the potential revenue streams from biochar and electricity sales. This analysis suggested that while the current market prices for biochar do not offer substantial financial returns, exploring high-value applications for biochar could significantly improve the economic viability of the PYROCO™ process.
- The key economic drivers identified include the biochar sale price, the calorific value of biosolids, and the GP:PYROCO ratio. These factors influenced the NPV, indicating potential areas for strategic improvements to enhance economic outcomes.
- The results of the current techno-economic analysis should be used primarily for benchmarking purposes. For more definitive investment decisions, a more detailed study should be carried out involving technology providers to obtain detailed capital and operating cost estimates through a competitive tendering process.

Supplementary Materials: The following supporting information can be downloaded at: https://www.mdpi.com/article/10.3390/w16233399/s1. Figure S1: Continuous milling and sieving setup used in pilot trials. Table S1: Proximate and ultimate analysis of biosolids feedstocks used in Aspen plus modelling. Table S2: Capital cost estimates for the PYROCO plant. Table S3: Parameters considered for economic analysis. Table S4: Pharmaceuticals and estrogens biosolids and biochar samples. Table S5: Micro-plastics in biosolids and biochar samples. Table S6: Siloxanes in biosolids and biochar samples. Table S7: PAHs in biosolids and biochar. Table S8. PAHs in scrubber water. Table S9: Metals in scrubber water. Table S10: Halides and organics in scrubber water.

Author Contributions: Conceptualization, N.R., S.P., G.V. and K.S.; methodology, N.R., S.P., I.G.H., G.V., S.A. and K.S.; software, N.R., S.P., I.A.-W. and S.A.; validation, I.G.H., G.V., A.K.V., A.S. and K.S; formal analysis, N.R., S.P., I.G.H. and S.A.; investigation, N.R., S.P., I.G.H., G.V. and A.K.V.; resources, S.P., A.S., D.B. and K.S.; data curation, N.R., I.G.H. and I.A.-W.; writing—original draft preparation, N.R., S.P. and I.G.H.; writing—review and editing, G.V., I.A.-W., S.A., A.K.V., A.S. and K.S.; visualisation, N.R., S.P., I.G.H., I.A.-W. and S.A.; supervision, A.S., D.B. and K.S.; project administration, S.P. and K.S.; funding acquisition, K.S. All authors have read and agreed to the published version of the manuscript.

Funding: Funding for the PYROCO™ Mark 1 pilot plant development and trials was received from South East Water, Intelligent Water Networks, Greater Western Water, and RMIT University, Australia.

Data Availability Statement: The original contributions presented in this study are included in the article/Supplementary Material, and further inquiries can be directed to the corresponding author/s.

Acknowledgments: The authors wish to acknowledge the support of technical operators who participated in the PYROCO™ pilot plant trials. The use of the Scanning Electron Microscope instrument in the RMIT Microanalysis and Microscopy Facility based at RMIT University Australia is acknowledged.

Conflicts of Interest: Authors Aravind Surapaneni and David Bergmann were employed by South East Water interested in biosolids management via pyrolysis technique. The remaining authors declare that the research was conducted in the absence of any commercial or financial relationships that could be construed as a potential conflict of interest.

References

1. Statistics, A.B. Australian Biosolids Statistics. Guidelines 2023. Available online: https://www.biosolids.com.au/guidelines/australian-biosolids-statistics/ (accessed on 27 October 2024).
2. Tezel, U.; Tandukar, M.; Pavlostathis, S.G. Anaerobic Biotreatment of Municipal Sewage Sludge. In *Comprehensive Biotechnology*; Georgia Institute of Technology: Atlanta, GA, USA, 2011; pp. 447–461.
3. Wang, H.; Brown, S.L.; Magesan, G.N.; Slade, A.H.; Quintern, M.; Clinton, P.W.; Payn, T.W. Technological options for the management of biosolids. *Environ. Sci. Pollut. Res. Int.* **2008**, *15*, 308–317. [CrossRef] [PubMed]
4. Agrafioti, E.; Bouras, G.; Kalderis, D.; Diamadopoulos, E. Biochar production by sewage sludge pyrolysis. *J. Anal. Appl. Pyrolysis* **2013**, *101*, 72–78. [CrossRef]
5. Barry, D.; Barbiero, C.; Briens, C.; Berruti, F. Pyrolysis as an economical and ecological treatment option for municipal sewage sludge. *Biomass Bioenergy* **2019**, *122*, 472–480. [CrossRef]
6. Liu, Z.; McNamara, P.; Zitomer, D. Autocatalytic Pyrolysis of Wastewater Biosolids for Product Upgrading. *Environ. Sci. Technol.* **2017**, *51*, 9808–9816. [CrossRef] [PubMed]
7. Lu, H.; Zhang, W.; Wang, S.; Zhuang, L.; Yang, Y.; Qiu, R. Characterisation of sewage sludge-derived biochars from different feedstocks and pyrolysis temperatures. *J. Anal. Appl. Pyrolysis* **2013**, *102*, 137–143. [CrossRef]
8. Patel, S.; Kundu, S.; Paz-Ferreiro, J.; Surapaneni, A.; Fouche, L.; Halder, P.; Setiawan, A.; Shah, K. Transformation of biosolids to biochar: A case study. *Environ. Prog. Sustain. Energy* **2018**, *38*, 13113. [CrossRef]
9. Paz-Ferreiro, J.; Nieto, A.; Méndez, A.; Askeland, M.P.J.; Gascó, G. Biochar from Biosolids Pyrolysis: A Review. *Int. J. Environ. Res. Public Health* **2018**, *15*, 956. [CrossRef]
10. Sánchez, M.; Martínez, O.; Gómez, X.; Morán, A. Pyrolysis of mixtures of sewage sludge and manure: A comparison of the results obtained in the laboratory (semi-pilot) and in a pilot plant. *Waste Manag.* **2007**, *27*, 1328–1334. [CrossRef]
11. Agarwal, M.; Tardio, J.; Mohan, S.V. Effect of pyrolysis parameters on yield and composition of gaseous products from activated sludge: Towards sustainable biorefinery. *Biomass Convers. Biorefinery* **2014**, *5*, 227–235. [CrossRef]
12. Brown, J.N. Development of a lab-scale auger reactor forbiomass fast pyrolysis and process optimizationusing response surface methodology. In *Mechanical Engineering; Biorenewable Resources and Technology*; Iowa State University: Ames, IA, USA, 2009.
13. Khanmohammadi, Z.; Afyuni, M.; Mosaddeghi, M.R. Effect of pyrolysis temperature on chemical and physical properties of sewage sludge biochar. *Waste Manag. Res.* **2015**, *33*, 275–283. [CrossRef]
14. Ledakowicz, S.; Stolarek, P.; Malinowski, A.; Lepez, O. Thermochemical treatment of sewage sludge by integration of drying and pyrolysis/autogasification. *Renew. Sustain. Energy Rev.* **2019**, *104*, 319–327. [CrossRef]
15. Gianico, A.; Braguglia, C.M.; Gallipoli, A.; Montecchio, D.; Mininni, G. Land Application of Biosolids in Europe: Possibilities, Con-Straints and Future Perspectives. *Water* **2021**, *13*, 103. [CrossRef]
16. Australian and New Zeland Biosolids Partnership. Land Application of BiosolidsFact Sheet. Available online: https://www.biosolids.com.au/wp-content/uploads/Land-application-of-biosolids.pdf (accessed on 27 October 2024).
17. Yang, Y.; Meehan, B.; Shah, K.; Surapaneni, A.; Hughes, J.; Fouché, L.; Paz-Ferreiro, J. Physicochemical Properties of Biochars Produced from Biosolids in Victoria, Australia. *Int. J. Environ. Res. Public Health* **2018**, *15*, 1459. [CrossRef]
18. de Figueiredo, C.C.; Pinheiro, T.D.; de Oliveira, L.E.Z.; de Araujo, A.S.; Coser, T.R.; Paz-Ferreiro, J. Direct and residual effect of biochar derived from biosolids on soil phosphorus pools: A four-year field assessment. *Sci. Total Environ.* **2020**, *739*, 140013. [CrossRef] [PubMed]
19. Nicomel, N.R.; Li, L.Y.; Du Laing, G. Biosolids-based activated carbon for enhanced copper removal from citric-acid-rich aqueous media. *Environ. Sci. Pollut. Res.* **2022**, *29*, 74742–74755. [CrossRef] [PubMed]
20. Mong, G.R.; Chong, C.T.; Chong, W.W.F.; Ng, J.-H.; Ong, H.C.; Ashokkumar, V.; Tran, M.-V.; Karmakar, S.; Goh, B.H.H.; Yasin, M.F.M. Progress and challenges in sustainable pyrolysis technology: Reactors, feedstocks and products. *Fuel* **2022**, *324*, 124777. [CrossRef]
21. Idris, J.; Shirai, Y.; Andou, Y.; Mohd Ali, A.A.; Othman, M.R.; Ibrahim, I.; Yamamoto, A.; Yasuda, N.; Hassan, M.A. Successful scaling-up of self-sustained pyrolysis of oil palm biomass under pool-type reactor. *Waste Manag. Res.* **2016**, *34*, 176–180. [CrossRef] [PubMed]

22. Thoma, E.D.; Wright, R.S.; George, I.; Krause, M.; Presezzi, D.; Villa, V.; Preston, W.; Deshmukh, P.; Kauppi, P.; Zemek, P.G. Pyrolysis processing of PFAS-impacted biosolids, a pilot study. *J. Air Waste Manag. Assoc.* **2022**, *72*, 309–318. [CrossRef]
23. Kundu, S.; Patel, S.; Halder, P.; Patel, T.; Marzbali, M.H.; Pramanik, B.K.; Paz-Ferreiro, J.; de Figueiredo, C.C.; Bergmann, D.; Surapaneni, A.; et al. Removal of PFASs from biosolids using a semi-pilot scale pyrolysis reactor and the application of biosolids derived biochar for the removal of PFASs from contaminated water. *Environ. Sci. Water Res. Technol.* **2021**, *7*, 638–649. [CrossRef]
24. Liu, Z.; Singer, S.; Zitomer, D.; McNamara, P. Sub-Pilot-Scale Autocatalytic Pyrolysis of Wastewater Biosolids for Enhanced Energy Recovery. *Catalysts* **2018**, *8*, 524. [CrossRef]
25. Sarvi, M.; Kainulainen, A.; Malk, V.; Kaseva, J.; Rasa, K. Industrial pilot scale slow pyrolysis reduces the content of organic contaminants in sewage sludge. *Waste Manag.* **2023**, *171*, 95–104. [CrossRef] [PubMed]
26. Schlederer, F.; Martín-Hernández, E.; Vaneeckhaute, C. Ensuring safety standards in sewage sludge-derived biochar: Impact of pyrolysis process temperature and carrier gas on micropollutant removal. *J. Environ. Manag.* **2024**, *352*, 119964. [CrossRef] [PubMed]
27. Campuzano, F.; Brown, R.C.; Martínez, J.D. Auger reactors for pyrolysis of biomass and wastes. *Renew. Sustain. Energy Rev.* **2019**, *102*, 372–409. [CrossRef]
28. Chen, J.C.; Grace, J.R.; Golriz, M.R. Heat transfer in fluidised beds: Design methods. *Powder Technol.* **2005**, *150*, 123–132. [CrossRef]
29. Flegkas, S.; Birkelbach, F.; Winter, F.; Groenewold, H.; Werner, A. Profitability Analysis and Capital Cost Estimation of a Thermochemical Energy Storage System Utilising Fluidised Bed Reactors and the Reaction System $MgO/Mg(OH)_2$. *Energies* **2019**, *12*, 4788. [CrossRef]
30. Onarheim, K.; Solantausta, Y.; Lehto, J. Process Simulation Development of Fast Pyrolysis of Wood Using Aspen Plus. *Energy Fuels* **2014**, *29*, 205–217. [CrossRef]
31. Li, T.Y.; Xiang, H.; Yang, Y.; Wang, J.; Yildiz, G. Prediction of char production from slow pyrolysis of lignocellulosic biomass using multiple nonlinear regression and artificial neural network. *J. Anal. Appl. Pyrolysis* **2021**, *159*, 105286. [CrossRef]
32. Craig, K.R.; Mann, M.K. *Cost and Performance Analysis of Biomass-Based Integrated Gasification Combined-Cycle (BIGCC) Power Systems;* National Renewable Energy Laboratory: Golden, CO, USA, 1996.
33. PLANT COST INDEX. 2020. Available online: https://www.chemengonline.com/site/plant-cost-index/ (accessed on 27 October 2024).
34. Nieder-Heitmann, M.; Savadkouhi, S.S.; Venderbosch, R.; Leijenhorst, E.; van der Pol, E.; Vleeming, H. Technoeconomic Feasibility of a Sunflower Husk Fast Pyrolysis Value Chain for the Production of Advanced Biofuels. *Bioenergy Biofuels* **2022**, *36*, 13084–13093. [CrossRef]
35. Kazi, F.K.; Fortman, J.A.; Anex, R.P.; Hsu, D.D.; Aden, A.; Dutta, A.; Kothandaraman, G. Techno-economic comparison of process technologies for biochemical ethanol production from corn stover. *Fuel* **2010**, *89*, S20–S28. [CrossRef]
36. Steve, T.W. Biosolids and Residuals Management. In *Municipal Technology Branc;* United States Environmental Protection Agency: Washington, DC, USA, 2000.
37. Patel, S.; Kundu, S.; Halder, P.; Veluswamy, G.; Pramanik, B.; Paz-Ferreiro, J.; Surapaneni, A.; Shah, K. Slow pyrolysis of biosolids in a bubbling fluidised bed reactor using biochar, activated char and lime. *J. Anal. Appl. Pyrolysis* **2019**, *144*, 104697. [CrossRef]
38. Isemin, R.L.; Kuzmin, S.N.; Viryasov, D.M. Co-Fluidization of Fine Particles and Straw Pellets at Room and Elevated Temperatures. *Int. J. Chem. Eng. Appl.* **2012**, *3*, 323–327. [CrossRef]
39. Mian, I.; Li, X.; Jian, Y.; Dacres, O.D.; Zhong, M.; Liu, J.; Ma, F.; Rahman, N. Kinetic study of biomass pellet pyrolysis by using distributed activation energy model and Coats Redfern methods and their comparison. *Bioresour. Technol.* **2019**, *294*, 122099. [CrossRef] [PubMed]
40. Paulionytė, J.; Vaiškūnaitė, R. Research on the physical and chemical properties of sewage treatment sludge biochar and its preparation for wastewater. *Moksl.–Liet. Ateitis/Sci.–Future Lith.* **2023**, *15*, 1–6. [CrossRef]
41. Phillips, C.L.; Meyer, K.M.; Garcia-Jaramillo, M.; Weidman, C.S.; Stewart, C.E.; Wanzek, T.; Grusak, M.A.; Watts, D.W.; Novak, J.; Trippe, K.M. Towards predicting biochar impacts on plant-available soil nitrogen content. *Biochar* **2022**, *4*, 9. [CrossRef]
42. Rathnayake, N.; Patel, S.; Halder, P.; Aktar, S.; Pazferreiro, J.; Sharma, A.; Surapaneni, A.; Shah, K. Co-pyrolysis of biosolids with alum sludge: Effect of temperature and mixing ratio on product properties. *J. Anal. Appl. Pyrolysis* **2022**, *163*, 105488. [CrossRef]
43. Rathnayake, N.; Patel, S.; Hakeem, I.G.; Pazferreiro, J.; Sharma, A.; Gupta, R.; Rees, C.; Bergmann, D.; Blackbeard, J.; Surapaneni, A.; et al. Co-pyrolysis of biosolids with lignocellulosic biomass: Effect of feedstock on product yield and composition. *Process Saf. Environ. Prot.* **2023**, *173*, 75–87. [CrossRef]
44. Fan, X.; Zhang, J.; Xie, Y.; Xu, D.; Liu, Y.; Liu, J.; Hou, J. Biochar produced from the co-pyrolysis of sewage sludge and waste tire for cadmium and tetracycline adsorption from water. *Water Sci. Technol.* **2021**, *83*, 1429–1445. [CrossRef]
45. Jin, J.; Wang, M.; Cao, Y.; Wu, S.; Liang, P.; Li, Y.; Zhang, J.; Zhang, J.; Wong, M.H.; Shan, S.; et al. Cumulative effects of bamboo sawdust addition on pyrolysis of sewage sludge: Biochar properties and environmental risk from metals. *Bioresour. Technol.* **2017**, *228*, 218–226. [CrossRef]
46. Biosolids Land Application. In *Guidelines for Environmental Management;* EPA Victoria: Southbank, Victoria, Australia, 2004.

47. HEPA. Draft PFAS National Environmental Management Plan, Version 3.0. 2022. Available online: https://consult.dcceew.gov.au/nemp-pfas (accessed on 27 October 2024).
48. Alves, O.; Calado, L.; Panizio, R.M.; Gonçalves, M.; Monteiro, E.; Brito, P. Techno-economic study for a gasification plant processing residues of sewage sludge and solid recovered fuels. *Waste Manag.* **2021**, *131*, 148–162. [CrossRef]

water

MDPI

Article

Designing, Modeling and Developing Scale Models for the Treatment of Water Contaminated with Cr (VI) through Bacterial Cellulose Biomass

Uriel Fernando Carreño Sayago *, Vladimir Ballesteros Ballesteros and Angelica Maria Lozano Aguilar

Faculty of Engineering and Basic Sciences, Fundación Universitaria los Libertadores, Bogotá 111221, Colombia; vladimir.ballesteros@libertadores.edu.co (V.B.B.); amlozanoa@libertadores.edu.co (A.M.L.A.)
* Correspondence: ufcarrenos@libertadores.edu.co

Abstract: The present research presents a method for scaling up a continuous treatment system with bacterial cellulose biomass for the removal of contaminants on a large industrial scale from effluents loaded with chromium (VI). This consisted of a laboratory-scale modeling process of the chromium (VI) adsorption processes, which would provide the necessary parameters to build a system on an industrial scale. The research also involved designing, modeling and developing scale models for the treatment of water contaminated with chromium (VI) through bacterial cellulose biomass. The results of the model indicated the specific route for the construction of a treatment system on an industrial scale, with the experimental data adjusted to achieve this objective. The pilot scale prototype was built using 450 g of biomass, including elution processes, with the data obtained from the aforementioned processes. In general, the excellent efficiency of the two models at different scales, together with the excellent elution results, suggests that this prototype could be presented to polluting industries for the treatment of water from different industrial effluents, being an advanced biotechnology for the treatment of industrial wastewater.

Keywords: cellulose bacterial; chromium; isoterm; kinetics

Citation: Sayago, U.F.C.; Ballesteros, V.B.; Aguilar, A.M.L. Designing, Modeling and Developing Scale Models for the Treatment of Water Contaminated with Cr (VI) through Bacterial Cellulose Biomass. *Water* **2024**, *16*, 2524. https://doi.org/10.3390/w16172524

Academic Editors: Issam A. Al-Khatib, Rehab O. Abdel Rahman, Tsuyoshi Imai and Yung-Tse Hung

Received: 28 June 2024
Revised: 26 August 2024
Accepted: 28 August 2024
Published: 5 September 2024

1. Introduction

Caring for water is a fundamental and intrinsic aspect of human existence. The conservation and reuse of water are crucial for the sustainability of our planet and the well-being of future generations. To achieve this, it is essential to implement effective treatment systems for industrial effluents to prevent irreversible damage to water systems, such as wetlands and rivers. Pollutants that cause these impacts are various heavy metals, which cause serious damage to the ecosystem in both animals and aquatic plants due to their bioaccumulation [1–4]. Therefore, it is of the utmost importance to be able to treat water contaminated with heavy metals in order to mitigate their increasingly harmful impacts. An unconventional form of treatment is through chemical adsorption processes with biomass, in which cationic exchanges occur between their multiple functional groups (OH) and the metals present in the water [5–8]. This is a biochemical technology [9]. The use of biomass capable of chemiadsorbing metals in an efficient, economical and easy way is a method in development [10,11].

These types of technologies are viable due to the ease of producing or obtaining them, as waste products are usually developed productively in the laboratory. This is exemplified by bacterial cellulose. This polysaccharide is produced from basic sources such as red tea, sugar and kombucha, where kombucha is a symbiotic colony formed by several species of bacteria (*Acetobacter xylinum*, *Acetobacter aceti*, *Acetobacter pasteurianus*, and *Gluconobacter bluconicum*) and yeasts (*Brettanomyces*, *Brettanomyces bruxellensis*, and *Brettanomyces intermedius*) [12–15]. Bacterial cellulose meets the three criteria previously described: it is easy to produce, economical, and, above all, it is effective in removing heavy metals from water due to its multiple hydroxyl groups (OH), which are responsible for chemisorbing these contaminants [16–18].

A number of investigations have been conducted worldwide that have employed bacterial cellulose in the removal of heavy metals, including Cu (II) [19], Pb (II) [20], Cd (III) [21], and Cr (VI) [22]. However, the implementation of these investigations on a larger scale has not yet commenced, with the majority remaining within the laboratory setting. One method for scaling this type of treatment is through the use of external-layer (K_f) and internal-layer (K_s) adsorption models, which are employed in mathematical modeling processes with the objective of parameterizing designs for the implementation of these treatment systems [23–28]. K_f is employed as a design and scaling parameter, as it relates the particle density, porosity and ideal diameter, in conjunction with the behavior of its isotherm. K_s is then identified as a parameter that relates the adsorption capacities, speeds and ideal treatment flow rates. This is to facilitate the design process for large-scale treatment systems [29–35]. By obtaining the design variables through laboratory-scale experimental processes, similarities in the contact time of the adsorbent and the hydrodynamic characteristics between the adsorption systems could be obtained, thus enabling the development of pilot-scale processes. It is feasible to use data at laboratory scale to calculate parameters and model the efficiency of larger column systems using vertical column systems [36,37]. Pilot-scale trials have been conducted using biomasses to treat this type of water, including waste cellulose biomass from the aquatic plant *Eichhornia crassipes* [38–41]. But currently, a treatment system based on extraparticle and intraparticle diffusion models with bacterial cellulose biomass has not been built; due to this, it is necessary to develop treatment systems on a real industrial scale with bacterial cellulose material through adsorption models in order to identify the optimal parameters for their implementation. In light of this, this project was initiated with the objective of designing, modeling and developing scale models for the treatment of water contaminated with Cr (VI) through bacterial cellulose biomass.

2. Materials and Methods

2.1. Production of Bacterial Cellulose (BC)

The production of bacterial cellulose BC was undertaken using a tea and sugar culture medium obtained from the bioprocess laboratory at the University Foundation Los Libertadores. A BC film was produced and subsequently cut into 20 mm × 20 mm × 5 mm pieces. These pieces were then placed in 8 L glass containers containing 6 L of sterilized water, 30 g of sugar, and 20 g of red tea. To produce wet bacterial cellulose films, 600 mL of kombucha sediment (approximately 200 mL per sheet of kombucha tea, with a total of 3 sheets) and 1.5 g of commercial yeast of (*saccharomyces cerevisiae)* were added. Samples of pH and temperature were taken in situ for internal control. After three weeks, the films were dried at 70 °C for 48 h to remove moisture (see Figure 1). The films were ground until they had a diameter of 0.216 mm. This process yielded 70 g of bacterial cellulose in two weeks.

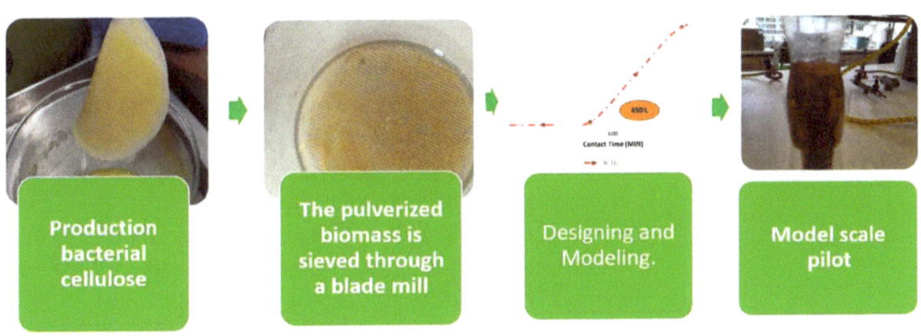

Figure 1. Production of cellulose bacterial.

2.2. The Research Parameters

The research parameters included initial chromium concentrations of 50, 100, 200, 300, and 600 mg (chromium/L). Samples were taken at each time interval, and the residual chromium concentration was analyzed. Samples of 20 µL were obtained and subsequently taken to the centrifuge (KASAI MIKRO 200, Hettich, Föhrenstr Germany). In the present investigation, the tests were conducted under neutral pH conditions. Given that the pH was neutral, the adsorption process was favored in this type of biomass.

Aliquots of the reaction mixture were analyzed for residual chromium concentration using a (UV84 Hettich, Föhrenstr Germany). The tests were conducted in accordance with the standard methods outlined by the American Public Health Association (APHA) for standard tests, specifically the Standard Methods for the Examination of Water and Wastewater. All procedures for determining chromium (VI) levels in water were carried out in accordance with the APHA procedure outlined in method 3500-Cr. The study's measurement uncertainty indicates that heavy element measurements, specifically Cr (VI), can be performed with an uncertainty level of approximately 3.95%.

Batch adsorption experiments were conducted in a 100mL glass vessel with constant stirring (IKA Ks 4000 shaker, Hettich, Föhrenstr Germany) at 20 °C and 150 rpm. Data were recorded every 20 min until 180 min had elapsed. The sample size was 20 µm. All experiments were performed in triplicate, and the final values were averaged.

The amount of Cr (VI) residue was estimated by means of the diphenylcarbazide method. For this, a phosphate-buffered solution was prepared by adjusting it to a pH equal to 2 with (H_3PO_4) to 90%. In an eppendorf tube, 200 µL of 0.5% diphenylcarbazide (P/V acetone) was added, along with 900 µL of phosphate buffer and 100 µL of residual sample. The absorbance was measured at 540 nm, whereupon it was transferred to an adsorption cell.

The adsorption capacity was determined with a suspension of 0.3 g of biomass in 100 mL of Cr (VI) solution for 140 min at 200 rpm, taking samples every 20 min, before determining the residue and discarding the sediment. All procedures were performed in duplicate.

An Evolution 300 spectrophotometer was used to monitor changes in absorbance during the determination of chromium in water and substrates. The procedures followed the standard methods for the examination of water and wastewater as outlined by the American Public Health Association (APHA).

2.3. Desorption–Adsorption

Following the Cr (VI) adsorption process, the chromium-loaded biomass underwent an elution process. After each biomass was subjected to the adsorption process, it was washed with distilled water. The elution process was then carried out in an Erlenmeyer flask at 25 °C for 24 h with constant stirring, using 20 mL of 20 mL of EDTA. The biomass was subsequently separated using a filter [31].

2.4. Column Design and Experiments

Two treatment systems were developed using recyclable PET bottles of varying sizes. The smaller-scale treatment system (laboratory) had a biomass diameter of 2.3 cm and a biomass length between the two compartments of 25 cm. Through experimental tests on this, a larger-scale system was developed. The larger-scale treatment system (pilot) was 5 cm in diameter with 75 cm of biomass height. Both systems had meshes that separated the compartments. The laboratory-scale system comprised 45 g of bacterial cellulose biomass, which reached a height of 25 cm and had a volume of 50 mL (with an area of 2.33 cm^2).

Subsequently, the column was scaled up to a larger laboratory scale, in accordance with the criteria of geometric, kinematic and dynamic similarity [33]. The design parameters of the two treatment systems were employed to ascertain the relationship between the heights and diameters of the biomasses, with a view to determining the optimal relationship criterion between the biomasses. The relationship between the densities of the biomasses was established as the primary similarity criterion through the identification of the most suitable particle diameter for the removal and elution processes. A pilot-scale prototype was constructed using 450 g of biomass, including the elution processes, with the data obtained from the aforementioned processes.

The biomasses of the bacterial cellulose samples were measured at the following diameters: 0.212 mm, 0.315 mm and 0.318 mm.

The three diameters resulted in three different biomass density ratios on a laboratory scale. Subsequent evaluations were conducted, and the diffusion constants K_f were established. The intra-particle diffusion constant K_s was established, allowing for the establishment of the relationship between height and contact area for the treatment of 50 liters of wastewater contaminated with 1000 g/L of chromium.

In accordance with the remaining criteria, the physical properties of the fluid and the linear velocity (v, cm min^{-1}) were maintained at a constant value in both columns, thus ensuring the same mass transfer and hydrodynamic conditions.

The initial experiments were conducted on a laboratory scale with a flow rate of 20 mL/min. The biomass particle diameter was 0.212 mm, 0.315 mm and 0.318 mm, with the objective of evaluating three different types of biomass densities. The initial concentration of Cr (VI) was 500, 300, 200 and 100 mg/L. The sampling period was set at 15 min, and the experiments were conducted in a downward flow configuration, with three repetitions.

2.5. Development of Adsorption Models

The dynamics of the adsorption process are reflected in Equation (1)

$$m_A \frac{\partial q_A}{\partial t} = -Vl \frac{\partial c}{\partial t} \tag{1}$$

The development of adsorption models will be discussed. In Equation (1), m_A is the mass of the adsorbent, which in our case is bacterial cellulose, and Vl is the objective parameter, which is the volume of water to be treated. The equation links the change in adsorption capacity q_A in relation to the time spent between the biomass and the contaminant.

This model establishes that in external conditions, the contaminant adsorption process is carried out, where it has a concentration gradient (δ). Therefore, as long as equilibrium conditions are not established, the concentration in the adsorbent particle will be less than that in the fluid to be treated. This gradient extends around the outer layer. To establish this dynamism, the mass transfer equation for external diffusion or film is used, which can be derived from Fick's law.

$$N_F = Dl \frac{\partial C}{\delta} \tag{2}$$

The flux, N_f, is given in g/(m^{2*} s), while the diffusion coefficient in the contaminant phase, Dl, is expressed in m^2/s. Integrating the linear gradient δ within the boundary layer leads to the following equation:

$$N_F = K_f(co - c) \tag{3}$$

The mass transfer coefficient of the film, K_f, is the amount of contaminant chemisorbed in the fluid that has already been treated at this moment, and that is adsorbed in the biomass per unit of time.

This parameter is dependent on the density of the adsorbent and its porosity. It directly relates to the contaminant and the chemiadsorption attraction between heavy metals and the functional groups of the biomass [29–31]. The relationship between functional groups and the metal to be removed from the water is directly proportional to the density of the particle. Therefore, a model must be designed to comply with national or international discharge regulations, or even reach 0.

$$n_F = \frac{N_F}{As} \tag{4}$$

The total external surface of all adsorbent particles within the designed bioreactor is also the external area of the volume of biomass occupied in the treatment. This is calculated using the following equation:

$$As = \frac{3Va(1-\varepsilon)}{rp} \tag{5}$$

where Va is the volume occupied by the biomass in the reactor. The total volume of the reactor (including spaces between pores and air) is denoted by ε. This is the relationship between the densities of the particle and the density of the biomass occupied in the reactor [32–34]. The radius of the biomass occupied in the reactor is denoted by rp. The relationship between the densities is a design parameter that is fundamental to the assembly and commissioning processes of adsorbent systems. The relationship between the volume of the liquid to be treated and the total volume of the reactor is:

$$\varepsilon = \frac{Vl}{Va} \tag{6}$$

The relationship between the volume of the reactor (Vl) and the volume of the adsorbent (Va) is given by the following equation:

$$\frac{Va}{Vr} = 1-\varepsilon \tag{7}$$

The volume of the liquid and the volume of the adsorbent is given by the following equation:

$$\frac{Va}{Vl} = \frac{1-\varepsilon}{\varepsilon} \tag{8}$$

$$\frac{Va}{Vl} = pp\frac{1-\varepsilon}{\varepsilon} \tag{9}$$

$$\varepsilon = 1-\frac{pb}{pp} \tag{10}$$

This equation will be fundamental due to the relationship between the general biomass and its small particles. The simplest particle will have a direct relationship with the contaminant coupled with the biomass, whereby the more space there is between particles, the better it will be for the treatment. *pp* is determined by:

$$\rho p = \frac{mp}{Vp} \tag{11}$$

The mass of the microparticle (*mp*) is its weight, and the volume of the microparticle *vp* is obtained by:

$$Vp = \frac{4\pi r^3}{3} \tag{12}$$

The radius of the tiny particle (*r*) is a dependent variable in this research. Its diameter is obtained in meshes and classified by size. Returning to Equations (1) and (4), the relationship and differences that exist in the time of the concentrations is given by:

$$\frac{\partial C}{\partial t} = k_F * A_M\frac{M_A}{vL}(c-cs) \tag{13}$$

where A_M is the total surface area related to the mass of adsorbent available in the reactor (A_M = As/mA); replacing Equations (5) and (10) in Equation (13) leaves the design equation unchanged:

$$\frac{\partial C}{(c-cs)} = k_F * A_M\frac{M_A}{vL}\partial t \tag{14}$$

Integrating Equation (14),

$$Ln\frac{Co}{(cs)} = -k_F * A_M\frac{M_A}{vL}t \quad (15)$$

The natural logarithm of the initial concentration and final concentration of Cr (VI) (designing the final parameter of discharges), together with the surface area (As) of contact, the volume of treated water, and the objective parameter (Vl), the diffusion constant (K_f) of Cr (VI) in the biomass, are found by graphing this term. This parameter summarizes the design behavior of the following:

Surface area that the biomass will occupy in the reactor;

Mass necessary to meet the objective of treating the contaminant;

The target treatment volume;

The target concentration (mg/l) of the contaminant.

These are fundamental parameters when designing and implementing water treatment systems contaminated with heavy metals. The diffusion constant (K_f) summarizes these parameters. The other part of Equation (1), which takes into account intraparticle diffusion, is given by:

$$\frac{\partial q}{\partial t} = K_s pp(qs - q) \quad (16)$$

The density of the particle plays a fundamental role in the change of adsorption capacities over time, and the parameter K_s is obtained. This constant summarizes, like K_f, the design parameters. In order to establish the behavior of the adsorption capacity in saturation, the present investigation employs the isotherm equations, which are the Langmuir and Freundlich isotherms. Equations (15) and (16) will be grouped.

$$ks(qs - q) = -k_F * A_M\frac{M_A}{vL} \quad (17)$$

In numerous studies of heavy metal adsorption, the principal focus has been the adsorption isotherm and its variables. Nonetheless, fixed bed or packed bed adsorbers' design necessitates both the adsorption isotherm and the complete adsorption pace.

If it is the Freundlich Equation (18):

$$qs = K(Cs)^n \quad (18)$$

The resulting expression with Freundlich modeling is Equation (19).

$$ks\left(K(Cs)^n - q\right) = -k_F * A_M\frac{M_A}{vL}t \quad (19)$$

Equation (19) does not obtain the intraparticle diffusion constant when adjusted to this Freundlich isotherm

$$k_S = \frac{3M_A k_F A_M}{\rho p(KCs^n - q)VL} \quad (20)$$

But if it is the Langmuir Equation (21):

$$q_S = \frac{qmBCs}{1 + BCs} \quad (21)$$

where B is the Langmuir parameter, if there is a fit to the Langmuir isotherm, this equation must be used to determine the K_S. If it is the Langmuir Equation (22):

$$\frac{KsqmBCs}{1 + BCs} - Ksq = A_M\frac{M_A}{vL}t \quad (22)$$

The resulting expression with Langmuir modeling is Equation (23).

$$k_S = \frac{3M_A k_F BCs^2 + 3M_A k_F Cs - k_F Cs + k_F Cs + 3M_A k_F Cs}{(ppqmBCs - qBCs - q)vL} \tag{23}$$

The intraparticle diffusion constant will be established through Equations (20) and (23), which will group design parameters such as:

- Biomass to use.
- Target volume.
- Target concentrations.
- Particle density.
- Maximum adsorption capacities.
- Constant K_f.

In order to ensure adequate staging and to obtain a full-scale treatment system, it is necessary to resolve various design parameters. These include determining the mass required to achieve a specific treatment objective or establishing the necessary treatment volume and adsorption capacity, taking into account all relevant variables. In the present investigation, Equation (24) was derived, which is ideal for establishing the biomass required for the effective treatment of wastewater contaminated with heavy metals.

$$M_A = \frac{(ppqmBCs - qBCs - q)vLk_S}{3(k_F BCs^2 + k_F Cs)} \tag{24}$$

3. Results

3.1. Result of Removal of Cr (VI)

Figure 2 illustrates the removal processes by the bacterial cellulose biomass with three different density models. The diameters represented each of the biomasses, which were BC (1): 0.212 mm, BC (2): 0.315 mm and BC (3): 0.318 mm. All experiments were performed in triplicate, with the average product and its margin of error shown in Figure 2.

All biomasses demonstrated excellent removal of Cr (VI) ions, reaching initial equilibrium levels. All achieved removal times above 450 min of treatment. In treatment processes at this laboratory scale, similar results were recorded in [42–47].

The BC(1) biomass was ideal, achieving better treatment intensity due to a breaking point of around 550 min. It was able to treat 5 L of wastewater with Cr (VI) ions. The biomass with BC (2) was able to treat 4.7 L of water with a break point of 500 min, while BC (3) spent approximately 435 min treating 4.65 L of water. The 0.0212 particle demonstrated superior performance, likely due to its intimate interaction with the Cr (VI) contaminant. This particle diameter has been recorded in some investigations as 10 microns. Table 1 presents the final results of the analysis of the relationship between densities.

Table 1. Results of the analysis of the relationship between densities.

Diameters	M (g)	Volume Mass (Vb)	Density Mas (pb)	Ma Particle (mp)	Volume Particle (Vp)	Density Particle (*pp*)	ε
BC (1)	50	80	0.62	0.01	0.005	0.59	0.68
BC (2)	50	85	0.58	0.019	0.044	0.43	0.33
BC (3)	50	90	0.55	0.028	0.039	0.71	0.23

Equations (3)–(12) were used to obtain the relationships between densities (ε), between the biomass in general and the tiny particle. This relationship is linked to the removals recorded in Figure 1, where the BC (1) biomass obtained the best treatment records. This is due to the compactness of the biomass density in general and the particle diameter, which is ideal for the direct relationship with the contaminant Cr (VI) [48,49].

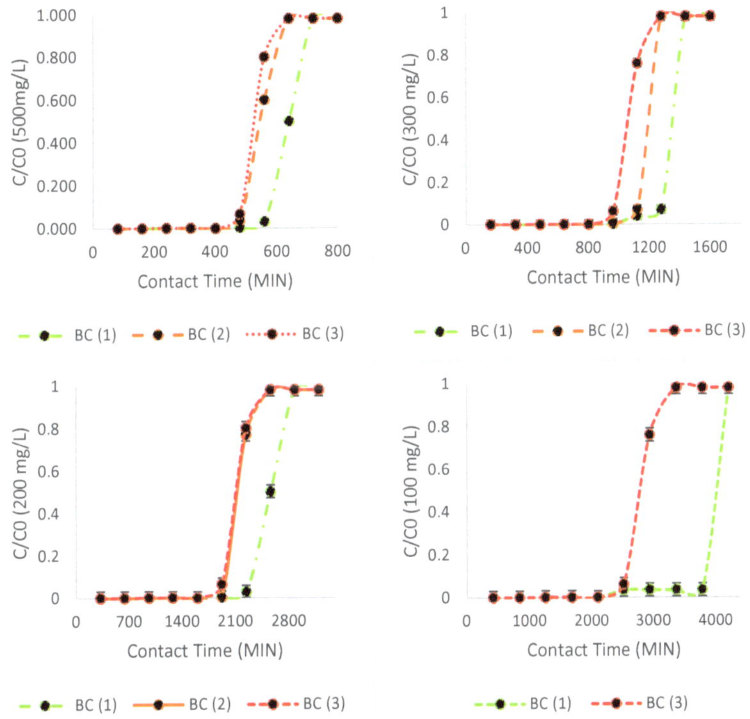

Figure 2. Removal processes by the bacterial cellulose.

The next step was to implement dynamic modeling at the laboratory scale. The three models were implemented using Equation (15) to establish the degree of incidence and design criteria for the different variables and design parameters. The initial and final concentrations of Cr (VI) were graphed with their natural logarithms, relating the rupture time and volume of treated water. For the case of BC (1), this was 5000 mL of water. Table 2 shows a summary of the variables in Equation (15). For illustrative purposes, let us consider the treatment with BC (1) as an example. In this case, Equation (15) becomes:

$$Ln\frac{1000}{(1)} = -k_F * \frac{61.5 * 50}{50 * 5000}550$$

where this equation can be solved to establish K_f or graph and through this the influences between the different variables can be observed, as shown in Figure 3.

Table 2. Summary of parameters given.

Biomass	As cm^2	K_f (cm/min)	Volume Goal (L)	Time Break (min)
BC (1)	61.5	0.99	5	550
BC (2)	61.5	0.895	4.7	500
BC (3)	61.5	0.889	4.65	435

The kinetic curve can be plotted according to Equation (3) in order to estimate K_f (Figure 3) in the which has the representative red color due to the breaking point. The two methods for establishing K_f were employed to complete the various processes, which are outlined in Table 2.

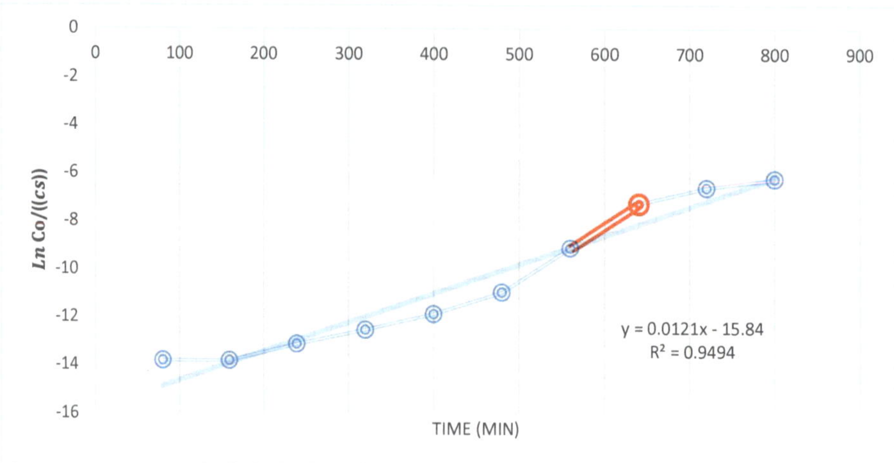

Figure 3. Plotting for find the K_f.

K_f is an adsorption rate that correlates with ideal diameters and densities in the removal processes of heavy metals in water. The K_f value for BC (1) was 0.99 cm/min, indicating that this biomass has an adsorption capacity for Cr (VI) ions. However, the K_f values for BC (2) and BC (3) were lower, and therefore, these biomass samples were not considered important in the design of treatment systems on a larger scale. In treatment systems employing plant biomass, the K_f for the Cr (VI) treatment was found to be less than 0.88 [50], indicating that the biomass of bacterial cellulose is greater under different circumstances of its particle sizes. K_f is the design and scaling parameter because it relates the particle density, porosity and ideal diameter, coupled with the behavior of its isotherm. Consequently, the K_s will be found to complement the scaling process in the design of a large-scale treatment system.

Subsequently, analysis of the removals must be carried out to establish how the isotherms in which the removals of Cr (VI) by the different biomasses occur behave. Figure 4 depicts the isotherms of each of the biomasses.

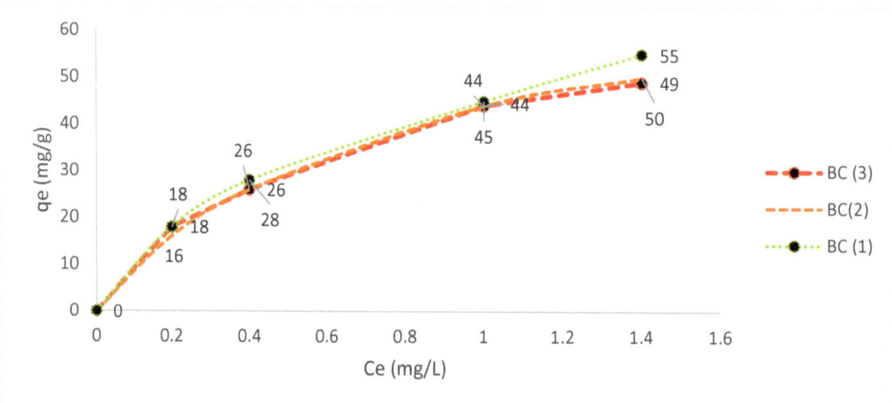

Figure 4. Isotherms of each biomass.

Figure 4 illustrates the relationship between bacterial cellulose and adsorption capacities (q_e) at different equilibrium concentrations (Ce) obtained in each of the initial concentrations of 400, 200, 100 and 50 mg/L of Cr (VI). It shows that adsorption capacities of more than 55 mg/g per biomass can be achieved, with the BC biomass (1) reporting the best results, at around 65 mg/g. The active sites (OH) in the bacterial cellulose biomass are filled by Cr (VI) ions, which is characteristic of this biomass. It is also homogenous in adsorption processes and adjusts to the Langmuir isotherm. It should be noted that this isotherm assumes that all active sites on the surface are energetically homogeneous. A representative fit of 0.99 R^2 can be seen in Table 3.

Table 3. Isotherm representatives.

	Isotherm	**Constant**	$\mathbf{R^2}$
BC(1)	Langmuir	B = 0.03; q_m; 65	0.99
	Freundlich	K = 0.17	0.91
	Isoterm	**Constant**	$\mathbf{R^2}$
BC (2)	Langmuir	B = 0.028; q_m; 60	0.99
	Freundlich	K = 0.11	0.92
	Isoterm	**Constant**	$\mathbf{R^2}$
BC(3)	Langmuir	B = 0.027; q_m; 58	0.98
	Freundlich	K = 0.18	0.96

For all biomasses, Langmuir isotherms were obtained due to the homogeneity of active sites in these; therefore, the models of Equation (23) were used to obtain the intra-particle design variable K_s, as shown in Table 4. Similarly to the variable K_f, the constant K_s plays a pivotal role in the design and scaling of the treatment system. This is because the intraparticle diffusion constant contains intertwined design parameters such as density variables and the influence of the contaminant.

Table 4. Isotherm equation representatives.

Biomass	$\mathbf{K_s}$ **(1/s)**	**Isotherm Equation**	$\mathbf{Q_m}$	**pp**
BC (1)	0.018	Langmuir	65	2
BC (2)	0.020	Langmuir	60	1.5
BC (3)	0.022	Langmuir	55	1.4

To ascertain the impact of the EDTA eluent on recycling ratios and biomasses, a series of elution and recycling cycles were conducted, as illustrated in Figure 5. The design variables were established between densities after each elution process and subsequent reuse. Figure 5 illustrates the characteristics and elution processes of each biomass, along with their respective removal percentages, with its margin of error product of the arithmetic average.

It is evident that each elution process yielded a notable performance, with similar yields observed across the different cycles. All biomasses were tested with up to four elution and reuse cycles, from (4) cycles of Elutions/Recycling. The biomass with the best performance was BC (1), although the other biomasses also demonstrated promising results in the elution and reuse processes. The results of the investigations by [51,52] indicate that bacterial cellulose performs optimally with the EDTA eluent. In each adsorption and desorption process of each biomass, the same analyses were carried out to determine the influence of the EDTA eluent on the biomass, along with all the parameters previously detailed. Table 5 shows the elutions and the behavior of all biomasses.

The elutions did not compromise the behavior of their isotherms, but as the elutions were carried out, they approached the Freundlich isotherms and moved away from the

Langmuir isotherm. This was due to the loss of adsorption capacities, in turn due to the gradual wear on the bacterial cellulose biomass caused by the same eluent. The K_s value was established using the equation developed by Freundlich.

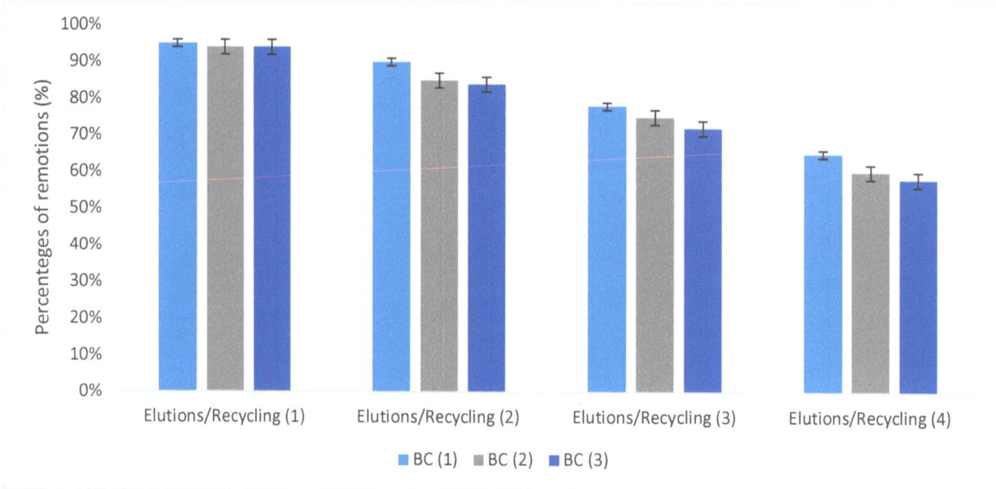

Figure 5. Cycles and reutilizations.

Table 5. The elutions and behavior of all biomasses.

Biomass	Elutions	K_f (cm/min)	Volume Goal (L)	Time Break (min)	K_s (1/s)	Isotherm Equation	Qm
BC (1)	1	0.98	4.5	500	0.017	Langmuir	58
	2	0.98	4.4	450	0.018	Langmuir	55
	3	0.96	4.4	440	0.019	Langmuir	52
	4	0.88	4.0	410	0.020	Langmuir	50
BC (2)	1	0.90	4.1	458	0.020	Langmuir	56
	2	0.90	4.0	450	0.021	Langmuir	54
	3	0.89	3.5	410	0.022	Langmuir	50
	4	0.80	3.0	390	0.022	Freundlich	44
BC (3)	1	0.90	4.1	450	0.022	Langmuir	55
	2	0.88	3.8	440	0.023	Langmuir	50
	3	0.88	3.5	380	0.023	Langmuir	45
	4	0.82	3.1	360	0.024	Freundlich	40

3.2. Redesign of Process Treatment

The redesign conditions were established by considering the similarity relationship between the contact area, biomass and volume of water transferred, as well as the elution processes. The best conditions of K_f and K_s were then established; Figure 6 shows the scaling of both process.

In the Figure 6, in the which observed the characteristics of the biomass of bacterial cellulose, with chemical bonds in the adsorption of Cr (VI) [16]. The biomass to be treated, which was around 50 L of water, was calculated using Equation (15), which solved the mass of the equation. This was undertaken because we had K_f and K_s.

The values of K_f and K_s in the pilot-scale prototype were found to be of a similar magnitude to those found at the laboratory scale, due to similar conditions. The treatment was found to be very similar, with an inlet concentration of 500 mg/L and a flow rate that was the same.

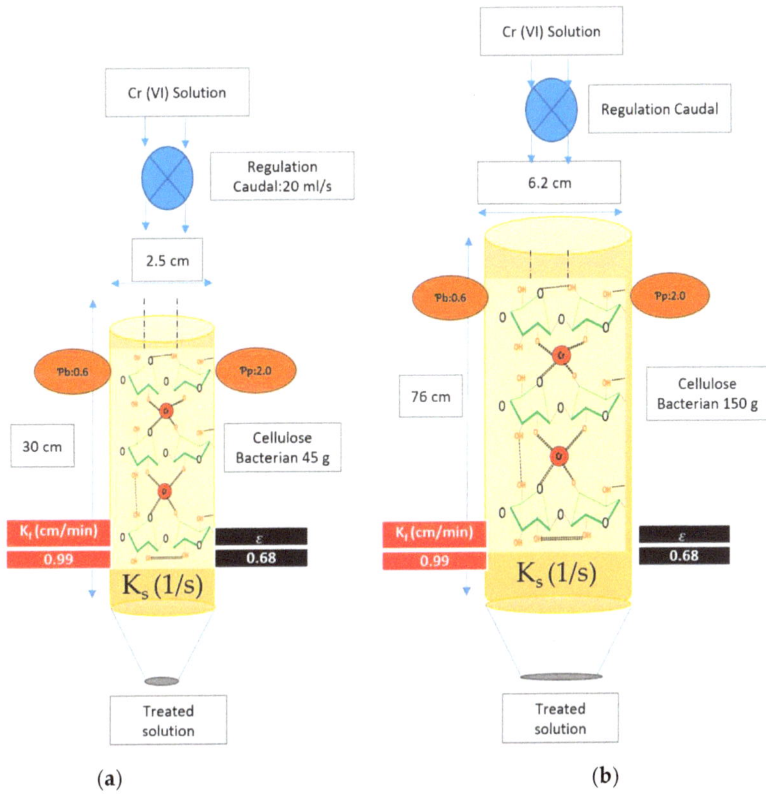

Figure 6. Description of both processes. (**a**) shows the system to laboratory scale, and (**b**) shows the system to pilot scale.

The extension was carried out using the representative models of K_f and K_s, due to similar conditions in the treatment and relationship between contaminant and bacterial cellulose biomass. This proved to be an appropriate procedure in the treatment of 500 mg/L of Cr (VI) and around 50 l of water. Equation (23) was used, where all the biomass information was available and the Table 6, shows the parameters of scaling.

$$M_A = \frac{(\text{ppqmBCs} - \text{qBCs} - \text{q})vLk_S}{3(k_F\text{BCs}^2 + k_F\text{Cs})}$$

In the Table 6, is observed that, a pilot-scale prototype was constructed using 450 g of biomass, including the elution processes, with the data obtained from the aforementioned processes. The flow rate used was 200. Figure 7 shows the process of treatment.

Table 6. Parameters of scaling.

Diameters	M (g)	Volume Mass (Vb)	Density Mass (pb)	Caudal mL/min	Volume Treat (L)	Elutions
BC (1)	50	80	0.62	20	5	4
Scaling	450	500	0.58	200	550	3

The device created in this research carried out an effective treatment of 450 L of water contaminated with 1000 mg/L of Cr (VI) with four elution cycles under the same conditions as in the laboratory-scale treatment. The pilot-scale prototype was constructed using 450 g of biomass, including the elution processes, with the data obtained from the aforementioned processes.

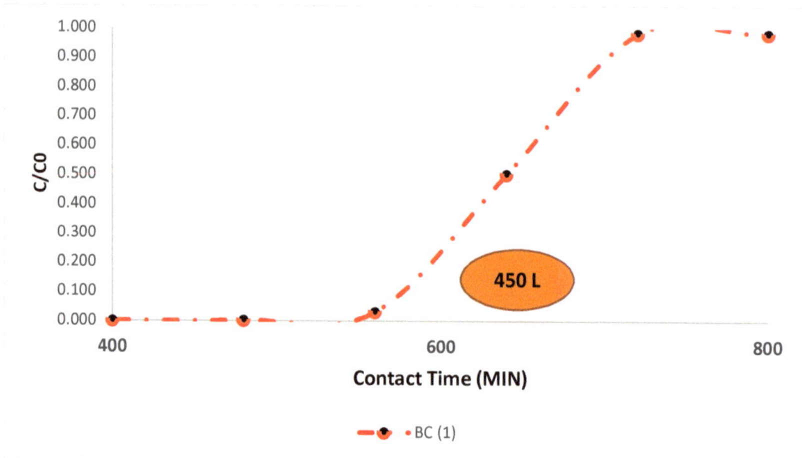

Figure 7. Process of treatment.

4. Conclusions

The production of bacterial cellulose in a laboratory setting represents an economical and straightforward technique for the development of treatment systems for industrial wastewater contaminated with heavy metals. The high level of adsorption of this material, with a particle diameter of 0.212 mm, was confirmed due to the high level of contact between the contaminant and this biomass. Based on this information and mathematical modeling of extra-particle and intra-particle diffusion, together with representative isotherms, a treatment system was developed at a scale appropriate for the treatment of an effluent from a company that contaminates with Cr (VI). The results of the model indicated the specific route for the construction of a treatment system at an industrial scale, with the experimental data adjusted to achieve this objective. A pilot-scale prototype was constructed utilizing 450 g of biomass, incorporating elution processes and incorporating the data obtained from the aforementioned processes. The excellent efficiency of the two models at different scales, together with the excellent elution results where the design variables were implemented in the EDTA elution process, providing more complete adsorption capabilities, suggests that this prototype could be presented to polluting industries for the treatment of waters coming from different industrial effluents. This represents an advanced biotechnology for the treatment of industrial wastewater. It is anticipated that bacterial cellulose biomass will be applicable for the removal of diverse effluents contaminated with heavy metals and implemented in wastewater-polluting industries, representing a transformative model of advanced biotechnology in the treatment of industrial wastewater.

Author Contributions: Methodology, U.F.C.S., V.B.B. and A.M.L.A.; Software, U.F.C.S.; Validation, U.F.C.S.; Formal analysis, U.F.C.S.; Investigation, U.F.C.S. and V.B.B.; Resources, U.F.C.S. and V.B.B.; Data curation, U.F.C.S.; Writing—original draft, U.F.C.S.; Writing—review & editing, U.F.C.S.; Supervision, U.F.C.S. All authors have read and agreed to the published version of the manuscript.

Funding: The university Los Libertadores is the company that contributed to development of this article and related processes.

Institutional Review Board Statement: Not applicable.

Informed Consent Statement: Not applicable.

Data Availability Statement: The data presented in this study is available on request from the author.

Conflicts of Interest: The authors declare no conflict of interest.

References

1. Sayago, U.F.C.; Castro, Y.P.; Rivera, L.R.C.; Mariaca, A.G. Estimation of equilibrium times and maximum capacity of adsorption of heavy metals by *E. crassipes* (review). *Environ. Monit. Assess.* **2020**, *192*, 141. [CrossRef] [PubMed]
2. Chaurasia, P.K.; Nagraj; Sharma, N.; Kumari, S.; Yadav, M.; Singh, S.; Mani, A.; Yadava, S.; Bharati, S.L. Fungal assisted bio-treatment of environmental pollutants with comprehensive emphasis on noxious heavy metals: Recent updates. *Biotechnol. Bioeng.* **2023**, *120*, 57–81. [CrossRef] [PubMed]
3. Carreño Sayago, U.F. Diseño y evaluación de un biosistema de tratamiento a escala piloto de aguas de curtiembres a través de la *Eichhornia crassipes*. *Rev. Colomb. De Biotecnol.* **2016**, *18*, 74–81. [CrossRef]
4. Almeida, J.C.; Cardoso, C.E.; Tavares, D.S.; Freitas, R.; Trindade, T.; Vale, C.; Pereira, E. Chromium removal from contaminated waters using nanomaterials–a review. *TrAC Trends Anal. Chem.* **2019**, *118*, 277–291. [CrossRef]
5. Hu, H.; Tang, C.S.; Shen, Z.; Pan, X.; Gu, K.; Fan, X.; Lv, C.; Mu, W.; Shi, B. Enhancing lead immobilization by biochar: Creation of "surface barrier" via bio-treatment. *Chemosphere* **2023**, *327*, 138477. [CrossRef]
6. Arumugam, M.; Gopinath, S.C.; Anbu, P.; Packirisamy, V.; Yaakub, A.R.W.; Wu, Y.S. Efficient Copper Adsorption from Wastewater Using Silica Nanoparticles Derived from Incinerated Coconut Shell Ash. *BioNanoScience* **2024**, 1–11. [CrossRef]
7. Blaga, A.C.; Zaharia, C.; Suteu, D. Polysaccharides as support for microbial biomass-based adsorbents with applications in removal of heavy metals and dyes. *Polymers* **2021**, *13*, 2893. [CrossRef]
8. Chakraborty, R.; Asthana, A.; Singh, A.K.; Jain, B.; Susan, A.B.H. Adsorption of heavy metal ions by various low-cost adsorbents: A review. *Int. J. Environ. Anal. Chem.* **2022**, *102*, 342–379. [CrossRef]
9. Sayago, U.F.C.; Ballesteros Ballesteros, V. Recent Advances in the Treatment of Industrial Wastewater from Different Celluloses in Continuous Systems. *Polymers* **2023**, *15*, 3996. [CrossRef]
10. Sayago, U.F.C. Design and Development of a Pilot-Scale Industrial Wastewater Treatment System with Plant Biomass and EDTA. *Water* **2023**, *15*, 3484. [CrossRef]
11. Sayago, U.F.C.; Ballesteros Ballesteros, V. Development of a treatment for water contaminated with Cr (VI) using cellulose xanthogenate from *E. crassipes* on a pilot scale. *Sci. Rep.* **2023**, *13*, 1970. [CrossRef] [PubMed]
12. Laavanya, D.; Shirkole, S.; Balasubramanian, P. Current challenges, applications and future perspectives of SCOBY cellulose of Kombucha fermentation. *J. Clean. Prod.* **2021**, *295*, 126454. [CrossRef]
13. AL-Kalifawi, E.J.; Hassan, I.A. Factors Influence on the yield of bacterial cellulose of Kombucha (Khubdat Humza). *Baghdad Sci. J.* **2014**, *11*, 1420–1428. [CrossRef]
14. Domskiene, J.; Sederaviciute, F.; Simonaityte, J. Kombucha bacterial cellulose for sustainable fashion. *Int. J. Cloth. Sci. Technol.* **2019**, *31*, 644–652. [CrossRef]
15. Zhu, C.; Li, F.; Zhou, X.; Lin, L.; Zhang, T. Kombucha-synthesized bacterial cellulose: Preparation, characterization, and biocompatibility evaluation. *J. Biomed. Mater. Res. Part A* **2014**, *102*, 1548–1557. [CrossRef]
16. Sayago, U.F.C.; Castro, Y.P. Development of a composite material between bacterial cellulose and E crassipes, for the treatment of water contaminated by chromium (VI). *Int. J. Environ. Sci. Technol.* **2022**, *19*, 6285–6298. [CrossRef]
17. Chen, X.; Cui, J.; Xu, X.; Sun, B.; Zhang, L.; Dong, W.; Chen, C.; Sun, D. Bacterial cellulose/attapulgite magnetic composites as an efficient adsorbent for heavy metal ions and dye treatment. *Carbohydr. Polym.* **2020**, *229*, 115512. [CrossRef]
18. Li, H.; Wang, Y.; Ye, M.; Zhang, X.; Zhang, H.; Wang, G.; Zhang, Y. Hierarchically porous poly (amidoxime)/bacterial cellulose composite aerogel for highly efficient scavenging of heavy metals. *J. Colloid Interface Sci.* **2021**, *600*, 752–763. [CrossRef]
19. Oktaviani, O.; Yunus, A.L.; Nuryanthi, N.; Syahputra, A.R.; Puspitasari, T.; Indriyati, I.; Safitri, A.T.; Mellawati, J. Adsorption of Cu (II) and Pb (II) onto bacterial cellulose-co-polyacrylamide: A study of equilibrium adsorption isotherm. *AIP Conf. Proc.* **2023**, *2568*, 030005.
20. Song, S.; Liu, Z.; Zhang, J.; Jiao, C.; Ding, L.; Yang, S. Synthesis and adsorption properties of novel bacterial cellulose/graphene oxide/attapulgite materials for Cu and Pb Ions in aqueous solutions. *Materials* **2020**, *13*, 3703. [CrossRef]
21. Mir, I.S.; Riaz, A.; Roy, J.S.; Fréchette, J.; Morency, S.; Gomes, O.P.; Dumée, L.F.; Greener, J.; Messaddeq, Y. Removal of cadmium and chromium heavy metals from aqueous medium using composite bacterial cellulose membrane. *Chem. Eng. J.* **2024**, *490*, 151665. [CrossRef]
22. Hosseini, H.; Mousavi, S.M. Bacterial cellulose/polyaniline nanocomposite aerogels as novel bioadsorbents for removal of hexavalent chromium: Experimental and simulation study. *J. Clean. Prod.* **2021**, *278*, 123817. [CrossRef]
23. Sayago, U.F.C. Design and development of a biotreatment of *E. crassipes* for the decontamination of water with Chromium (VI). *Sci. Rep.* **2021**, *11*, 9326. [CrossRef] [PubMed]
24. Malik, D.S.; Jain, C.K.; Yadav, A.K. Heavy metal removal by fixed-bed column—A review. *ChemBioEng Rev.* **2018**, *5*, 173–179. [CrossRef]
25. Abdolali, A.; Ngo, H.H.; Guo, W.; Zhou, J.L.; Zhang, J.; Liang, S.; Chang, S.W.; Nguyen, D.D.; Liu, Y. Application of a breakthrough biosorbent for removing heavy metals from synthetic and real wastewaters in a lab-scale continuous fixed-bed column. *Bioresour. Technol.* **2017**, *229*, 78–87. [CrossRef]

26. Dissanayake, D.M.R.E.A.; Chathuranga, P.K.D.; Perera, P.I.; Vithanage, M.; Iqbal, M.C.M. Modeling of Pb (II) adsorption by a fixed-bed column. *Bioremediation J.* **2016**, *20*, 194–208. [CrossRef]
27. Ahmad, M.; Lubis, N.M.; Usama, M.; Ahmad, J.; Al-Wabel, M.I.; Al-Swadi, H.A.; Rafique, M.I.; Al-Farraj, A.S. Scavenging microplastics and heavy metals from water using jujube waste-derived biochar in fixed-bed column trials. *Environ. Pollut.* **2023**, *335*, 122319. [CrossRef]
28. Bringas, A.; Bringas, E.; Ibañez, R.; San-Román, M.F. Fixed-bed columns mathematical modeling for selective nickel and copper recovery from industrial spent acids by chelating resins. *Sep. Purif. Technol.* **2023**, *313*, 123457. [CrossRef]
29. Ahmad, A.; Singh, A.P.; Khan, N.; Chowdhary, P.; Giri, B.S.; Varjani, S.; Chaturvedi, P. Bio-composite of Fe-sludge biochar immobilized with *Bacillus* Sp. in packed column for bio-adsorption of Methylene blue in a hybrid treatment system: Isotherm and kinetic evaluation. *Environ. Technol. Innov.* **2021**, *23*, 101734. [CrossRef]
30. Allahkarami, E.; Soleimanpour Moghadam, N.; Jamrotbe, B.; Azadmehr, A. Competitive adsorption of Ni (II) and Cu (II) ions from aqueous solution by vermiculite-alginate composite: Batch and fixed-bed column studies. *J. Dispers. Sci. Technol.* **2023**, *44*, 1402–1412. [CrossRef]
31. Dinesha, B.L.; Hiregoudar, S.; Nidoni, U.; Ramappa, K.T.; Dandekar, A.T.; Ganachari, S.V. Adsorption modelling and fixed-bed column study on milk processing industry wastewater treatment using chitosan zinc-oxide nano-adsorbent–coated sand filter bed. *Environ. Sci. Pollut. Res.* **2023**, *30*, 37547–37569. [CrossRef] [PubMed]
32. Abin-Bazaine, A.A.; Olmos-Marquez, M.A.; Campos-Trujillo, A. A Fixed-Bed Column Sorption: Breakthrough Curves Modeling. In *Sorption—New Perspectives and Applications [Working Title]*; IntechOpen: Rijeka, Croatia, 2024.
33. Kavand, M.; Fakoor, E.; Mahzoon, S.; Soleimani, M. An improved film–pore–surface diffusion model in the fixed-bed column adsorption for heavy metal ions: Single and multi-component systems. *Process Saf. Environ. Prot.* **2018**, *113*, 330–342. [CrossRef]
34. Qu, J.; Song, T.; Liang, J.; Bai, X.; Li, Y.; Wei, Y.; Huang, S.; Dong, L.; Jin, Y.U. Adsorption of lead (II) from aqueous solution by modified Auricularia matrix waste: A fixed-bed column study. *Ecotoxicol. Environ. Saf.* **2019**, *169*, 722–729. [CrossRef]
35. Song, T.; Liang, J.; Bai, X.; Li, Y.; Wei, Y.; Huang, S.; Dong, L.; Qu, J.; Jin, Y. Biosorption of cadmium ions from aqueous solution by modified Auricularia Auricular matrix waste. *J. Mol. Liq.* **2017**, *241*, 1023–1031. [CrossRef]
36. Abbasi, M.; Safari, E.; Baghdadi, M.; Janmohammadi, M. Enhanced adsorption of heavy metals in groundwater using sand columns enriched with graphene oxide: Lab-scale experiments and process modeling. *J. Water Process Eng.* **2021**, *40*, 101961. [CrossRef]
37. Olatunji, M.A.; Salam, K.A.; Evuti, A.M. Continuous removal of Pb (II) and Cu (II) ions from synthetic aqueous solutions in a fixed-bed packed column with surfactant-modified activated carbon. *Sep. Sci. Technol.* **2024**, *59*, 561–579. [CrossRef]
38. Park, S.S.; Park, Y.; Repo, E.; Shin, H.S.; Hwang, Y. Three-dimensionally printed scaffold coated with graphene oxide for enhanced heavy metal adsorption: Batch and fixed-bed column studies. *J. Water Process Eng.* **2024**, *57*, 104658. [CrossRef]
39. Apiratikul, R. Application of analytical solution of advection-dispersion-reaction model to predict the breakthrough curve and mass transfer zone for the biosorption of heavy metal ion in a fixed bed column. *Process Saf. Environ. Prot.* **2020**, *137*, 58–65. [CrossRef]
40. Khalfa, L.; Sdiri, A.; Bagane, M.; Cervera, M.L. A calcined clay fixed bed adsorption studies for the removal of heavy metals from aqueous solutions. *J. Clean. Prod.* **2021**, *278*, 123935. [CrossRef]
41. Yusuf, M.; Song, K.; Li, L. Fixed bed column and artificial neural network model to predict heavy metals adsorption dynamic on surfactant decorated graphene. *Colloids Surf. A Physicochem. Eng. Asp.* **2020**, *585*, 124076. [CrossRef]
42. Renu; Agarwal, M.; Singh, K.; Gupta, R.; Dohare, R.K. Continuous fixed-bed adsorption of heavy metals using biodegradable adsorbent: Modeling and experimental study. *J. Environ. Eng.* **2020**, *146*, 04019110. [CrossRef]
43. Ma, B.; Zhu, J.; Sun, B.; Chen, C.; Sun, D. Influence of pyrolysis temperature on characteristics and Cr (VI) adsorption performance of carbonaceous nanofibers derived from bacterial cellulose. *Chemosphere* **2022**, *291*, 132976. [CrossRef]
44. Li, Z.Y.; Dong, J.J.; Azi, F.; Feng, X.; Ge, Z.W.; Yang, S.; Sun, Y.-X.; Guan, X.-Q.; Dong, M.S. Mechanism of Cr (VI) removal by polyphenols-rich bacterial cellulose gel produced from fermented wine pomace. *Npj Clean Water* **2024**, *7*, 21. [CrossRef]
45. Lu, M.; Xu, Y.; Guan, X.; Wei, D. Preliminary research on Cr (VI) removal by bacterial cellulose. *J. Wuhan Univ. Technol.-Mater. Sci. Ed.* **2012**, *27*, 572–575. [CrossRef]
46. Carreño Sayago, U.F.; Piñeros Castro, Y.; Conde Rivera, L.R. Design of a fixed-bed column with vegetal biomass and its recycling for Cr (VI) treatment. *Recycling* **2022**, *7*, 71. [CrossRef]
47. Obernberger, I.; Thek, G. Physical characterisation and chemical composition of densified biomass fuels with regard to their combustion behaviour. *Biomass Bioenergy* **2004**, *27*, 653–669. [CrossRef]
48. Znad, H.; Awual, M.R.; Martini, S. The utilization of algae and seaweed biomass for bioremediation of heavy metal-contaminated wastewater. *Molecules* **2022**, *27*, 1275. [CrossRef]
49. Shanmugaprakash, M.; Sivakumar, V.; Manimaran, M.; Aravind, J. Batch and dynamics modeling of the biosorption of Cr (VI) from aqueous solutions by solid biomass waste from the biodiesel production. *Environ. Prog. Sustain. Energy* **2014**, *33*, 342–352. [CrossRef]
50. Verma, D.K.; Hasan, S.H.; Ranjan, D.; Banik, R.M. Modified biomass of *Phanerochaete chrysosporium* immobilized on luffa sponge for biosorption of hexavalent chromium. *Int. J. Environ. Sci. Technol.* **2014**, *11*, 1927–1938. [CrossRef]

51. Mohite, B.V.; Patil, S.V. Bacterial cellulose of *Gluconoacetobacter hansenii* as a potential bioadsorption agent for its green environment applications. *J. Biomater. Sci. Polym. Ed.* **2020**, *25*, 2053–2065. [CrossRef]
52. Sayago, U.F.C.; Ballesteros, V.A.B. The Design of a Process for Adsorbing and Eluting Chromium (VI) Using Fixed-Bed Columns of *E. crassipes* with Sodium Tripolyphosphate (TPP). *Water* **2024**, *16*, 952. [CrossRef]

water

MDPI

Article

Microbial Selection for the Densification of Activated Sludge Treating Variable and High-Strength Industrial Wastewater

Mukhtiar Ahmed, Dorothee Goettert, Catharina Vanherck, Koen Goossens and Jan Dries *

Research Group BioWAVE, Biochemical Wastewater Valorization & Engineering, Faculty of Applied Engineering, University of Antwerp, Campus Groenenborger, Groenenborgerlaan 171, 2020 Antwerp, Belgium
* Correspondence: jan.dries2@uantwerpen.be

Abstract: This study investigates the densification/granulation of activated sludge with poor settleability, treating real industrial wastewater from a tank truck cleaning company. The wastewater is low in nutrients, acidic in nature, and high and variable in chemical oxygen demand (COD, ranging from 2770 mg·L^{-1} to 14,050 mg·L^{-1}). A microbial selection strategy was applied to promote slow-growing glycogen-accumulating microorganisms (GAO) by the implementation of an anaerobic feast/aerobic famine strategy in a sequencing batch reactor (SBR). After 60 to 70 days, the uptake of carbon during the anaerobic phase exceeded 80%, the sludge morphology improved, and the sludge volume index (SVI) dropped below 50 mL·g^{-1}. 16S rRNA gene sequencing showed the enrichment of the GAOs *Defluviicoccus* and *Candidatus* Competibacter. Stable sludge densification was maintained when using a constant organic loading rate (OLR) of 0.85 ± 0.05 gCOD·(L·d)$^{-1}$, but the sludge quality deteriorated when switching to a variable OLR. In view of the integration of densified/granular sludge in a membrane bioreactor configuration, the filtration properties of the densified SBR sludge were compared to the seed sludge from the full-scale plant. The densified sludge showed a significantly lower resistance due to pore blockage and a significantly higher sustainable flux (45 vs. 15 L·(m^2·h)$^{-1}$).

Keywords: aerobic granular sludge (AGS); anaerobic feeding; bulking sludge; filamentous microorganisms; membrane bioreactor (MBR); organic loading rate (OLR)

Citation: Ahmed, M.; Goettert, D.; Vanherck, C.; Goossens, K.; Dries, J. Microbial Selection for the Densification of Activated Sludge Treating Variable and High-Strength Industrial Wastewater. *Water* **2024**, *16*, 2087. https://doi.org/10.3390/w16152087

Academic Editors: Yung-Tse Hung, Issam A. Al-Khatib, Rehab O. Abdel Rahman and Tsuyoshi Imai

Received: 27 June 2024
Revised: 17 July 2024
Accepted: 19 July 2024
Published: 24 July 2024

1. Introduction

The activated sludge (AS) process is widely used in the biological treatment of domestic and industrial wastewater for the removal of organic matter (C) and nutrients (mainly, nitrogen (N) and phosphorous (P)) [1]. The AS process consists of two tanks, namely the aeration tank, where the biological reactions take place, followed by the secondary settling tank, where sludge is separated from the effluent by sedimentation. After settling, the effluent is discharged, and sludge is recycled back into aeration tank. Sludge separation from the effluent represents a significant challenge in AS processes and affects about 50% of biological wastewater treatment plants worldwide [2]. In Flanders (Belgium), one-third of biological wastewater treatment plants are facing problems due to sludge bulking [3]. The reasons for sludge bulking are nutrient deficiency, limited oxygen supply, low organic load rate (OLR), and low pH (<6.5). Besides these, the slow feeding of readily biodegradable COD (rbCOD) under aerobic conditions also negatively impacts sludge settling [4].

To control sludge bulking, both unspecific (curative) and specific (selector-based) preventive strategies are applied. Unspecific methods involve chemicals such as chlorine (Cl_2), ozone (O_3), hydrogen peroxide (H_2O_2), iron, or aluminum salts to eliminate filamentous microorganisms extending from the flocs [5]. However, they have drawbacks like being temporary solutions and often negatively affecting nitrifiers, denitrifiers, and polyphosphate-accumulating organisms (PAOs) [6,7]. A selector is a small tank placed upstream from the main aeration tank that receives recycled sludge and influent wastewater, resulting in a high substrate concentration. The principle of aerobic selectors is based on the premise that creating a high food-to-microorganism ratio (F/M) to low F/M ratio gradient generates a 'feast'-to-'famine' regime. The high substrate availability during the feast phase promotes preferential growth and simultaneous substrate storage by floc-forming microorganisms. The stored substrate is then used for growth in the famine phase when the F/M ratio decreases, granting these organisms a competitive advantage [8]. Aerobic selectors need to be operated at high F/M ratios of 3–12 gCOD·$(gVSS·d)^{-1}$ [9]. The quality of the organic matter also plays a role in the effectiveness of aerobic selectors, and the presence of high amounts of particulate organic matter (POM) can be harmful to its performance. Anaerobic selectors select for floc-forming bacteria that store rbCOD and use the stored substrate for growth in the subsequent aerobic phase. Filamentous bacteria lacking substrate storage capacity under anaerobic conditions are then outcompeted [10].

The anaerobic feast and aerobic famine regimes favor the selection of slow-growing microorganisms such as PAOs and glycogen-accumulating organisms (GAOs), which are the main contributors to granulation [11–13]. The growth of PAOs and GAOs is enhanced by the availability of rbCOD in the influent [14]. In the anaerobic selector, the sludge in suspension contacts the influent, where rbCOD is fully utilized and converted to polyhydroxyalkoate (PHA). In the subsequent aerobic phase, these PHAs are oxidized for microbial growth [15]. The densification of the sludge is not only based on the anaerobic uptake of rbCOD, but it should also be concentrated on the best settling fraction of the sludge. In the bottom feeding strategy, both conditions are fulfilled, as there is a higher probability that the best-settling sludge will remain on the sludge bed. This process design results in well-settling dense sludge, which is further improved by selective discharge of the worst-settling sludge. The leakage of rbCOD into the aerobic phase, either directly or through the hydrolysis of POM, favors the growth of ordinary heterotrophic organisms (OHOs), which is detrimental to granulation. According to one study, a 20% leakage of rbCOD into the aerobic phase has no detrimental effect on the settling of granular sludge [16].

Aerobic granular sludge (AGS) is an innovative technology for industrial and domestic wastewater treatment. AGS has a dense, compact, and spherical structure and high biomass retention as compared to conventional floccular AS [17]. Also, AGS has easy dewaterability, which improves the quality of the effluent. These properties give a higher settling velocity (18–40 mph), which eliminates the requirement of a secondary clarifier tank [18]. Due to the compact size, the AGS reactor needs 25–30% less energy and 50–75% lower land footprint. These microenvironments help diverse microbial communities to coexist in a single granule, which makes a granule a hub of mini-ecosystems. This diversity of microorganisms leads to the removal of organic matter, nitrogen, and phosphorus in a single AGS reactor [19]. In AS, the COD is oxidized directly, while, in AGS, the COD is first converted to storage polymers and then oxidized. In addition, due to the large granule size, aerobic and anoxic zones coexist, which allows for simultaneous nitrification and denitrification.

There are many factors that affect the cultivation of AGS, such as the feeding strategy, the type of substrate, and the organic loading rate (OLR). The OLR, which depends on the strength of wastewater, ranges from 0.6 to 24 gCOD·$(L·d)^{-1}$ for AGS cultivation [20]. A high OLR results quickly in large but unstable granules, while a low OLR slowly forms small and compact granules, sometimes failing too [21]. When the OLR is changed during a reactor operation, an imbalance between the feast and the famine phase may occur, which gives an opportunity to filamentous microorganisms to proliferate at the cost of slow-growing microorganisms, leading to worsening of the sludge structure [22]. The change in the microbial community thus influences the granule stability, as well as the performance of the reactor. In a previous study by Adav et al. [23], the stepwise increase of the OLR from 9 gCOD·$(L·d)^{-1}$ to 12 gCOD·$(L·d)^{-1}$ led to a loss of microbial diversity, and a high OLR functional microbial community was selected.

Over the last decade, membrane bioreactors (MBRs) have gained attention due to their small footprint, excellent effluent properties, the ability to withstand high OLR, and the application of a high sludge concentration [24]. In MBRs, activated sludge processes are combined with membrane filtration to separate the effluent from sludge. However, membrane biofouling is the main barrier for the wide application of MBR technology, as this increases its operational and maintenance costs [25,26]. The sludge properties that cause biofouling are the mixed liquor suspended solids (MLSSs) concentration, the particle size of flocs, the soluble microbial products (SMPs), and the extracellular polymers substances (EPSs) [27].

To mitigate the fouling problem of conventional MBRs, the aerobic granular sludge membrane bioreactor (AGMBR) concept has been proposed as an alternative. AGMBR combines AGS and MBR for municipal and industrial wastewater treatment [28]. Due to its granular and larger structure, AGS is believed to reduce pore blockage and to form a loose cake layer, ultimately leading to increased permeability [29]. One study showed that, under the same conditions, conventional MBRs have 27-times higher fouling rates than AGMBRs [30]. The reason for the better filtration properties of AGS is due to its higher hydrophobicity than conventional AS [28]. In addition, AGMBR can be operated at higher sludge concentrations (up to 20 $g \cdot L^{-1}$) [31].

Using anaerobic feast/aerobic famine conditions, AGS have been cultured in SBR for the treatment of different industrial WW, such as the tank truck cleaning industry [32], dyeing industry [33], potato processing industry [34], and brewery industry [35]. However, to the best of our knowledge, there is no study available that shows aerobic granulation with the application of a microbial selection strategy only when treating high strength and variable industrial WW and how this strategy impacts the sludge filtration properties in a MBR.

This study investigates the densification of sludge with poor settleability, originating from a full-scale tank truck cleaning company (TTC) usually transporting chocolates and beverages. The wastewater originating from the TTC is low in nutrients, acidic in nature, and high and variable in COD. In addition, the evolution of the filterability of the sludge was determined to study the effect of the densification strategy on the filtration properties.

2. Methods

2.1. Laboratory-Scale SBR Reactor Set Up

A lab-scale SBR with a working volume of 10 L and height-to-diameter ratio = 1 was operated for 165 days at room temperature (20 ± 3 °C). The SBR reactor was equipped with a mechanical stirrer (RZR2020, Heidolph Instruments, Schwabach, Germany) to keep the sludge in suspension, and its speed was 120 revolutions per minute (rpm). A feeding pump (ES-B16VC-3, IWAKI, Tokyo, Japan), a discharge valve (Eriks RX10.X33.S00, ERIKS, Berchem, Belgium), a 13-cm diameter aeration disc (Aquadistri), and an aeration pump (Super Fish Koi Flow 60, Aquadistri, Klundert, The Netherlands) were also used for the stable operation of the reactor. Process operation was controlled by a Siemens PLC and a custom-built LABVIEW™ program (National instruments, Austin, TX, USA). A luminescent dissolved oxygen (LDO) sensor (Hach, Loveland, CO, USA) was used, and the DO level was kept between 1 and 4 $mgO_2 \cdot L^{-1}$ during the aerobic phase.

The seed-activated sludge for the SBR reactor was taken from a full-scale TTC plant. At the start of the experiment, when the sludge properties such as settling, dewaterability, and filterability were not good, the anaerobic feeding step (of 300 min) was followed by an (unfed) anaerobic reaction step (of 180 min) and an aerobic reaction step (of 860 min). The hydraulic retention time (HRT) varied between 4 and 9 days, depending upon the COD of the influent. After day 55, when the sludge properties improved, the anaerobic reaction was reduced to 60 min, while the aerobic reaction was increased to 1020 min. From days 69 to 130, a constant OLR was applied. The OLR was kept constant at a value of approximately 0.85 ± 0.05 $gCOD \cdot (L \cdot d)^{-1}$ by adjusting the volume fed to the reactor. The volume of wastewater fed varied between 1.1 $L \cdot d^{-1}$ and 2.4 $L \cdot d^{-1}$, depending on the influent COD. From day 130 onward, a variable OLR was applied to investigate its effect on the sludge characteristics and the reactor performance. At this stage, a fixed volume of wastewater was fed (2 L·d^{-1}). The 24-h SBR reactor cycle ended with settling (55 min) and discharge (4 min). The sludge retention time (SRT) was held constant at approximately 30 days by wasting mixed sludge daily from the reactor. The reactor was fed with the real TTC wastewater.

2.2. Set Up of Membrane Filtration Unit

A lab-scale membrane microfiltration unit was used to quantify the sludge filtration properties. A Kubota A4 size membrane (Type 510 cartridge, Osaka, Japan, with a surface area of 1040 cm^2 and average pore size of 0.2 μm) was used in a submerged design inside a Plexiglas cascade. An air pump (Aqua Forte V-30, SIBO BV, Doornhoek, The Netherlands) was used to supply air through a bubble aerator at the bottom of the reactor. To recycle the filtered effluent, a peristaltic pump (Heidolph 5101, Schwabach, Germany) was used. To control the filtration process, a custom-built LABVIEW™ program (National Instruments) was used. Different fluxes were applied stepwise for sludge filtration, ranging from 5 L·(m^2·h)$^{-1}$ to 55 L·(m^2·h)$^{-1}$ with an air flowrate between 3.5 and 4 m·s^{-1}. The flux of each magnitude was run twice with equal time intervals to check for irreversible fouling. The average transmembrane pressure (TMP_{avg}), permeability, fouling rate (FR), and sustainable flux were calculated as mentioned by Tsertou et al. [36]. FR is an increase in the TMP as a function of time. It is calculated as the slope of TMP to time in minutes.

A clean water filtration test was carried out before (CW1) and after (CW2) the sludge filtration test. The CW filtration tests were carried out at a single flux rate of 55 L·(m^2·h)$^{-1}$ for 10 min, while the sludge filtration test was carried out at different flux ranges from 5 to 55 L·(m^2·h)$^{-1}$ (Figure 1). During the sludge filtration test, the flux was increased in steps of 10 min each. After each step, a relaxation time of 5 min was allowed to remove the reversible cake layer by air flushing. At the end of the sludge filtration test, the membrane was gently washed with demineralized water to remove the cake layer. By manually removing the cake layer, the cake layer resistance is assumed to be 0. In our experiment, variable flux was used to investigate its effect on TMP change. The sludge filtration allows the estimation of the different contributors to the total membrane resistance, based on the clean water tests CW1 and CW2.

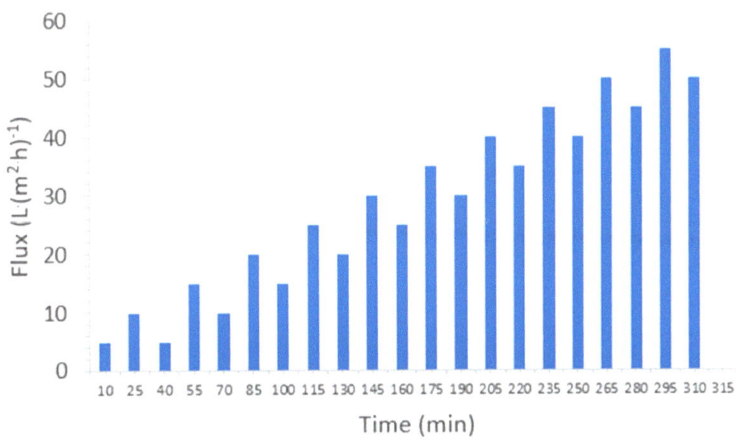

Figure 1. Flux profile for sludge filtration.

To determine the membrane resistance, Darcy's Law [37] was used:

$$Rt = Rm + Rf = Rm + Rp + Rc = Rm + Rp + 0$$

$$Rt = Rm + Rp$$

$$Rm = \frac{TMPcw1}{Flux \times \mu}$$

$$Rt = \frac{TMPcw2}{Flux \times \mu}$$

$$Rp = Rt - Rm$$

where Rt is the total resistance, Rm is the membrane resistance, Rf is the fouling resistance, Rp is the resistance by pore blocking, Rc is the cake layer resistance, and μ is the viscosity of the demineralized water (μ = 0.001 Pa·s). As the cake layer was removed manually, Rc is assumed to be 0.

2.3. Analytical Techniques

For the measurement of the influent and effluent parameters in the low range (Test 0-36 COD LR 150) for soluble COD (sCOD) and high range (Test 0-38 COD HR 1500) for total COD, a volatile fatty acid (VFA) test kit (Test-050 Ref: 985050), total nitrogen (Test-083 Ref: 985083), and total phosphorus (Test-080 Ref: 985080) test kits from Macherey-Nagel (Duren, Germany) were used. A Sievers InnovOx laboratory total carbon analyzer (Veolia, Aubervilliers, France) was used to measure the DOC. Nutrients were measured using an AQ300 discrete analyzer (SEAL Analytical, Inc., Southampton, UK). A pH meter of Hanna Instruments (HI 991003, Temse, Belgium) was used for pH measurement and a turbidimeter (Hach, Germany) for turbidity. All samples (except total COD) were filtered on 1.2-μm microfiber glass filters (VWR International, Leuven, Belgium) prior to analyses.

For the sludge analyses, biomass concentrations such as mixed liquor (volatile) suspended solids (ML(V)SSs) and the (diluted) sludge volume index ((D)SVI) were measured according to the standard method (APHA 1998). For the DSVI, the dilution factor was 2 times with the effluent. Sludge morphology was analyzed using an Olympus microscope (CX 43, Tokyo, Japan) with a bright field and phase contrast. The capillary suction time (CST) of the sludge was determined using a type 304B (Essex, UK).

2.4. In Situ DOC Substrate Removal

In situ DOC removal experiments were carried out regularly to determine the DOC removal during the anaerobic feeding and reaction step. Samples were taken at the start of the anaerobic feeding step and just before the end of the anaerobic phase. All sludge samples were filtered by fiberglass filter before analysis. The following formula was used to calculate the DOC removal during the anaerobic step:

$$\%\ \text{Anaerobic DOC removal} = 100 \times \frac{DOCs - DOCe}{DOCs - DOCc}$$

DOCs is DOC right after feeding, DOCe is the DOCs at the end of the anaerobic step, and DOCc is the DOCs at the end of the SBR cycle.

2.5. DNA Extraction and Bacteria 16S rRNA Gene Sequencing

To extract DNA from the nine activated sludge samples, 0.50 mL of each sample was processed using the FastDNA Spin Kit for Soil (MP Biomedicals, Santa Ana, CA, USA), following the manufacturer's instructions. The frozen samples were then shipped to DNAsense ApS (Aalborg, Denmark) for library preparation, sequencing, and bioinformatics processing. The gene amplicon sequencing targeted the bacterial 16S rRNA gene variable regions 1–8 (bV18-A); after which, a taxonomic classification was conducted against the MiDAS 5.2 database. A custom DNAsense ApS (Aalbor, Denmark) protocol was used to prepare amplicon libraries for the 16S rRNA gene variable regions 1–8 (bV18-A) sequencing. Up to 25 ng of extracted DNA served as the basis for PCR amplification, each sample receiving custom 24 nt barcode sequence forward and reverse primers. The resulting amplicon libraries were purified using the standard protocol for CleanNGS SPRI beads (CleanNA, Waddinxveen, The Netherlands) and then prepared using the SQK-LSK114 kit (Oxford Nanopore Technologies, Oxford, UK), according to the manufacturer's instructions, in addition to a few modifications. Next, the sequencing library was loaded onto a PromethION R10.4.1 flow cell and sequenced using MinKNOW 23.11.4 software (Oxford Nanopore Technologies, UK). Reads were base-called and demultiplexed with MinKNOW Dorado 7.2.13. The project also utilized DNASense's custom bioinformatics workflow (version MGA_ONT_DS240221) for bioinformatics processing. The sequencing reads in the fastq files were first filtered for length and quality and then mapped to the QIIME-formatted MiDAS version 5.2 database. Further bioinformatics processing was conducted using RStudio IDE (2023.12.1.402), running R version 4.3.2 (2023-10-31) and employing R packages such as ampvis2 (2.8.7).

3. Results and Discussion

3.1. Characteristics of the Industrial Wastewater

The TTC deals with the cleaning of tanks that transport various types of brewery and chocolate, generating highly variable wastewater. The lab-scale SBR was fed with 18 different batches of influent wastewater with COD values ranging from 14,050 mg·L^{-1} to 2770 mg·L^{-1} (Figure 2). The average total COD was 5967 mg·L^{-1} (standard deviation (SD) = 2437 mg·L^{-1}, %CV = 42%), and the average soluble COD (sCOD) was 5135 mg·L^{-1} (SD = 2122 mg·L^{-1}, %CV = 42%). On average, the sCOD was 86% of the total COD. The average ratio of COD to dissolved organic carbon (DOC) was 3.2. The influent wastewater was acidic, with an average pH of 3.22. Therefore, the pH was adjusted to 7 by using 1 M NaOH. A COD:N:P ratio of 100:3:0.5 was applied to provide nutrients by adding NH_4Cl and K_2HPO_2.

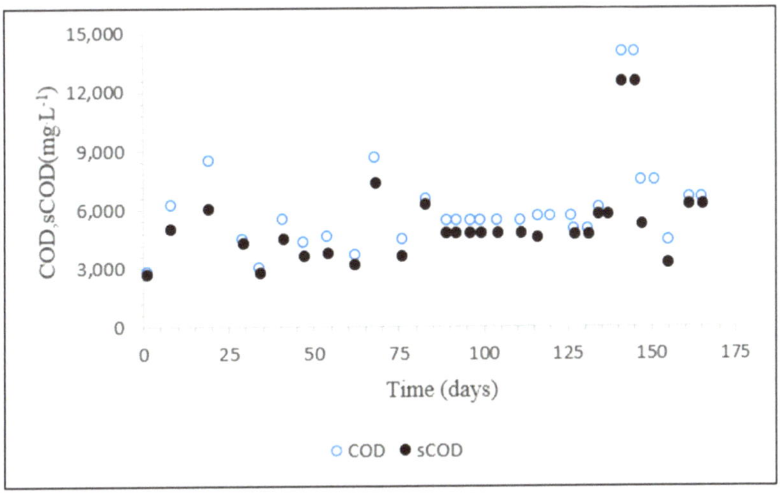

Figure 2. COD values of all batches of influent wastewater collected from the full-scale TTC company.

Figure 3 shows the evolution of the OLR values. The OLR was not constant when the reactor was started with the seed sludge. The initial MLSS values of the sludge were high, and sludge washout was observed due to poor sludge settling. When the sludge settling improved on day 69 (SVI_{30} = 113 mL·g^{-1}), a constant OLR of 0.85 ± 0.05 gCOD·(L·d)$^{-1}$ was applied by varying the volume of the influent feed. After day 130, a variable OLR was applied by a constant daily feed of 2 L influent into the reactor.

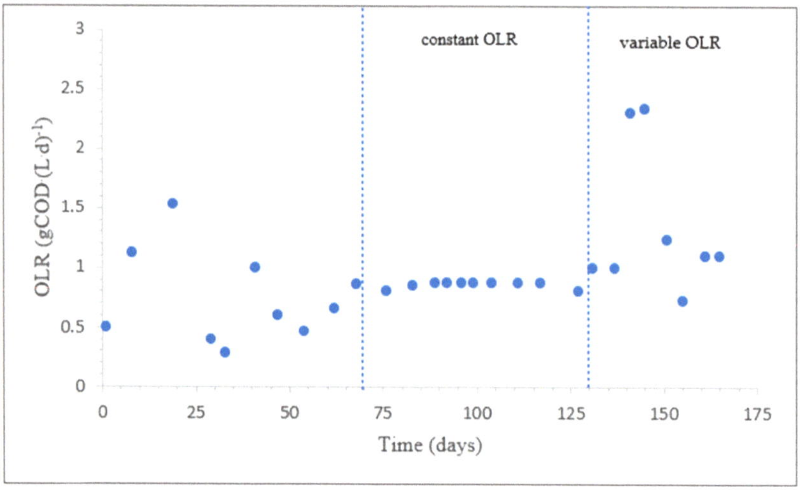

Figure 3. Evolution of the OLR values.

3.2. *In Situ DOC Removal*

The purpose of the anaerobic feeding and reaction step is to allow uptake of the readily biodegradable organic carbon by the substrate storing organism. During feeding, the high concentration of substrate in the feed penetrates the entire depth of the sludge. Under anaerobic conditions, this readily biodegradable COD is converted into storage polymers such as PHA and glycogen by PAOs and GAOs [38]. When oxygen is supplied to the microorganisms in the subsequent aerobic phase, the internally stored polymers are slowly consumed for the production of a new biomass. As the growth of heterotrophic microorganisms on storage polymers is low compared to their growth rate on readily biodegradable carbon substrate, stable and dense granulation is possible. If the readily biodegradable substrate would not be converted to storage polymers in the anaerobic phase, the granule structure will be fluffy, and filamentous bacteria may be stimulated [21].

At the beginning of the experiment, the total anaerobic phase lasted 480 min, and the anaerobic DOC removal was 47% (Figure 4), which means more than half of the biodegradable substrate entered the aerobic phase. At this stage, the sludge was highly flocculated, and its sedimentation was poor (SVI_{30} = 155 mL·g^{-1}). After a few days, the anaerobic DOC uptake decreased to 30% and then further to 24%. During this period, a decrease in MLSS was observed due to sludge washout. On day 35, the anaerobic DOC removal increased to 63%. On day 42, the DOC removal decreased slightly to 54%. However, the DOC removal efficiency soon reached 73% and then increased steadily to 94% on day 69. At this time, the filamentous microorganisms were almost gone from the sludge, the floc size increased, and the settling characteristics improved (see below). From day 69 to the end, the DOC removal efficiency during the anaerobic phase was over 84%. A constant OLR of 0.85 ± 0.05 gCOD·(L·d)$^{-1}$ was applied from day 69 until day 130. During the constant OLR period, an average of 7.71 mg·L^{-1} of DOC leaked into the aerobic phase. After day 130, when a variable OLR was applied, an average of 11.45 mg·L^{-1} of DOC leaked into the anaerobic phase.

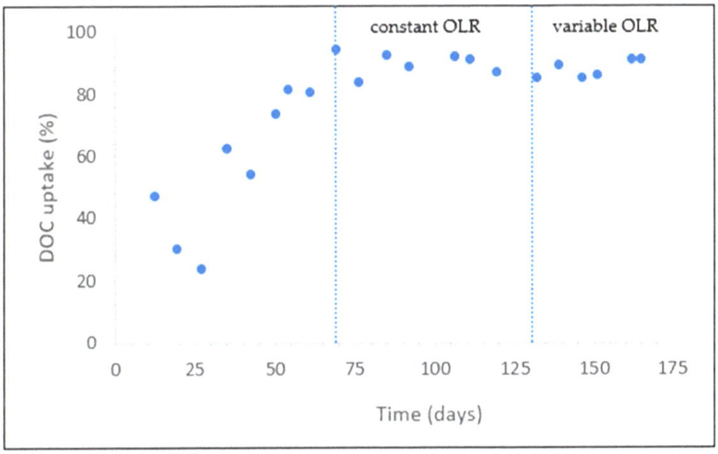

Figure 4. Anaerobic DOC uptake.

3.3. Sludge Properties

3.3.1. Sludge Concentration ML(V)SS

The MLSS of the seed sludge was 11.7 g·L^{-1} and MLVSS was 9.6 g·L^{-1} (Figure 5). On day 4, the MLSS decreased to 6.65 g·L^{-1}. The sudden decrease in the MLSS is attributed to washout of loose sludge flocs due to bad settling. The decline in MLSS continued until day 44, when it became stable. At this point, a SRT of 30 days was applied. From day 44 to day 98, the MLSS values were between 6 g·L^{-1} and 8.6 g·L^{-1}. On day 140, the MLSS reached 10.4 g·L^{-1} when a high COD of 14,050 mg·L^{-1} was fed [39]. From day 145, the SRT was lowered to 25 days to reduce the sludge concentration for proper aeration. In Figure 5 the evolution of the ML(V)SS is given.

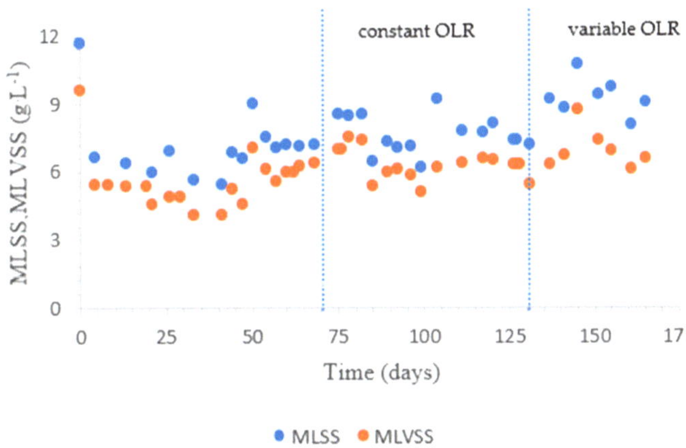

Figure 5. Evolution of the ML(V)SS values.

3.3.2. Settleability

The change in settleability of the sludge is shown in Figure 6. The SVI_{30} of the seed sludge was 147 mL·g^{-1}, but due to the high sludge concentration, this means it did not settle at all. The DSVI was 295 mL·g^{-1}. From day 40, the sludge settleability improved gradually until day 70. On day 89, the SVI_{30} reached the value of 26 mL·g^{-1}. The sludge settleability was stable until day 130 but then deteriorated when the reactor operation switched to the application of a variable OLR. Bad settling was caused by filamentous bulking, which continued to increase until day 166, as evidenced from microscopic analyses. Filamentous bulking resulted in poor settleability of the sludge (SVI_{30} values above 120 $mL \cdot g^{-1}$) [40].

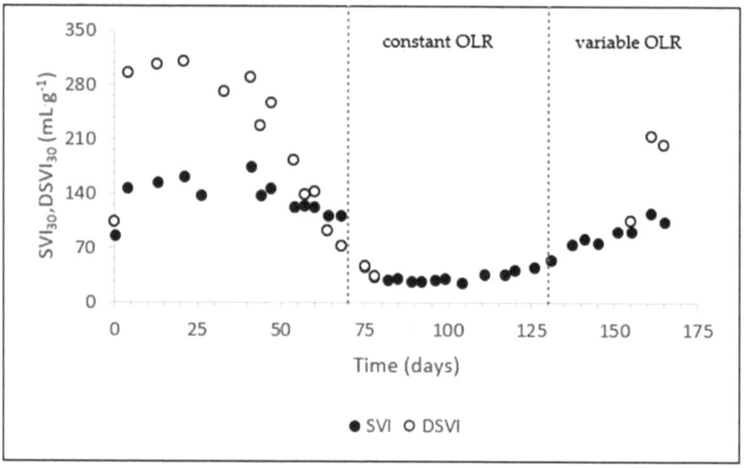

Figure 6. The (D)SVI of the sludge.

3.3.3. CST

The CST of the sludge was measured regularly to characterize its dewaterability. The CST of the seed sludge was infinite. On day 4, the CST dropped to 543 s, and it decreased further to 48 s on day 13 and to 10 s on day 57. Afterwards, the CST value was almost stable at 11 s until the end of the study. A study by Wang et al. [41] showed that a larger floc size and lower amount of SMP are the main reasons for a low CST. It should be noted that a high CST is correlated to poor filtration properties and could result in elevated membrane fouling [42].

3.3.4. Microscopy

Microscopic analysis (Figure 7) was carried out to monitor the development of the sludge morphology. The structure of the flocs is an indicator of undesirable sludge characteristics. The seed sludge came from a plant where carbohydrate-rich influent is treated. The seed sludge on day 0 was free of filamentous microorganisms but had very small flocs. The complete absence of filamentous bacteria results in small, weak, and dispersed flocs [43]. On day 27, some filamentous growth appeared, which may aid in the granulation process [44]. On day 54, the flocs increased in size with the presence of a few filaments. On day 78, the filaments disappeared completely, and the flocs increased in size and were more compact in shape. During this period, the anaerobic DOC removal increased to 94% (Figure 4), indicating that the substrate had completely penetrated the flocs, resulting in smooth flocs. In addition, the settling characteristics of the sludge greatly improved during this time (SVI_{30} = 26 $mL \cdot g^{-1}$), as compact flocs have a high settling velocity. On day 117, some filaments reappeared on the flocs, and the size of the flocs also increased with the improved settling. On day 127, the OLR was changed from constant to variable mode. On day 151, after switching to a variable OLR operation, filaments dominated the sludge, and the settling capability deteriorated again [25]. The reason for the dominance of filamentous organisms during the variable OLR period was probably the availability of rbCOD and hydrolyzed products in the aerobic phase, as a higher amount of DOC (11.45 $mg \cdot L^{-1}$) leakage was observed in this period. The increased DOC leakage into the aerobic phase has harmful effects on densification [45,46].

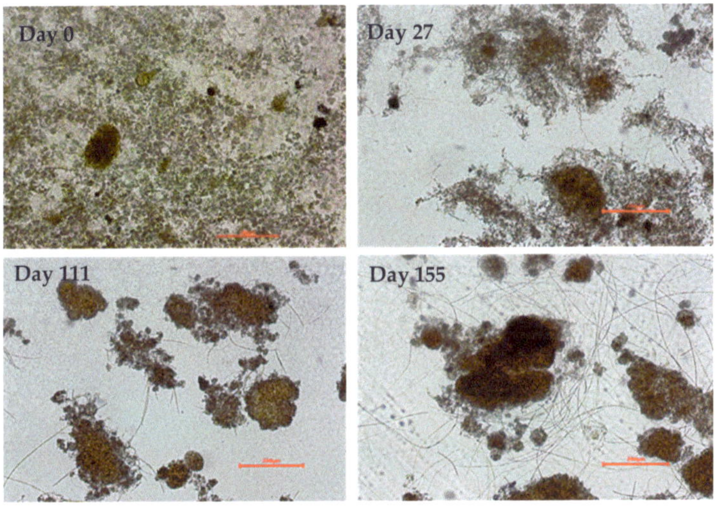

Figure 7. Microscopic evaluation of the sludge for floc size and filamentous bacteria (scale bar = 200 µm).

3.3.5. Microbial Community

Microorganisms from the phylum *Bacteriodota* and *Patescibacter* were in abundance in the seed sludge. The most abundant genus was *Ca.* Epiflobacter from *Bacteriodota* and *midas_g_8390* from *Patescibacteria*, with an abundance of 16.26%. The role of *midas_g_8390* is still unclear, while *Ca.* Epiflobacter degrades proteins, polysaccharides, and complex molecules [47]. At the start, the flocs were small, and the sludge was facing a viscous bulking problem with no settling and infinite CST. On day 19, microorganisms from the phylum *Bacteriodota* were still predominant, with the OHO *Terrimonas* high in abundance, reaching 37.5%. The values of the CST, SVI, and anaerobic uptake were not good at this time. From day 60, the sludge morphology showed significant densification, the anaerobic DOC uptake was high, and the CST reached low values. During this time, the GAO *Candidatus* Competibacter became the most abundant GAO, followed by the genus *Defluviicoccus*. *Candidatus* Competibacter secretes EPS to form compact AGS [48,49]. The substrate of the second-most abundant GAO *Defluviicoccus* is propionate and butyrate, and the formation of alginate-like extracellular polymers are linked to *Defluviicoccus* [50,51]. The overall abundance of GAOs reached 8% on day 82. The flocs become denser and larger in size without any filaments. The overall abundance of the PAOs and GAOs reached 12% on day 117. The PAO *Candidatus* Accumulibacter appeared during the variable OLR phase and reached a maximum value of 0.23% on day 161. A study conducted by Caluwe et al. [52] to form hybrid granular sludge using microbial selection found that the PAO abundance was never more than 1%, and *Defluviicoccous* and *Candidatus* Competibacter were the common genera of the GAOs. These GAOs can store VFA and convert them into PHA under anaerobic conditions. The storage by GAOs of rbCOD under anaerobic conditions and preventing its leakage to the aerobic phase hinders filamentous organism growth. Another study conducted by Tsertou et al. [11] also found that *Defluviicoccus* and *Candidatus* Competibacter were the most common GAOs. A study carried out by Burzio et al. [53] found that *Candidatus* Competibacter was more dominant in AGS while treating complex synthetic water. Li et al. [51] found that the *Defluviicoccus* and *Candidatus* Competibacter abundance ranged from 16.88% to 39.31% in a successful AGS cultivation. A study by Begum et al. [54] showed that a high abundance of *Defluviicoccus* resulted in the deterioration of P-removal in the enhanced biological phosphorous removal (EBPR) system. The abundance of filamentous organisms was low until variable OLR was applied from day 130 onwards. The filamentous organisms *Thiothrix* and *Leptothrix* from the phylum proteobacteria appeared in the sludge and also affected the sludge morphology

(Figure 6). *Thiothrix* is a common filamentous organism in WWTPs and its presence negatively correlates with settling. On day 161, the abundance of filamentous organisms reached 11%. The increase in abundance of filamentous MOs was related to the increase in SVI values, but the CST and the anaerobic DOC uptake values were not affected. In Table 1a–c, the abundances of GAOs, PAOs, and filamentous microorganisms are listed.

Table 1. (**a**) Abundance of genera of the GAOs found in the SBR (%). (**b**) Abundance of genera of the PAOs found in the SBR (%). (**c**) Abundance of genera of the filamentous microorganisms found in the SBR (%).

(a)								
Genus/Day	0	19	60	82	104	117	141	161
Defluviicoccus; Proteobacteria	1.489	0.098	1.850	3.021	2.123	5.837	2.034	1.260
*Candidatus*_Competibacter; *Proteobacteria*	0.405	0.000	6.018	4.401	3.934	4.173	3.437	1.680
Propionivibrio; Proteobacteria	1.963	1.188	0.032	0.217	0.121	0.088	0.170	0.024
Micropruina; Actinobacteriota	0.000	0.000	0.025	0.000	0.056	0.173	0.493	0.104
(b)								
Genus/Day	0	19	60	82	104	117	141	161
Candidatus Accumulibacter; *Proteobacteria*	0.00	0.00	0.00	0.09	0.00	0.00	0.19	0.23
Tetrasphaera; Actinobacteriota	0.02	0.00	0.08	0.00	0.08	0.05	0.03	0.00
(c)								
Genus/Day	0	19	60	82	104	117	141	161
Thiothrix; Proteobacteria	0.00	0.00	0.00	0.07	0.01	0.06	6.73	9.81
Ca_Villigracilis; Chloroflexi	0.00	0.00	0.13	0.33	0.27	0.80	0.22	0.29
Lewinella; Bacteroidota	0.05	0.04	0.07	0.08	0.08	0.14	0.03	0.06
Kouleothrix; Chloroflexi	0.00	0.00	0.02	0.00	0.04	0.04	0.28	0.10
Erysipelothrix; Firmicutes	0.35	0.01	0.00	0.00	0.02	0.00	0.00	0.00

3.4. Membrane Filtration Experiments

The membrane filtration process operates in two modes: the constant flux and the constant TMP modes. In the constant flux mode, the TMP rises as fouling occurs to maintain a constant flux. In the constant TMP mode, the permeability declines as a result of fouling to maintain a constant TMP. In our experiments, the constant flux mode is used for the filtration tests. For SBR sludge, the TMP_{avg} increased with the increasing flux as the fouling potential increased. The minimum TMP_{avg} of 0.9 mbar was observed at a flux of 5 $L{\cdot}(m^2{\cdot}h)^{-1}$, and the maximum TMP_{avg} was observed at a flux of 55 $L{\cdot}(m^2{\cdot}h)^{-1}$, which was 29.4 mbar. The permeability first decreased and then increased at a flux rate of 15 $L{\cdot}(m^2{\cdot}h)^{-1}$. After the flux of 15 $L{\cdot}(m^2{\cdot}h)^{-1}$ was reached, the permeability decreased continuously and reached a minimum value at the flux of 55 $L{\cdot}(m^2{\cdot}h)^{-1}$. The permeability decreased as the flux rate increased and as the chances of fouling increased. When fouling occurs, the available pores for permeation become limited, which reduces the overall permeability [55]. For SBR sludge, the contribution of irreversible fouling to the initial permeability loss was almost 20%. During the relaxation step, a fraction of the cake layer was removed, which helped in regaining the permeability. At higher flux rates, the cake layers were probably too difficult to be removed completely during relaxation, leading to continuous permeability losses [56]. The sustainable flux ($SF_{0.5}$) is the flux at which the fouling rate (FR) remains below the value of 0.5 mbar·min^{-1}, as shown by Stes et al. [57]. The concept of sustainable flux is suitable to determine the practical and economic feasibility of the filtration processes. In SBR sludge filtration, the FR crossed the value of 0.5 mbar·min^{-1} at the flux of 50 L·(m^2·h)$^{-1}$. The $SF_{0.5}$ value for the lab-scale SBR sludge was therefore 45 L·(m^2·h)$^{-1}$. The flocculent sludge from the full-scale TTC treatment plant showed very high values of TMP_{avg} and FR and low values for permeability as compared to the lab-scale SBR cultivated sludge. First, the TMP_{avg} increased steadily up to 10.1 mbar until a flux of 20 L·(m^2·h)$^{-1}$. The FR value at a flux of 20 L·(m^2·h)$^{-1}$ was 0.52 mbar·min^{-1}, which showed that the $SF_{0.5}$ for the full-scale sludge was 15 L·(m^2·h)$^{-1}$ [58]. For the TTC sludge, the contribution of irreversible fouling was 40% after the initial permeability loss. In SBR sludge filtration, the average membrane resistance due to pore blockage (Rp) was 5.3×10^{10} m^{-1}, which was 2.1 times lower than the Rp of the TTC sludge. During the filtration of the full-scale sludge, the main reason for fouling was the cake layer with the contribution of pore blocking, while, during filtration of the lab-scale SBR sludge, pore blocking was the sole factor for fouling [59]. The comparison of TMP_{avg}, permeability and FR of the seed sludge from full-scale and SBR cultivated sludge are shown in Figure 8a,b.

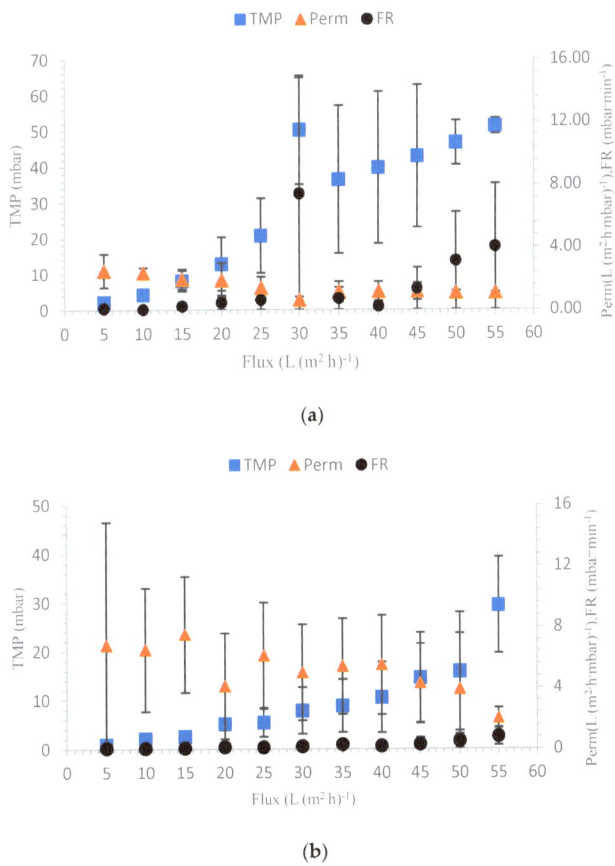

Figure 8. (**a**) TTC sludge filtration characteristics. (**b**) SBR sludge filtration characteristics.

The findings of Tsertou et al. [36] confirm our results. Their study was to check the performance of AGS and conventional AS in membrane filtration, and it was shown that the $SF_{0.5}$ value for AGS was higher than for activated sludge. For AGS, the FR never exceeded 0.1 mbar·min^{-1} at a maximum flux of 55 L·(m^2·h)$^{-1}$, while the $SF_{0.5}$ for conventional sludge was less than 35 L·(m^2·h)$^{-1}$. The permeability of AGS was also more than twice that of conventional AS. The high $SF_{0.5}$ value indicates less fouling problems.

Truong et al. [60] also carried out a comparative study of AGS and AS. They found that the Rc of AS was 4–13 times higher than AGMBR, resulting in two to three-times higher Rt for AS than AGMBR. Iorhemen et al. [29] also compared the filtration results of AGS and AS. They observed that the TMP increased more rapidly in AS than in AGS. The reasons for the rapid increase in TMP for AS were that AS was easily deposited on the membrane surface, resulting in a dense cake layer and pore blockage. For AGS, the large and rigid structure of AGS causes shear on the membrane surface, which mitigates the fouling. Wang et al. [37] conducted filtration experiments on AGS and AS. They found that fouling was higher in AS than in AGS because of the larger AGS particle size. In a study by Zhang et al. [24], it was concluded that pore blocking was the main form of the fouling in the AGMBR.

3.5. Unsuccessful Granulation

The objective of this study was to form AGS starting from floccular sludge using microbial selection only. The results showed that the applied strategy resulted in a more densified sludge as compared to the initial full-scale sludge, but full granulation was not achieved. AGS has some unique features, like a SVI_{30} of less than 50 mL·g^{-1}, SVI_{10}/SVI_{30} value of approximately 1, and a minimum size of 200 μm. In our experiment, the lowest SVI_{30} value was 26 mL·g^{-1}, the SVI_{10}/SVI_{30} was 1.16, and the DV_{50} of the flocs reached 254 μm, but its shape was never spherical. There are two possible reasons for the incomplete granulation in our experiment. These reasons are:

1. High settling time: The settling of the reactor in our experiment was 55 min. For AGS, the settling time should be less, so the loose flocs are washed out, and the best settling flocs remains in the reactor. In AGS formation, typically, a short settling of 2–10 min is applied [20].
2. High particulate COD (pCOD): A high amount of pCOD in the influent is not good for granulation. In our experiment, the average pCOD value was 832 mg·L^{-1}. In pCOD, the most challenging part is the slowly biodegradable fraction, which is hydrolyzed to rbCOD in the reactor. The leakage of rbCOD to the aerobic phase favors OHO, and it is detrimental to granulation [48].

4. Conclusions

The aim of this study was to improve the settleability and filterability of a seed sludge originating from a TTC company treating highly variable influent wastewater. A slow anaerobic feeding strategy was adopted to improve the sludge characteristics and pollutant removal by the microbial selection of slow-growing organisms such as GAOs. Slow anaerobic feeding, sludge densification, improved settling, and better membrane filtration properties were achieved when applying a constant OLR. The application of variable OLR was detrimental to sludge settling, as filamentous organisms appeared. For the densification of the sludge, a stable OLR is required. The strategy of microbial selection to form aerobic granules can be further improved by implementing an adequate influent pretreatment to remove particulates in the influent.

Author Contributions: Conceptualization, M.A. and J.D.; methodology, D.G.; software, D.G.; validation, M.A., D.G. and J.D.; formal analysis, C.V.; investigation, M.A.; resources, K.G.; data curation, M.A.; writing—original draft preparation, M.A.; writing—review and editing, M.A.; visualization, M.A.; supervision, J.D.; project administration, J.D.; funding acquisition, M.A. All authors have read and agreed to the published version of the manuscript.

Funding: This research was funded by the Higher Education Commission (HEC) of Pakistan.

Data Availability Statement: All relevant data are included in the paper.

Conflicts of Interest: The authors declare no conflicts of interests.

References

1. Fan, N.; Wang, R.; Qi, R.; Gao, Y.; Rossetti, S.; Tandoi, V.; Yang, M. Control strategy for filamentous sludge bulking: Bench-scale test and full-scale application. *Chemosphere* **2018**, *210*, 709–716. [CrossRef] [PubMed]
2. Tandoi, V.; Rossetti, S.; Wanner, J. (Eds.) *Activated Sludge Separation Problems: Theory, Control Measures, Practical Experiences*; IWA Publishing: Melbourne, Australia, 2017. [CrossRef]
3. Cornelissen, R.; Van Dyck, T.; Dries, J.; Ockier, P.; Smets, I.; Van den Broeck, R.; Van Hulle, S.; Feyaerts, M. Application of online instrumentation in industrial wastewater treatment plants—A survey in Flanders, Belgium. *Water Sci. Technol.* **2018**, *78*, 957–967. [CrossRef] [PubMed]
4. van Loosdrecht, M.C.M.; Pot, M.A.; Heijnen, J.J. Importance of bacterial storage polymers in bioprocesses. *Water Sci. Technol.* **1997**, *35*, 41–47. [CrossRef]
5. Guo, J.; Peng, Y.; Wang, Z.; Yuan, Z.; Yang, X.; Wang, S. Control filamentous bulking caused by chlorine-resistant Type 021N bacteria through adding a biocide CTAB. *Water Res.* **2012**, *46*, 6531–6542. [CrossRef] [PubMed]
6. Wang, P.; Yu, Z.; Qi, R.; Zhang, H. Detailed comparison of bacterial communities during seasonal sludge bulking in a municipal wastewater treatment plant. *Water Res.* **2016**, *105*, 157–166. [CrossRef] [PubMed]
7. Zerva, I.; Remmas, N.; Kagalou, I.; Melidis, P.; Ariantsi, M.; Sylaios, G.; Ntougias, S. Effect of Chlorination on Microbiological Quality of Effluent of a Full-Scale Wastewater Treatment Plant. *Life* **2021**, *11*, 68. [CrossRef] [PubMed]
8. Ferreira, V.; Martins, C.; Pereira, M.O.; Nicolau, A. Use of an aerobic selector to overcome filamentous bulking in an activated sludge wastewater treatment plant. *Environ. Technol.* **2014**, *35*, 1525–1531. [CrossRef] [PubMed]
9. Jenkins, D.; Richard, M.G.; Daigger, G.T. *Manual on the Causes and Control of Activated Sludge Bulking, Foaming, and Other Solids Separation Problems*; CRC Press: Boca Raton, FL, USA, 2003.
10. Cao, Y.S.; Teo, K.H.; Yuen, W.A.; Long, W.Y.; Seah, B. Performance analysis of anoxic selector in upgrading activated sludge process in tropical climate. *Water Sci. Technol.* **2005**, *52*, 27–37. [CrossRef] [PubMed]
11. Tsertou, E.; Caluwé, M.; Goossens, K.; Dobbeleers, T.; Dockx, L.; Poelmans, S.; Suazo, K.S.; Dries, J. Is building up substrate during anaerobic feeding necessary for granulation? *Water Sci. Technol.* **2022**, *86*, 763–776. [CrossRef] [PubMed]
12. Caluwé, M.; Goossens, K.; Seguel Suazo, K.; Tsertou, E.; Dries, J. Granulation strategies applied to industrial wastewater treatment: From lab to full-scale. *Water Sci. Technol.* **2022**, *85*, 2761–2771. [CrossRef]
13. Stes, H.; Aerts, S.; Caluwé, M.; Dobbeleers, T.; Wuyts, S.; Kiekens, F.; D'aes, J.; De Langhe, P.; Dries, J. Formation of aerobic granular sludge and the influence of the pH on sludge characteristics in a SBR fed with brewery/bottling plant wastewater. *Water Sci. Technol.* **2018**, *77*, 2253–2264. [CrossRef] [PubMed]
14. Ali, M.; Wang, Z.; Salam, K.W.; Hari, A.R.; Pronk, M.; van Loosdrecht, M.C.M.; Saikaly, P.E. Importance of Species Sorting and Immigration on the Bacterial Assembly of Different-Sized Aggregates in a Full-Scale Aerobic Granular Sludge Plant. *Environ. Sci. Technol.* **2019**, *53*, 8291–8301. [CrossRef]
15. Martins, A.M.P.; Heijnen, J.J.; van Loosdrecht, M.C.M. Bulking sludge in biological nutrient removal systems. *Biotechnol. Bioeng.* **2004**, *86*, 125–135. [CrossRef] [PubMed]
16. Haaksman, V.; Schouteren, M.; van Loosdrecht, M.; Pronk, M. Impact of the anaerobic feeding mode on substrate distribution in aerobic granular sludge. *Water Res.* **2023**, *233*, 119803. [CrossRef]
17. Iorhemen, O.T.; Hamza, R.A.; Zaghloul, M.S.; Tay, J.H. Simultaneous organics and nutrients removal in side-stream aerobic granular sludge membrane bioreactor (AGMBR). *J. Water Process Eng.* **2018**, *21*, 127–132. [CrossRef]
18. Ni, B.-J.; Xie, W.-M.; Liu, S.-G.; Yu, H.-Q.; Wang, Y.-Z.; Wang, G.; Dai, X.-L. Granulation of activated sludge in a pilot-scale sequencing batch reactor for the treatment of low-strength municipal wastewater. *Water Res.* **2009**, *43*, 751–761. [CrossRef]
19. Nancharaiah, Y.; Reddy, G.K.K. Aerobic granular sludge technology: Mechanisms of granulation and biotechnological applications. *Bioresour. Technol.* **2018**, *247*, 1128–1143. [CrossRef]
20. Show, K.-Y.; Lee, D.-J.; Tay, J.-H. Aerobic granulation: Advances and challenges. *Appl. Biochem. Biotechnol.* **2012**, *167*, 1622–1640. [CrossRef]
21. de Kreuk, M.; van Loosdrecht, M. Selection of slow growing organisms as a means for improving aerobic granular sludge stability. *Water Sci. Technol.* **2004**, *49*, 9–17. [CrossRef]
22. Corsino, S.F.; Di Trapani, D.; Torregrossa, M.; Viviani, G. Aerobic granular sludge treating high strength citrus wastewater: Analysis of pH and organic loading rate effect on kinetics, performance and stability. *J. Environ. Manag.* **2018**, *214*, 23–35. [CrossRef]
23. Adav, S.S.; Lee, D.-J.; Lai, J.-Y. Functional consortium from aerobic granules under high organic loading rates. *Bioresour. Technol.* **2009**, *100*, 3465–3470. [CrossRef] [PubMed]
24. Zhang, W.; Jiang, F. Membrane fouling in aerobic granular sludge (AGS)-membrane bioreactor (MBR): Effect of AGS size. *Water Res.* **2019**, *157*, 445–453. [CrossRef] [PubMed]

25. Gurung, K.; Ncibi, M.C.; Fontmorin, J.-M.; Särkkä, H.; Sillanpää, M. Incorporating Submerged MBR in Conventional Activated Sludge Process for Municipal Wastewater Treatment: A Feasibility and Performance Assessment. *J. Membr. Sci. Technol.* **2016**, *6*, 3. [CrossRef]
26. Le-Clech, P. Membrane bioreactors and their uses in wastewater treatments. *Appl. Microbiol. Biotechnol.* **2010**, *88*, 1253–1260. [CrossRef] [PubMed]
27. Du, X.; Shi, Y.; Jegatheesan, V.; Haq, I.U. A Review on the Mechanism, Impacts and Control Methods of Membrane Fouling in MBR System. *Membranes* **2020**, *10*, 24. [CrossRef] [PubMed]
28. Zhang, B.; Huang, D.; Shen, Y.; Yin, W.; Gao, X.; Zhang, B.; Shi, W. Treatment of municipal wastewater with aerobic granular sludge membrane bioreactor (AGMBR): Performance and membrane fouling. *J. Clean. Prod.* **2020**, *273*, 123124. [CrossRef]
29. Iorhemen, O.T.; Hamza, R.A.; Zaghloul, M.S.; Tay, J.H. Aerobic granular sludge membrane bioreactor (AGMBR): Extracellular polymeric substances (EPS) analysis. *Water Res.* **2019**, *156*, 305–314. [CrossRef] [PubMed]
30. Tay, J.H.; Yang, P.; Zhuang, W.Q.; Tay, S.T.L.; Pan, Z.H. Reactor performance and membrane filtration in aerobic granular sludge membrane bioreactor. *J. Membr. Sci.* **2007**, *304*, 24–32. [CrossRef]
31. Tu, X.; Zhang, S.; Xu, L.; Zhang, M.; Zhu, J. Performance and fouling characteristics in a membrane sequence batch reactor (MSBR) system coupled with aerobic granular sludge. *Desalination* **2010**, *261*, 191–196. [CrossRef]
32. Caluwé, M.; Dobbeleers, T.; Daens, D.; Geuens, L.; Blust, R.; Dries, J. SBR treatment of tank truck cleaning wastewater: Sludge characteristics, chemical and ecotoxicological effluent quality. *Environ. Technol.* **2018**, *39*, 2524–2533. [CrossRef]
33. Manavi, N.; Kazemi, A.S.; Bonakdarpour, B. The development of aerobic granules from conventional activated sludge under anaerobic-aerobic cycles and their adaptation for treatment of dyeing wastewater. *Chem. Eng. J.* **2017**, *312*, 375–384. [CrossRef]
34. Dobbeleers, T.; Daens, D.; Miele, S.; D'aes, J.; Caluwé, M.; Geuens, L.; Dries, J. Performance of aerobic nitrite granules treating an anaerobic pre-treated wastewater originating from the potato industry. *Bioresour. Technol.* **2017**, *226*, 211–219. [CrossRef] [PubMed]
35. Corsino, S.F.; di Biase, A.; Devlin, T.R.; Munz, G.; Torregrossa, M.; Oleszkiewicz, J.A. Effect of extended famine conditions on aerobic granular sludge stability in the treatment of brewery wastewater. *Bioresour. Technol.* **2017**, *226*, 150–157. [CrossRef] [PubMed]
36. Tsertou, E.; Caluwé, M.; Goossens, K.; Seguel Suazo, K.; Dries, J. Performance of an aerobic granular sludge membrane filtration in a full-scale industrial plant. *Water Sci. Technol.* **2023**, *87*, 3002–3016. [CrossRef] [PubMed]
37. Wang, Y.; Zhong, C.; Huang, D.; Wang, Y.; Zhu, J. The membrane fouling characteristics of MBRs with different aerobic granular sludges at high flux. *Bioresour. Technol.* **2013**, *136*, 488–495. [CrossRef]
38. Pronk, M.; Abbas, B.; Al-zuhairy, S.H.K.; Kraan, R.; Kleerebezem, R.; van Loosdrecht, M.C.M. Effect and behaviour of different substrates in relation to the formation of aerobic granular sludge. *Appl. Microbiol. Biotechnol.* **2015**, *99*, 5257–5268. [CrossRef] [PubMed]
39. Toja Ortega, S.; Pronk, M.; de Kreuk, M.K. Effect of an Increased Particulate COD Load on the Aerobic Granular Sludge Process: A Full Scale Study. *Processes* **2021**, *9*, 1472. [CrossRef]
40. Bagheri, M.; Mirbagheri, S.A.; Bagheri, Z.; Kamarkhani, A.M. Modeling and optimization of activated sludge bulking for a real wastewater treatment plant using hybrid artificial neural networks-genetic algorithm approach. *Process Saf. Environ. Prot.* **2015**, *95*, 12–25. [CrossRef]
41. Wang, Q.; Wang, Z.; Wu, Z.; Han, X. Sludge reduction and process performance in a submerged membrane bioreactor with aquatic worms. *Chem. Eng. J.* **2011**, *172*, 929–935. [CrossRef]
42. Karczmarczyk, A.; Kowalik, W. Combination of Microscopic Tests of the Activated Sludge and Effluent Quality for More Efficient On-Site Treatment. *Water* **2022**, *14*, 489. [CrossRef]
43. Layer, M.; Adler, A.; Reynaert, E.; Hernandez, A.; Pagni, M.; Morgenroth, E.; Holliger, C.; Derlon, N. Organic substrate diffusibility governs microbial community composition, nutrient removal performance and kinetics of granulation of aerobic granular sludge. *Water Res. X* **2019**, *4*, 100033. [CrossRef] [PubMed]
44. Müller, E.; Lind, G.; Lemmer, H.; Wilderer, P.A. Population structure and chemical EPS analyses of activated sludge and scum. *Acta Hydrochim. Hydrobiol.* **2005**, *33*, 189–196. [CrossRef]
45. Kondrotaite, Z.; Valk, L.C.; Petriglieri, F.; Singleton, C.; Nierychlo, M.; Dueholm, M.K.D.; Nielsen, P.H. Diversity and Ecophysiology of the Genus OLB8 and Other Abundant Uncultured *Saprospiraceae* Genera in Global Wastewater Treatment Systems. *Front. Microbiol.* **2022**, *13*, 917553. [CrossRef] [PubMed]
46. Lemaire, R.; Webb, R.I.; Yuan, Z. Micro-scale observations of the structure of aerobic microbial granules used for the treatment of nutrient-rich industrial wastewater. *ISME J.* **2008**, *2*, 528–541. [CrossRef] [PubMed]
47. Seviour, T.W.; Lambert, L.K.; Pijuan, M.; Yuan, Z. Selectively inducing the synthesis of a key structural exopolysaccharide in aerobic granules by enriching for Candidatus "Competibacter phosphatis". *Appl. Microbiol. Biotechnol.* **2011**, *92*, 1297–1305. [CrossRef] [PubMed]
48. Oehmen, A.; Yuan, Z.; Blackall, L.L.; Keller, J. Comparison of acetate and propionate uptake by polyphosphate accumulating organisms and glycogen accumulating organisms. *Biotechnol. Bioeng.* **2005**, *91*, 162–168. [CrossRef] [PubMed]
49. Begum, S.A.; Batista, J.R. Impact of butyrate on microbial selection in enhanced biological phosphorus removal systems. *Environ. Technol.* **2014**, *35*, 2961–2972. [CrossRef] [PubMed]

50. Schambeck, C.M.; Magnus, B.S.; de Souza, L.C.R.; Leite, W.R.M.; Derlon, N.; Guimarães, L.B.; da Costa, R.H.R. Biopolymers recovery: Dynamics and characterization of alginate-like exopolymers in an aerobic granular sludge system treating municipal wastewater without sludge inoculum. *J. Environ. Manag.* **2020**, *263*, 110394. [CrossRef]
51. Li, D.; Yang, J.; Li, Y.; Zhang, J. Research on rapid cultivation of aerobic granular sludge (AGS) with different feast-famine strategies in continuous flow reactor and achieving high-level denitrification via utilization of soluble microbial product (SMP). *Sci. Total Environ.* **2021**, *786*, 147237. [CrossRef]
52. Caluwé, M.; Dobbeleers, T.; D'aes, J.; Miele, S.; Akkermans, V.; Daens, D.; Geuens, L.; Kiekens, F.; Blust, R.; Dries, J. Formation of aerobic granular sludge during the treatment of petrochemical wastewater. *Bioresour. Technol.* **2017**, *238*, 559–567. [CrossRef] [PubMed]
53. Burzio, C.; Ekholm, J.; Modin, O.; Falås, P.; Svahn, O.; Persson, F.; van Erp, T.; Gustavsson, D.J.I.; Wilén, B.-M. Removal of organic micropollutants from municipal wastewater by aerobic granular sludge and conventional activated sludge. *J. Hazard. Mater.* **2022**, *438*, 129528. [CrossRef] [PubMed]
54. Begum, S.A.; Batista, J.R. Microbial selection on enhanced biological phosphorus removal systems fed exclusively with glucose. *World J. Microbiol. Biotechnol.* **2012**, *28*, 2181–2193. [CrossRef] [PubMed]
55. Fernandes, C.S.; Bilad, M.R.; Nordin, N.A.H.M. Silica incorporated membrane for wastewater based filtration. *AIP Conf. Proc.* **2017**, *1891*, 020041. [CrossRef]
56. De Vleeschauwer, F.; Caluwé, M.; Dobbeleers, T.; Stes, H.; Dockx, L.; Kiekens, F.; Copot, C.; Dries, J. A dynamically controlled anaerobic/aerobic granular sludge reactor efficiently treats brewery/bottling wastewater. *Water Sci. Technol.* **2021**, *84*, 3515–3527. [CrossRef] [PubMed]
57. Stes, H.; Caluwé, M.; Dockx, L.; Cornelissen, R.; Langhe, P.; Smets, I.; Dries, J. Cultivation of aerobic granular sludge for the treatment of food-processing wastewater and the impact on membrane filtration properties. *Water Sci. Technol.* **2020**, *83*, 39–51. [CrossRef] [PubMed]
58. Kim, H.-G.; Jang, H.-N.; Kim, H.-M.; Lee, D.-S.; Chung, T.-H. Effects of the sludge reduction system in MBR on the membrane permeability. *Desalination* **2010**, *250*, 601–604. [CrossRef]
59. Li, J.; Liu, Y.; Li, X.; Cheng, F. Reactor performance and membrane fouling of a novel submerged aerobic granular sludge membrane bioreactor during long-term operation. *J. Water Reuse Desalination* **2019**, *9*, 1–9. [CrossRef]
60. Truong, H.T.B.; Bui, H.M. Potential of aerobic granular sludge membrane bioreactor (AGMBR) in wastewater treatment. *Bioengineered* **2023**, *14*, 2260139. [CrossRef]

water

Article

Construction of Genetically Engineered *Escherichia coli* Cell Factory for Enhanced Cadmium Bioaccumulation in Wastewater

Lingna Tian [1,2], Daiwei Wang [2], Yueying Liu [2], Mingjie Wei [2], Xuexue Han [2], Xiaomei Sun [1], Liang Yin [2] and Guanghong Luo [1,2,*]

[1] College of Resources and Environmental Sciences, Gansu Agricultural University, Lanzhou 730000, China; 17339853564@163.com (L.T.); sunxm@gsau.edu.cn (X.S.)

[2] Gansu Microalgae Engineering Technology Research Center, Hexi University, Zhangye 734000, China; wangdaiwei1216@163.com (D.W.); xiaoliu9616@163.com (Y.L.); 15318662213@163.com (M.W.); 17793638830@163.com (X.H.); yinl03@163.com (L.Y.)

* Correspondence: 13993693452@163.com

Abstract: The heavy metal cadmium poses severe threats to both ecosystems and human health. Utilizing genetic engineering to enhance the microbial capability for efficient cadmium accumulation has emerged as a pivotal research direction. This study constructed a genetically engineered bacterium capable of expressing multivalent phytochelatins with a self-assembly ability and explored its efficacy in cadmium adsorption. Molecular biology techniques were adopted to fuse the recombinant human ferritin (rHF) gene and the synthetic phytochelatin (EC) gene, known for its robust adsorption capacity for heavy metals. The expression vector was constructed. *Escherichia coli (E. coli)* served as the host cell to express multivalent nanochelator rHF-ECs tailored for high-efficiency heavy metal adsorption. The results reveal the successful soluble expression of the recombinant fusion protein in *E. coli* cells, forming self-assembled multivalent nanoparticles with a size of about 13 nm, and the target protein rHF-EC20 (monomer) could adsorb approximately 9.2 μmol of Cd^{2+} in vitro. Moreover, this recombinant strain demonstrated cadmium adsorption across a temperature range of 16–45 °C and a pH range of 5–9, with the optimal performance observed at pH 7.0 and 37 °C. Compared with the control strain, the recombinant strain BL21 (FLE), expressing nano-chelating peptides, achieves an adsorption rate of 80% for Cd^{2+} at 60 min, resulting in an approximately 18% increase in the Cd^{2+} enrichment efficiency. The maximum adsorption capability of cadmium reached 12.62 mg per gram of dry cell weight. This work indicated that the synthesis of multivalent chelating peptides in *E. coli* cells could efficiently enhance the bioaccumulation of the heavy metal cadmium, which renders novel avenues and methodologies for addressing cadmium pollution, offering promising prospects for environmental remediation.

Keywords: *Escherichia coli*; multivalent nanoparticles; cadmium adsorption; wastewater treatment

Citation: Tian, L.; Wang, D.; Liu, Y.; Wei, M.; Han, X.; Sun, X.; Yin, L.; Luo, G. Construction of Genetically Engineered *Escherichia coli* Cell Factory for Enhanced Cadmium Bioaccumulation in Wastewater. *Water* **2024**, *16*, 1759. https://doi.org/10.3390/w16131759

Academic Editors: Issam A. Al-Khatib, Rehab O. Abdel Rahman, Tsuyoshi Imai and Yung-Tse Hung

Received: 14 May 2024
Revised: 13 June 2024
Accepted: 18 June 2024
Published: 21 June 2024

1. Introduction

Cadmium, a natural heavy metal characterized by strong mobility, possesses a high toxicity to human health. The discharge of waste and sewage in industries such as electronics, chemicals, and photovoltaics as well as cadmium-laden dust and smoke deposition in the atmosphere have led to a substantial accumulation of cadmium in the environment. Moreover, as this metal traverses water bodies and food chains and accumulates, it further adversely affects the human respiratory, cardiovascular, gastrointestinal, reproductive, renal, and nervous systems, posing a grave threat to ecological balance and human health [1,2]. Therefore, efficiently reducing the content of cadmium in the living environment or eliminating its bioavailability has emerged as an imperative issue demanding human intervention.

In recent years, substantial research efforts have focused on the environmental remediation of cadmium pollution. Traditional physical and chemical treatment methods include ion exchange, dialysis, reverse osmosis, ozonation, nanofiltration, and chemical precipitation [3,4]. Meanwhile, there is growing interest in utilizing natural or modified biomass materials as well as carbonizing biomass into materials with adsorption properties for the adsorption and fixation of Cd^{2+} [5,6]. In recent years, hydrogel-based adsorbents have been designed and applied for the removal of heavy metals from water [7,8]. Microorganisms, pivotal in ecosystems, exhibit diverse mechanisms for cadmium adsorption and storage, including cell membrane/wall surface adsorption, intracellular aggregation, mineralization precipitation, and intracellular conversion [9,10]. Compared with conventional treatment methods, microorganisms offer advantages like a superior adsorption efficiency, absence of secondary pollution, and environmental compatibility [11–13]. Naturally occurring bacteria, fungi, seaweed, and algae have demonstrated the capability to adsorb and accumulate heavy metals [14–17]. Due to their small volume, ease of cultivation, and rapid reproduction, they have been extensively used for the remediation of heavy metal pollution in the environment. However, naturally isolated microorganisms usually possess drawbacks in their adsorption capacity and stress resistance. With the increasingly stringent requirements of national pollutant emission standards, challenges persist in the adaptability and efficiency of employing these microorganisms for heavy metal treatment. In response to this issue, recent research has focused on leveraging genetic engineering to enhance the efficiency of microbial cells in treating heavy metals [18–20]. This involves introducing dominant genes associated with the enrichment and degradation of heavy metals in microbial cells into recipient strains with a robust reproductive capacity and ease of manipulation. Candidate genes involved in the adsorption and transport of cadmium ions include cadmium transporters, metallothionein capable of binding to heavy metal ions, and synthetic phytochelatins (ECs). A previous study has verified the cadmium transport activities of mercury superfamily proteins (MerC, MerE, MerF, and MerT) [21]. Furthermore, the overexpression of these transporters has been shown to elevate the Cd uptake [22]. Metallothioneins (MTs) are intracellular proteins with a low molecular weight and high cysteine content that exhibit a robust adsorption capacity for heavy metals [23]. Organisms that overexpress metallothionein and biomaterials based on metallothionein have been extensively utilized in treating heavy metals [24–26]. Ying Cai et al. successfully expressed both the cadmium transporter protein and pea metallothionein in *E. coli* cells, resulting in a 50% increase in cadmium ion accumulation in stem cells per gram compared to wild strains [27]. ECs are heavy metal binding peptides that mimic phytochelatins. With a structure similar to that of the cell detoxifying agent glutathione, ECs contain repeating dipeptide units composed of glutamic acid and cysteine {(Glu-Cys) n Gly (n: repetition times)}. These two amino acids can efficiently chelate heavy metals through carboxyl and thiol groups [28]. Bae et al. fused and expressed synthetic chelating peptide ECs, which can adsorb heavy metals, with ice nucleation protein INPNC and successfully displayed ECs on the surface of *E. coli* and *Moraxella* sp. cell membranes for the treatment of Hg^{2+} in aquatic environments. The results show that the *Moraxella* sp. engineered bacterium has a better ability to adsorb Hg^{2+} than the *E. coli* engineered bacterium, which is ten times higher [29]. Yu et al. showed metallothionein on the surface of the strain *Alishewanella* sp. WH16-1-MT, which doubled the adsorption capacity of the strain for Cd^{2+}. The engineered strain was further used to remediate a cadmium-contaminated paddy soil, effectively improving the resistance of rice to cadmium and reducing the accumulation of cadmium in rice [30]. Moreover, Zhu et al. constructed a genetically engineered strain that could efficiently enrich heavy metals by displaying a fusion protein (SynHM) carrying a 6*His-tag, a cysteine-rich short peptide, and metallothionein on the surface of *E. coli* cells. The adsorption efficiency of this strain for Cd^{2+} can reach 50 mg/L [31].

In nature, a plethora of proteins can precisely self-assemble and form various nanostructures, such as nanowires and protein nanocages. Among these, due to its unique structure and stable properties, the human ferritin heavy chain (rHF) has been widely used as a structural unit in the design and synthesis of nanodevices, showcasing its multifaceted function. Human-derived ferritin rHF primarily facilitates the storage and dynamic regulation of iron elements within the human body, exhibiting highly conserved biochemical and structural characteristics [32]. The rHF subunit contains 180 amino acids, consisting of a 5-α-helix composition, spontaneously assembled into a 24-polymer nanocage structure in cells, with an outer diameter of 12 nm and an inner diameter of 8 nm. The N-terminus of each subunit is exposed on the outer side of the cage. The overall spatial structure presents a 4-3-2 symmetry [33]. It has been verified that ferritin has an extremely high thermal stability, with a Tm value of up to 85 °C. It can achieve soluble expression and self-assembly in hosts, making it an ideal self-assembled nanoelement [34]. The successful engineering of ferritin proteins as potential nanoreactors for synthesizing/incorporating nanoparticles and as nanocarriers for tumor detection and therapy has been documented [35,36].

E. coli stands as an ideal microorganism for wastewater treatment due to its rapid growth and clear genetic background, making it a current research hotspot to enrich heavy metals through genetic engineering modification. Upon the enrichment of heavy metals, achieving subsequent non-hazardous treatment is straightforward through solidification and stabilization, including the encapsulation of pollutants into inert substrate materials (such as cement and lime), organic binders like asphalt and other thermoplastic materials, and thermally hardened organic polymers such as urea, phenolic plastics, and epoxides. These approaches aim to convert pollutants into less soluble, less migratory, or less toxic forms, thereby abating the risk of harm to ecosystems [37–39]. The design concept and system are shown in Figure 1.

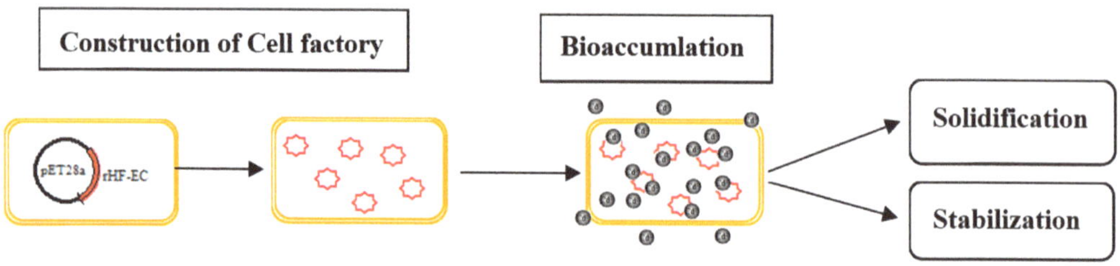

Figure 1. Design concept of cadmium treatment with *Escherichia coli* cell factory. Yellow boxes represent *E. coli* cells, the red circles represent nanochelator rHF-ECs, and black balls represent cadmium ions.

This study selected 20 repetitive units of artificial plant chelating peptide EC20 and the coding gene of human transferrin rHF for gene fusion through the peptide linker coding sequence. By overexpressing the fusion protein in *E. coli* cells, a 24-polymer multivalent, nano-chelating peptide self-assembled in the cells for heavy metal treatment. This recombinant strain was used to investigate the adsorption capacity and characteristics of Cd^{2+} in simulated wastewater, providing new research ideas and highly effective biomaterials for the bioremediation of cadmium-contaminated environments.

2. Materials and Methods

2.1. Material

2.1.1. Strain and Plasmid

The host bacteria *E. coli* DH5α and *E. coli* BL21 (DE3) and the plasmids pET28a and pET28a-rHF were preserved in the laboratory of this study. The plasmid pUC57-L-EC20 carried a linker peptide sequence (GGGSGGGS) and the fusion gene EC20. Endonucleases *Bam*H I and *Xho* I were introduced at both ends of the gene to recognize sites. The codon optimization and synthesis were fulfilled by Genscript Biotech Corporation (Nanjing, China).

2.1.2. Main Reagent

The plasmid extraction kit and DNA purification kit were purchased from Beijing Solarbio Biotechnology Co., Ltd. (Beijing, China); the protein gel electrophoresis kit was from Crystal Biology; PrimeSTAR Max DNA Polymerase, restriction endonuclease, and T4 DNA ligase came from Takara Bio (Beijing, China); and the cadmium detection kit was purchased from Henan Suijing Environmental Protection Technology Co., Ltd. (Luoyang, China). All other reagents were analytical grade.

2.1.3. Culture Medium

The LB medium for *E. coli* culture comprised the following: 0.5% yeast extract, 1% Tryptone, 1% NaCl, and a solid plate with 2% agar. The protein expression self-induction culture medium was prepared according to the formula of Gawron et al. [40], including Tryptone, 10 g/L; yeast extract, 5 g/L; $Na_2HPO_4 \cdot 12H_2O$, 9 g/L; KH_2PO_4, 3.4 g/L; NH_4Cl, 2.68 g/L; Na_2SO_4, 0.71 g/L; $MgSO_4$, 0.24 g/L; glycerol, 6 mL/L; glucose, 0.5 g/L; and α-D-Lactose, 2 g/L. When using the above culture media, kanamycin sulfate with a final mass concentration of 0.1 mg/L was added.

2.2. Method

2.2.1. Construction of Fusion Protein Gene Expression Vector

Based on the nucleotide sequence of rHF (Genbank: M97164.1), specific primers were designed: FP (5′-GATCCATATGACGACCGCGTCCACC-3′, underlined as *Nde* I restriction site) and RP (5′-TACCGGATCCGGGCGCTCCCATCTTGCG-3′, underlined as *Bam*H I restriction site). The PCR amplification was performed using the plasmid pET28a-rHF as a template. The PCR product was detected using 1% agarose gel electrophoresis, and the target gene rHF was recovered with a DNA purification kit. The fragment was digested with *Nde* I and *BamH* I and recovered for later use. Simultaneously, the plasmid pUC57-L-EC20 was digested by *BamH* I and *Xho* I. After the reaction, the L-EC20 gene fragment was recovered for later use. The digested products of the rHF gene and the L-EC20 gene obtained in the above steps were ligated to the pET-28a (+) vector double digested with *Nde* I and *Xho* I using T4 DNA ligase. The ligation products were transformed into *E. coli* DH5α competent cells, and monoclones were selected for colony PCR validation. Clones with preliminary validation were sent to Sangon Biotech (Shanghai) Co., Ltd. (Shanghai, China) for sequencing. The correctly sequenced recombinant plasmid vector was named pET28a-rHF-EC20.

2.2.2. Expression and Purification of Recombinant Fusion Protein

The correctly sequenced recombinant plasmid pET28a-rHF-EC20 was transformed into *E. coli* BL21 (DE3) competent cells to obtain the recombinant strain BL21 (FLE). Meanwhile, pET28a was transformed into the same strain to serve as the control strain BL21 (28a). Single colonies were picked from LB agar plates and inoculated into a 5 mL LB liquid medium containing 100 mg/mL kanamycin. The cultures were incubated overnight at 37 °C and 180 rpm to prepare seed solutions. These seed solutions were then inoculated at 5% (volume fraction) into a 300 mL auto-induction medium containing the same antibiotic at 37 °C for 16 h. Then, the bacterial cells were collected by centrifugation at 4 °C and 12,000 rpm for 5 min. Part of the bacterial cells was resuspended in a pre-cooled cell disruption buffer, followed by sonication under ice bath conditions for 30 min (working duration: 3 s; interval time: 3 s; power: 300 W) until the bacterial suspension became clear. The suspension was centrifuged again at 12,000 rpm for 30 min to collect the supernatant and precipitate for SDS-PAGE analysis. The supernatant passed through a 0.45 μm filter membrane, and the recombinant protein in the filtered supernatant was purified using a Ni-NTA column through affinity chromatography. The purified protein was concentrated and desalted to determine its concentration before being stored at −80 °C for future use.

2.2.3. Morphological Characterization of Recombinant Fusion Protein

The purified target protein was diluted to lower the salt ion concentration. Then, 10 μL of the protein solution was pipetted onto a carbon membrane and dried at room temperature. Next, it was negatively stained with 1% phosphotungstic acid solution (pH 7.0) for three minutes, followed by rinsing with deionized water 2–3 times. After drying at room temperature, the morphology of the protein was observed using the transmission electron microscope FEI Tecnai G2 spirit Biotwin (Thermo Fisher Scientific Inc., Waltham, MA, USA).

2.2.4. Adsorption of Recombinant Protein rHF-EC20 to Cadmium

A total of 2 mg of purified target protein rHF-EC20 was mixed with 100 mg of Ni-NTA agarose resin loaded with Ni^{2+} and reacted for 30 min to ensure full binding to the resin. The resin was washed with ddH_2O to remove unbound target protein. The eluate was collected after centrifugation, and the protein content was detected to calculate the amount of protein fixed to the resin. The immobilized target protein was mixed with 100 mg/L cadmium chloride and shaken at room temperature for 30 min. The supernatant was collected after centrifugation and washed once with a small amount of ddH_2O. The cadmium content was determined using a cadmium quantification detection kit. Finally, the adsorption capacity of the target protein to cadmium was calculated.

2.2.5. Adsorption of Recombinant Strain to Cadmium

Influence of Induction Time on Cadmium Adsorption by Recombinant Strain

Single colonies of the recombinant strain BL21 (FLE) and control strain BL21 (28a) were picked and inoculated into 5 mL LB liquid medium containing kanamycin and cultured overnight to prepare seed solutions. These seed solutions were then inoculated at 5% (volume fraction) into 200 mL of auto-induction medium supplemented with cadmium chloride with a final concentration of 50 mg/L. The cultures were incubated at 37 °C for 24 h, with continuous sampling every four hours to determine the OD and the adsorption rate of the strains to cadmium.

Optimal Adsorption pH and Temperature

(1) Optimal adsorption temperature: The recombinant strain BL21 (FLE), auto-induced for expression, was collected after centrifugation and resuspended in Tris-HCl (pH 7.0). The bacterium with a density of 3 g/L was added to cadmium chloride with a final concentration of 50 mg/L. The strains were separately placed on a shaker at 16, 25, 37, and 45 °C and incubated for two hours at 150 rpm. Samples were collected into nitric acid-soaked centrifuge tubes and centrifuged at 12,000 rpm for two minutes. The bacterial cells were dried, digested with nitric acid, and then quantitatively analyzed for their cadmium content using cadmium detection kits.

(2) Optimal adsorption pH: The recombinant strain BL21 (FLE) were collected after centrifugation and resuspended in Tris-HCl buffer at pH values of 5.0, 6.0, 7.0, 8.0, and 9.0 at the optimal temperature. The bacterial density was 3 g/L. After adding cadmium chloride with a final concentration of 50 mg/L, the strains were separately incubated at 150 r/min for two hours. The bacterial cells were collected, and a quantitative analysis was performed on the adsorbed cadmium following the above procedure. The adsorption rate A was calculated using the formula $A = Ce/C_0 \times 100\%$, where C_0 represents the initial concentration of cadmium chloride (in mg), and Ce denotes the cadmium ions adsorbed by the recombinant strains (in mg).

Time Adsorption Curve of Recombinant Strain

Under optimal adsorption conditions, the self-induced expression of recombinant strain BL21 (FLE) and control strain BL21 (28a) were separately placed in 50 mg/L cadmium chloride solutions and continuously cultured for 120 min on a shaker at 150 rpm. Samples were taken at 5, 10, 20, 40, 60, and 120 min time points for detection, and the adsorption rates at these time points were calculated.

Equilibrium Adsorption Capacity of Recombinant Strain to Cadmium

Under optimal adsorption conditions, the recombinant strain BL21 (FLE), auto-induced for expression, and control strain BL21 (28a) were separately placed in cadmium chloride solutions with final concentrations of 20, 40, 60, 80, 100, and 120 mg/L and continuously cultured for 120 min on a shaker at 150 r/min. Samples were taken to determine the adsorption quantity of cadmium. The typical Langmuir model was used to calculate the equilibrium enrichment amount (qm) of strains for Cd^{2+} using the formula q = qm Ce/(K + Ce), where Ce represents the equilibrium concentration of Cd^{2+} in the solution (mg/L), q is the enrichment amount of the strain for Cd^{2+} (mg/g), qm stands for the maximum enrichment capacity (mg/g), and K denotes the dissociation constant (mg/L). The values of qm and K can be obtained graphically with the rearranged form of the above formula: Ce/q = Ce/qm + K/qm.

2.3. Statistical Analyses

In this study, all experiments were performed in three replicates. All measurements were repeated three times, and the data were represented as the mean ± standard deviation (SD). SPSS V19.0 (IBM, New York, NY, USA) software was used to conduct analysis variance, and $p < 0.05$ stands for statistical significance.

3. Result and Analysis

3.1. Construction of Fusion Gene Expression Vector and Analysis of Fusion Protein Expression

The rHF and EC20 genes were enzymatically cleaved and ligated into the prokaryotic expression vector pET28a to obtain the recombinant plasmid pET28a-rHF-EC20. After sequencing verification, the plasmid was transformed into *E. coli* BL21 (DE3) to obtain the recombinant strain *E. coli* BL21 (FLE). The fusion gene rHF-EC20 has a full length of 748 bp, encoding 248 amino acids. With the N-terminal His-tag, the fusion protein has a total length of 266 amino acids, with a theoretical molecular weight of approximately 30.8 ku and an isoelectric point of approximately 4.9. After continuous cultivation of the recombinant strain *E. coli* BL21 (FLE) in the auto-induction medium at 37 °C for 16 h, the bacterial cells were collected using centrifugation. The cells were crushed using sonication and centrifuged to harvest the supernatant and precipitate. The recombinant protein was purified using a Ni-NTA column. The samples were separately taken for SDS-PAGE analysis. The results are shown in Figure 2. The target protein mainly exists in the supernatant, indicating the soluble expression of the recombinant protein in *E. coli*. The protein molecular weight is around 30 ku, consistent with the predicted theoretical value.

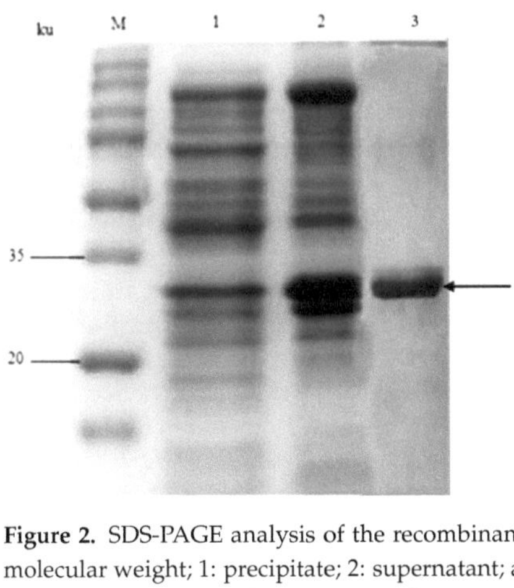

Figure 2. SDS-PAGE analysis of the recombinant fusion protein rHF-EC20. M: Protein standard molecular weight; 1: precipitate; 2: supernatant; and 3: purified protein; the band indicated by the black arrow represents the target recombinant protein rHF-EC20.

3.2. *Morphological Analysis of Fusion Protein*

The purified fusion protein was negatively stained with phosphotungstic acid, and its morphology was observed using transmission electron microscopy (TEM). As shown in Figure 3, uniformly sized nanoparticles were observed under the electron microscope, implying that the fusion protein rHF-EC20 auto-assembled into nanospheres in *E. coli*, with a size of approximately 13 nm, consistent with the expectation.

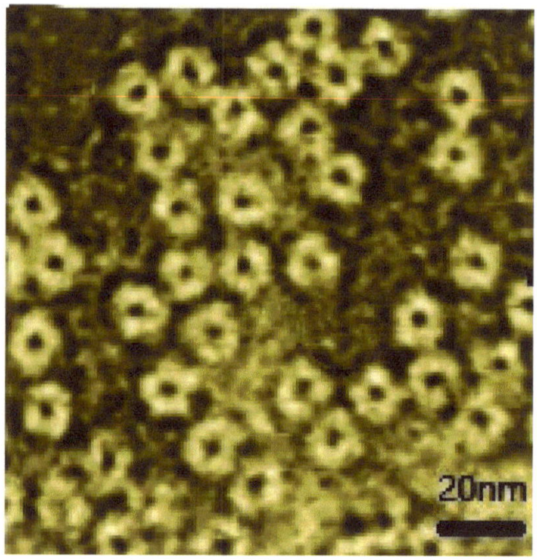

Figure 3. TEM analysis of the recombinant fusion protein rHF-EC20.

3.3. *Adsorption Capacity of Recombinant Fusion Protein rHF-EC20 to Cadmium*

The purified target protein was immobilized onto Ni-NTA agarose through a His-tag to treat cadmium chloride. It was calculated that 1 μmol of the target protein rHF-EC20 (monomer) could adsorb approximately 9.2 μmol of Cd^{2+}. Xu et al. conducted adsorption experiments on Cd^{2+} using the EC20 fusion protein in vitro. The results show that theoretically, one molecule of EC20 fusion protein can bind to ten molecules of Cd^{2+} [41]. The findings of this study agree with this, demonstrating the ability of the recombinant protein to bind Cd^{2+}.

3.4. *Adsorption Characteristics of Engineered Bacterium E. coli BL21 (FLE) to Cadmium*

3.4.1. Impact of Induction Time on Cadmium Adsorption by Engineered Strain

The recombinant engineered strain *E. coli* BL21 (FLE) and control strain *E. coli* BL21 (28a) were inoculated into auto-induction media containing cadmium and continuously cultured for 24 h. The adsorption characteristics of strains for cadmium were determined throughout the entire cultivation period. The results shown in Figure 4 reveal that the growth curve of the target strain is generally consistent with that of the control strain BL21. This verifies that under the same antibiotic and heavy metal stresses, the introduction of the target gene into BL21 does not significantly affect cell growth. During the initial 12 h of cultivation, there was no evident difference in the cadmium adsorption efficiency between the target and control strains. However, between 12 to 16 h, as the cell density increased, the target strain exhibited a higher cadmium adsorption efficiency than the control strain. After continuous cultivation for 16 h, *E. coli* BL21 (FLE) achieved an adsorption efficiency of 78.13% for Cd^{2+} in the solution, and the control strain was only 62.32%. After continuous cultivation for 20 h, the target strain *E. coli* BL21 (FLE) reached its maximum cadmium adsorption capacity. However, after 24 h, the adsorption efficiency of cadmium by the engineered strains decreased. This may be attributed to the excretion of heavy metal ions from the cells through specialized "pumps" or transporters, such as P-type ATPase and CDF (Cation Diffusion Facilitator) family proteins [42–44]. When the concentration of heavy metals in the cells of microorganisms reaches their tolerance capacity, the optimal detoxification method is to expel heavy metals from the cells.

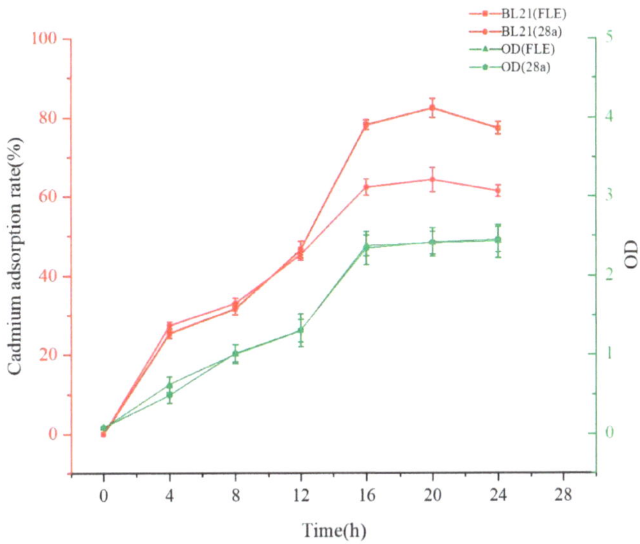

Figure 4. Curves of the growth of engineered strains and their adsorption to cadmium.

3.4.2. Optimal Adsorption Temperature of Engineered Bacterium BL21 (FLE)

The determination results of the optimal adsorption temperature of the recombinant engineered strain *E. coli* BL21 (FLE) for Cd^{2+} are shown in Figure 5. The optimal adsorption temperature for Cd^{2+} by the recombinant strain is 37 °C. After continuous adsorption for two hours, the adsorption rate reached 82.4%. Conversely, low temperatures are unfavorable for *E. coli* adsorption of Cd^{2+}. After continuous adsorption at 16 °C for two hours, the adsorption rate was merely 26.3%. By comparison, adsorption experiments at 45 °C result in a decline in the adsorption efficiency. The optimal growth temperature for *E. coli* is 37 °C, Prior research delineates heavy metal adsorption by microorganisms into stages encompassing heavy metal ion adsorption onto cell membrane surfaces, transmembrane transport, and intracellular detoxification [45]. A range of transport proteins and enzymes orchestrates the intricate process of heavy metal transport. Existing research indicates that temperature substantially influences gene expression and protein synthesis within cells. Microbial activities surge within an appropriate temperature range, enhancing both microbial metabolism and enzyme activity, thereby expediting the bioremediation process of heavy metals [46,47]. Conversely, low temperatures hamper the rapid synthesis of functional proteins and can impair the permeability of cell membranes. Additionally, higher environmental temperatures may disrupt the activities of metal transporters and related enzymes, potentially impeding the transport and adsorption of Cd^{2+} [48].

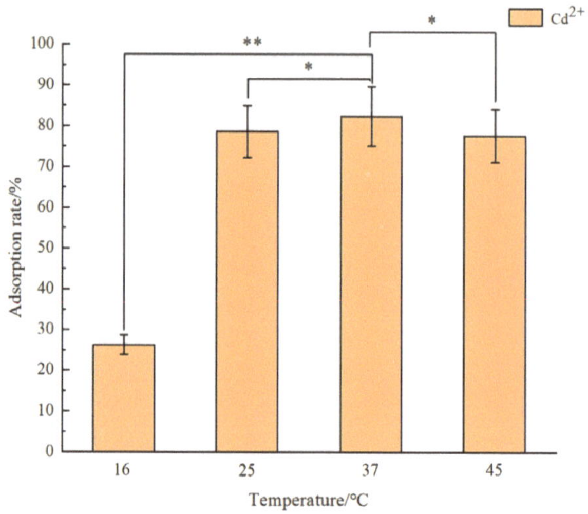

Figure 5. The optimal adsorption temperature of the engineered bacterium BL21 (FLE). Different asterisks mean significant difference between treatments (* $p < 0.05$, ** $p < 0.01$).

3.4.3. Optimal Adsorption pH of Engineered Bacterium *E. coli* BL21 (FLE)

The recombinant strain was subjected to different pH buffer solutions to determine its adsorption efficiency for Cd^{2+}. The results show that the recombinant strain *E. coli* BL21 (FLE) exhibits a high adsorption efficiency for Cd^{2+} in the pH range of 5–7. Specifically, treatment in a pH 7.0 solution for two hours results in the highest adsorption efficiency for Cd^{2+}, reaching 82.4%. In contrast, under pH 9.0 environmental conditions for two hours, the adsorption efficiency is only 69.2% (Figure 6). The pH exerts considerable influence on the redox and solubility of heavy metals. Under neutral and acidic conditions (pH < 7), negatively charged functional groups on the microbial cell surface, such as hydroxyl, phosphate, carboxyl, and amide groups, can readily bind with Cd^{2+}. In contrast, when the pH levels exceed 8, $Cd(OH)_2$ precipitates out of wastewater. Cadmium concentrations decrease rapidly, and the cellular uptake of Cd^{2+} also drops [7].

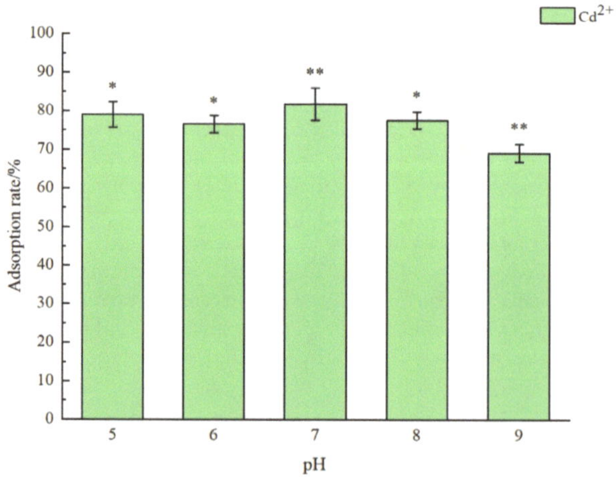

Figure 6. The optimum adsorption pH of engineered bacterium BL21 (FLE). Different asterisks mean significant difference between treatments (* $p < 0.05$, ** $p < 0.01$).

3.4.4. Evaluation of Adsorption Efficiency of Recombinant Strain to Cd^{2+}

Under the optimal adsorption conditions, the adsorption efficiencies of the recombinant genetically engineered strain BL21 (FLE) and control strain BL21 (28a) were determined at different time points (Figure 7a). Throughout the entire time process, the recombinant strain BL21 (FLE) consistently demonstrates a higher adsorption efficiency for Cd^{2+} compared to the control strain. The recombinant strain BL21 (FLE), expressing nano-chelating peptides, achieves an adsorption rate of 80% for Cd^{2+} at 60 min, whereas the control strain BL21 (28a) only reaches 62%, resulting in an approximately 18% increase in the Cd^{2+} enrichment efficiency. These results indicate that the overexpression of multivalent phytochelatins in *E. coli* cells effectively enhances the adsorption efficiency and tolerance of the strain to Cd^{2+}, thereby improving the adsorption efficiency of heavy metals. As shown in Figure 7b, the results of the kinetic study regarding BL21 (FLE) reveal that during the initial adsorption stage (0–30 min), *E. coli* exhibits a swift adsorption rate for Cd^{2+}. At this stage, heavy metal ions adhere to cell membrane surfaces primarily through processes like ion exchange and complexation with chemical groups in biomolecules on the membrane surfaces, featured by rapid and large-scale increases. Subsequently, the adsorption rate gradually decelerates, reaching equilibrium at 60 min. Some heavy metals bound to the membrane surfaces are transported into the cells via metal transporters. Within the cells, these metals interact with heavy metal chelating proteins and form precipitates within the cytoplasm, thereby mitigating the impact of soluble heavy metal ions on cells. This phase represents the rate-limiting step throughout the entire adsorption process [45].

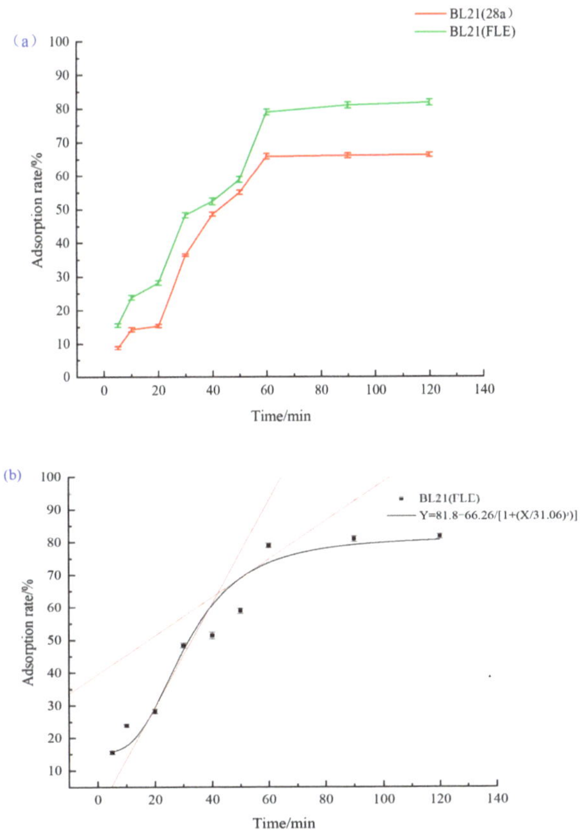

Figure 7. (**a**) The temporal profile curves of cadmium adsorption by the strains; (**b**) the kinetic profile curve of cadmium adsorption by the strain BL21 (FLE).

3.4.5. Equilibrium Adsorption Capacity of Recombinant Strain to Cd^{2+}

This experiment analyzed the enrichment behavior of the recombinant strain BL21 (FLE) and the control strain BL21 (28a) at different Cd^{2+} concentrations. As shown in Figure 8, the maximum enrichment of Cd^{2+} by the recombinant strain BL21 (FLE) and the control strain BL21 (28a) can be calculated based on the Langmuir model formula. The equilibrium enrichment of the control strain BL21 (28a) is 9.27 mg/g, and that of the recombinant strain BL21 (FLE) is 12.62 mg/g. Meanwhile, the equilibrium curve shows that in solutions with high concentrations of Cd^{2+}, the recombinant genetically engineered bacterium presents a higher resistance and adsorption capacity to cadmium. However, due to the absence of the secretion mechanism, proteins expressed in *E. coli* cells mainly reside in the cytoplasm. Therefore, the genetically engineered bacterium in this study aimed to promote heavy metal adsorption by elevating the expression levels of intracellular heavy metal chelating peptides. However, the results indicate that compared with the control strain, the engineered strain merely augments the cadmium adsorption efficiency by approximately 18%. This may stem from the ineffective transport of heavy metal ions into cells.

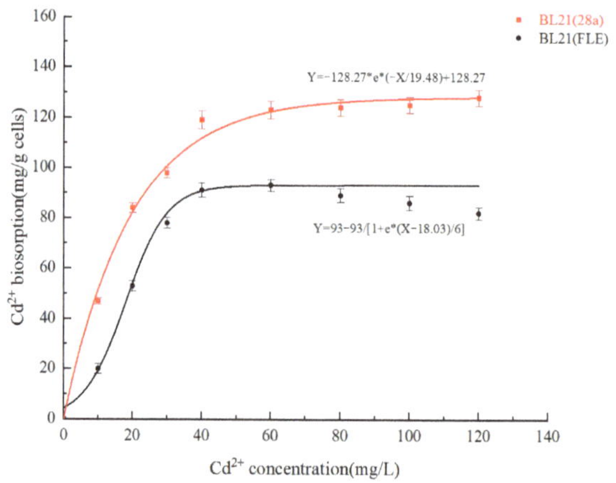

Figure 8. The equilibrium curve of cadmium adsorption by strains.

4. Conclusions

This work demonstrates the feasibility of employing genetically engineered *E. coli* cell factories to reinforce the efficient bioaccumulation of heavy metal cadmium via the overexpression of a fusion protein, rHF-EC. Moreover, the recombinant proteins can self-assemble into multivalent nanochelators with a size of about 13 nm within cells, effectively adsorbing heavy metal cadmium both intracellularly and extracellularly. The target protein rHF-EC20 (monomer) could adsorb approximately 9.2 μmol of Cd^{2+} in vitro. The adsorption capacity varies with temperature and pH. The engineered strain BL21 (FLE) exhibits a superior performance in cadmium biosorption at pH 7.0 and 37 °C. Meanwhile, a notable 18% increase in cadmium accumulation implies that the strain that produces rHF-ECs maintains significant advantages over the control host cells.

Considering the findings of the above research, promoting heavy metal adsorption by elevating the expression levels of intracellular heavy metal chelating peptides emerges as a promising strategy. Nonetheless, this approach faces limitations due to the ineffective transport of heavy metal ions into cells. Future research can focus on tethering chelating peptides to the surface of cell membranes and co-expressing cadmium transporters alongside this nanochelator within cells to further augment the adsorption and transport of heavy metals [49]. Such advancements hold the potential to expand the utility of genetically engineered *E. coli* in environmental pollution control [50].

Author Contributions: G.L. and L.Y. conceived the study. L.T., D.W., Y.L., X.H. and M.W. performed the experiments, L.Y. wrote the draft of the manuscript. X.S. and G.L. critically reviewed the full manuscript content. All authors have read and agreed to the published version of the manuscript.

Funding: This research was supported by Gansu Province Key R&D Plan under Grant No. 21YF5FA129; Gansu Province Key R&D Plan under Grant No. 22YF7FG188; Gansu Province Natural Science Foundation (No. 22JR11RG222) and Hexi University Research Initiation Project (No. KYQD2020018).

Data Availability Statement: The datasets used and/or analyzed during the current study are available from the corresponding author upon reasonable request.

Conflicts of Interest: The authors declare no conflicts of interest.

References

1. Pratush, A.; Kumar, A.; Hu, Z. Adverse effect of heavy metals (As, Pb, Hg, and Cr) on health and their bioremediation strategies: A review. *Int. Microbiol.* **2018**, *21*, 97–106. [CrossRef] [PubMed]
2. Fang, Y.; Sun, J.; Sun, M.; Shi, X.; He, Z.; Gong, Z.; Yao, M.; Sun, Y.; Xu, X.; Sui, H. Research progress on cadmium toxicity and prevention measures. *J. Toxicol.* **2022**, *36*, 517–520.
3. Barakat, M.A. New trends in removing heavy metals from industrial wastewater. *Arab. J. Chem.* **2011**, *4*, 361–377. [CrossRef]
4. Al-Rashdi, B.A.M.; Johnson, D.J.; Hilal, N. Removal of heavy metal ions by nano-filtration. *Desalination* **2013**, *315*, 2–17. [CrossRef]
5. Priyanka, R.; Sharma, A.C.; Sunil, K.S.; Geng, L.; Nasim, A.; Darren, M.; Benjamin, S.H. Nanocellulose from Spinifex as an Effective Adsorbent to Remove Cadmium (II) from Water. *ACS Sustain. Chem. Eng.* **2018**, *6*, 3279–3290.
6. Guo, Z.; Zhang, X.; Kang, Y.; Zhang, J. Biomass-Derived Carbon Sorbents for Cd (II) Removal: Activation and Adsorption Mechanism. *ACS Sustain. Chem. Eng.* **2017**, *5*, 4103–4109. [CrossRef]
7. Khan, S.A.; Siddiqui, M.F.; Khan, T.A. Ultrasonic-assisted synthesis of polyacrylamide/bentonite hydrogel nanocomposite for the sequestration of lead and cadmium from aqueous phase: Equilibrium, kinetics and thermodynamic studies. *Ultrason. Sonochem.* **2020**, *60*, 104761. [CrossRef] [PubMed]
8. Muya, F.N.; Sunday, C.E.; Baker, P.; Iwuoha, E. Environmental remediation of heavy metal ions from aqueous solution through hydrogel adsorption: A critical review. *Water Sci. Technol.* **2016**, *73*, 983–992. [CrossRef] [PubMed]
9. Zheng, Y.; Xiao, C.; Chi, R. Remediation of soil cadmium pollution by biomineralization using microbial-induced precipitation: A review. *World J. Microbiol. Biotechnol.* **2021**, *37*, 208–213. [CrossRef]
10. Saumya, A.; Ankur, S.; Vipin, K. Recent advancements in cadmium-microbe interactive relations and their application for environmental remediation: A mechanistic overview. *Environ. Sci. Pollut. Res. Int.* **2023**, *30*, 17009–17038.
11. Alabssawy, A.N.; Hashem, A.H. Bioremediation of hazardous heavy metals by marine microorganisms: A recent review. *Arch. Microbiol.* **2024**, *206*, 103. [CrossRef] [PubMed]
12. Roy, R.; Samanta, S.; Pandit, S.; Naaz, T.; Banerjee, S.; Rawat, J.M.; Chaubey, K.K.; Saha, R.P. An Overview of Bacteria-Mediated Heavy Metal Bioremediation Strategies. *Appl. Biochem. Biotechnol.* **2024**, *196*, 1712–1751. [CrossRef] [PubMed]
13. Joseph, L.; Jun, B.; Flora, J.; Park, C.; Yoon, Y. Removal of heavy metals from water sources in the developing world using low-cost materials: A review. *Chemosphere* **2019**, *229*, 142–159. [CrossRef] [PubMed]
14. Liu, S.; Zeng, G.; Niu, Q.; Liu, Y.; Zhou, L.; Jiang, L.; Tan, X.; Xu, P.; Zhang, C.; Cheng, M. Bioremediation mechanisms of combined pollution of PAHs and heavy metals by bacteria and fungi: A mini review. *Bioresour. Technol.* **2017**, *224*, 25–33. [CrossRef] [PubMed]
15. Znad, H.; Awual, M.R.; Martini, S. The Utilization of Algae and Seaweed Biomass for Bioremediation of Heavy Metal-Contaminated Wastewater. *Molecules* **2022**, *27*, 1275. [CrossRef] [PubMed]
16. Salama, E.S.; Roh, H.S.; Dev, S.; Khan, M.A.; Abou-Shanab, R.A.I.; Chang, S.W.; Jeon, B.H. Algae as a green technology for heavy metals removal from various wastewater. *World J. Microbiol. Biotechnol.* **2019**, *35*, 75. [CrossRef] [PubMed]
17. Arumugam, N.; Chelliapan, S.; Kamyab, H.; Thirugnana, S.; Othman, N.; Nasri, N.S. Treatment of Wastewater Using Seaweed: A Review. *Int. J. Environ. Res. Public Health* **2018**, *15*, 2851. [CrossRef] [PubMed]
18. Li, P.; Tao, H. Cell surface engineering of microorganisms towards adsorption of heavy metals. *Crit. Rev. Microbiol.* **2015**, *41*, 140–149. [CrossRef]
19. Hansda, A.; Kumar, V.; Anshumali, A. comparative review towards potential of microbial cells for heavy metal removal with emphasis on biosorption and bioaccumulation. *World J. Microbiol. Biotechnol.* **2016**, *32*, 170. [CrossRef]
20. Saravanan, A.; Kumar, P.S.; Ramesh, B.; Srinivasan, S. Removal of toxic heavy metals using genetically engineered microbes: Molecular tools, risk assessment and management strategies. *Chemosphere* **2022**, *298*, 134341. [CrossRef]
21. Thévenod, F.; Fels, J.; Lee, W.K.; Zarbock, R. Channels.transporters and receptors for cadmium and cadmium complexes in eukaryotic cells: Myths and facts. *Biometals* **2019**, *32*, 469–489. [CrossRef]
22. Chang, J.D.; Huang, S.; Konishi, N. Overexpression of the manganese/cadmium transporter OsNRAMP5 reduces cadmium accumulation in rice grain. *J. Exp. Bot.* **2020**, *71*, 5705–5715. [CrossRef] [PubMed]
23. Coyle, P.; Philcox, J.C.; Carey, L.C.; Rofe, A.M. Metallothionein: The multipurpose protein. *Cell. Mol. Life Sci.* **2002**, *59*, 627–647. [CrossRef]

24. Esser-Kahn, A.P.; Iavarone, A.T.; Francis, M.B. Metallothionein-cross-linked hydrogels for the selective removal of heavy metals from water. *J. Am. Chem. Soc.* **2008**, *130*, 15820–15822. [CrossRef] [PubMed]
25. Okasha, H.; Abdel-Motleb, A.; Abdel-Wareth, M.T.A. Metallothionein expression in Aspergillus exposed to environmentally relevant concentrations of heavy metals at different pH levels. *Environ. Sci. Pollut. Res. Int.* **2021**, *28*, 49936–49948. [CrossRef]
26. Nordberg, M.; Nordberg, G.F. Metallothionein and Cadmium Toxicology-Historical Review and Commentary. *Biomolecules* **2022**, *12*, 360. [CrossRef] [PubMed]
27. Cai, Y.; Zhao, X.; Deng, X. Bioaccumulation of heavy metal cadmium in wastewater by genetically engineered bacteria. *Technol. Water Treat.* **2006**, *32*, 26–29.
28. Bae, W.; Chen, W.; Mulchandania, A.; Mehra, R.K. Enhanced bioaccu-mulation of heavy metals by bacterial cells displaying synthetic phytochelatins. *Biotechnol. Bioeng.* **2000**, *70*, 518–524. [CrossRef]
29. Bae, W.; Mulchandania, A.; Chen, W. Cell surface display of synthetic phytochelatins using ice nucleation protein for enhanced heavy metal bioaccumulation. *J. Org. Chem.* **2002**, *88*, 223–227. [CrossRef]
30. Yu, Y.; Shi, K.; Li, X.; Luo, X.; Wang, M.; Li, L.; Wang, G.; Li, M. Reducing cadmium in rice using metallothionein surface-engineered bacteria WH16-1-MT. *Environ. Res.* **2022**, *203*, 111801. [CrossRef]
31. Zhu, N.; Zhang, B.; Yu, Q. Genetic engineering-facilitated co-assembly of synthetic bacterial cells and magnetic nanoparticles for efficient heavy metal removal. *ACS Appl. Mater. Interfaces* **2020**, *12*, 22948–22957. [CrossRef] [PubMed]
32. Theil, E.C. Ferritin: The protein nanocage and iron biomineral in health and in disease. *Inorg. Chem.* **2013**, *52*, 12223–12233. [CrossRef] [PubMed]
33. Chakraborti, S.; Chakrabarti, P. Self-Assembly of Ferritin: Structure, Biological Function and Potential Applications in Nanotechnology. *Adv. Exp. Med. Biol.* **2019**, *1174*, 313–329. [PubMed]
34. Wu, J.; Li, Y.; Wu, H.; Zhang, H.; Sha, X.; Ma, J.; Yang, R. The application of ferritin in transporting and binding diverse metal ions. *Food Chem.* **2024**, *439*, 138132. [CrossRef] [PubMed]
35. Lu, Z.; Wu, Z.; Yuan, X. Overexpression of human-derived soluble transferrin receptor sTfR antigen and preparation and application of polyclonal antibodies. *J. Biol.* **2023**, *40*, 111–115.
36. Zhao, X.; Zhou, Y.; Zhang, Y.; Zhang, Y. Ferritin: Significance in viral infections. *Rev. Med. Virol.* **2024**, *34*, e2531. [CrossRef] [PubMed]
37. Chen, Q.Y.; Tyrer, M.; Hills, C.D.; Yang, X.M.; Carey, P. Immobilisation of heavy metal in cement-based solidification/stabilisation: A review. *Waste Manag.* **2009**, *29*, 390–403. [CrossRef] [PubMed]
38. Malviya, R.; Chaudhary, R. Factors affecting hazardous waste solidification/stabilization: A review. *J. Hazard. Mater.* **2006**, *137*, 267–276. [CrossRef]
39. Malviya, R.; Chaudhary, R. Leaching behavior and immobilization of heavy metals in solidified/stabilized products. *J. Hazard. Mater.* **2006**, *137*, 207–217. [CrossRef]
40. Studier, F.W. Protein production by auto-induction in high density shaking cultures. *Protein Expr. Purif.* **2005**, *41*, 207–234. [CrossRef]
41. Xu, Z.; Bae, W.; Mulchandani, A.; Mehra, R.K.; Chen, W. Heavy metal removal by novel CBD-EC20 sorbents immobilized on cellulose. *Biomacromolecules* **2002**, *3*, 462–465. [CrossRef] [PubMed]
42. Drees, S.L.; Beyer, D.F.; Lenders-Lomscher, C.; Lübben, M. Distinct functions of serial metal-binding domains in the *Escherichia coli* P1B-ATPase CopA. *Mol. Microbiol.* **2015**, *97*, 423–438. [CrossRef] [PubMed]
43. Kolaj-Robin, O.; Russell, D.; Hayes, K.A.; Pembroke, J.T.; Soulimane, T. Cation Diffusion Facilitator family: Structure and function. *FEBS Lett.* **2015**, *589*, 1283–1295. [CrossRef] [PubMed]
44. Montanini, B.; Blaudez, D.; Jeandroz, S.; Sanders, D.; Chalot, M. Phylogenetic and functional analysis of the Cation Diffusion Facilitator (CDF) family: Improved signature and prediction of substrate specificity. *BMC Genom.* **2007**, *8*, 107. [CrossRef] [PubMed]
45. Avanbakht, V.; Alavi, S.A.; Zilouei, H. Mechanisms of heavy metal removal using microorganisms as biosorbent. *Water Sci. Technol.* **2014**, *69*, 1775–1787. [CrossRef] [PubMed]
46. Kuriki, Y. Response to temperature shifts of expression of the amp gene on pBR322 in Escherichia coli K-12. *J. Bacteriol.* **1987**, *169*, 2294–2297. [CrossRef] [PubMed]
47. Vortuba, J.; Pazlarova, J.; Dvorakova, M.; Vachova, L.; Strnadova, M.; Kucerova, H.; Vinter, V.; Zourabian, R.; Chaloupka, J. External factors involved in the regulation of synthesis of an extracellular proteinase in Bacillus megaterium: Effect of temperature. *Appl. Microbiol. Biotechnol.* **1991**, *35*, 352–357. [CrossRef] [PubMed]
48. Quinn, P.J. Effects of temperature on cell membranes. *Symp. Soc. Exp. Biol.* **1988**, *42*, 237–258.
49. Leong, Y.K.; Chang, J.S. Bioremediation of heavy metals using microalgae: Recent advances and mechanisms. *Bioresour. Technol.* **2020**, *303*, 122886. [CrossRef]
50. Wang, J.; Chen, C. Biosorbents for heavy metals removal and their future. *Biotechnol. Adv.* **2009**, *27*, 195–226. [CrossRef]

water

Article

The Design of a Process for Adsorbing and Eluting Chromium (VI) Using Fixed-Bed Columns of *E. crassipes* with Sodium Tripolyphosphate (TPP)

Uriel Fernando Carreño Sayago * **and Vladimir Alfonso Ballesteros Ballesteros**

Facultad de Ingenieria, Fundacion Universitaria los Libertadores, 111221 Bogotá, Colombia
* Correspondence: ufcarrenos@libertadores.edu.co

Abstract: Proper water resource management is a critical global objective, both privately and in business, due to the continuous deterioration of this valuable resource. Scientific research in environmental sciences has made significant progress in the development and achievements of treatment. The use of transformed *E. crassipes* biomass with sodium tripolyphosphate (TPP) can help to achieve this important goal. The objective of this study was to develop an experimental process for the continuous adsorption and elution of chromium (VI) using fixed-bed columns of *E. crassipes* biomass modified with sodium tripolyphosphate (TPP). Additionally, design tools were created, and economic viability was assessed by analyzing adsorption capacity indicators and unit production costs of different biomasses. Treatment systems were designed and constructed to remove chromium from tannery wastewater, ensuring that the levels were below the current environmental regulations of 0.05 mg/L Cr(VI). The biomass had an adsorption capacity of 98 mg/g and was produced at a low cost of 8.5 dollars. This resulted in an indicator of 11.5 g Cr(VI)/(USD) when combined with the elution processes. The proposed strategy, which utilizes entirely green technologies, enables the recovery and valorization of water resources. This makes it an effective tool for the circular economy.

Keywords: *E. crassipes*; treatment water; tripolyphosphate (TPP); chromium

Citation: Sayago, U.F.C.; Ballesteros, V.A.B. The Design of a Process for Adsorbing and Eluting Chromium (VI) Using Fixed-Bed Columns of *E. crassipes* with Sodium Tripolyphosphate (TPP). *Water* **2024**, *16*, 952. https://doi.org/10.3390/w16070952

Academic Editors: Issam A. Al-Khatib, Rehab O. Abdel Rahman, Tsuyoshi Imai and Yung-Tse Hung

Received: 12 February 2024
Revised: 13 March 2024
Accepted: 14 March 2024
Published: 26 March 2024

1. Introduction

Non-conventional, economical, and efficient treatment is a major focus of research centers in developing countries, where wastewater from rivers, wetlands, and other water sources is still contaminated with heavy metals, phenols, and dyes. For this reason, there is a need to find suitable ways to treat water in an efficient way in the industrial sector, where most of the pollution occurs. One suitable method of treating industrial wastewater is through fixed-bed column systems, coupled with a process of reutilizing waste biomasses [1–3].

Fixed-bed columns are those with a constant biomass, together with pollutant flow inputs, which are easy to implement, effective due to the chelating capacity of the biomass, and economical. In different research, such as [4–6], treatment systems have been designed for large water bodies due to their mass balance models and intra-particle and ex-particle diffusion among other models and isotherms [7,8].

Among the varieties of heavy metal, the most impactful on water resources is chromium, which is utilized across a variety of industries due to its efficacy in metal alloys and the tannery industry's leather preservation techniques. However, its excessive use has resulted in significant environmental, social, and health issues [9]. A notable instance of this environmental threat can be observed in the south of Bogotá, Colombia. Over 350 tanneries in this area utilize chromium (VI), which is discharged into the Tunjuelo River and deteriorates its quality, thereby diminishing the ecosystem services available to the communities adjacent to the river [10].

To carry out sustainable projects, various plant species' biomasses have been used due to their high elimination efficiency, natural availability, and profitability, which favor the chemisorption process [11–13]. Chemisorption is a widely used technique for treating industrial wastewater due to its low cost and ability to remove various contaminants, as well as the ease of regenerating the adsorbent [14–16].

The presence of hydroxyl (OH) and carboxylic (COOH) groups in cellulose leads to the removal of heavy metals by adsorption through cation exchange or chemical mechanisms [17,18]. This cellulose is found in considerable amounts within the aquatic plant *E. crassipes* [19–22], which is abundant in tropical environments due to its dry biomass resulting from regular cleaning in the wetlands, lagoons, and rivers of Bogotá D.C. [23,24].

Pilot-scale process trials offer assistance in designing treatment systems that meet discharge regulations [25–28]. To create greater consistency in biomass and improve the heavy metal chemisorption process, the adsorption of heavy metals on modified cellulose involving sodium tripolyphosphate (TPP) has been investigated [29–31]. Microspheres, comprising *E. crassipes* and sodium tripolyphosphate (TPP), were examined for their ability to adsorb lead (II). This novel adsorbent exhibited a maximum loading capacity of 312.5 mg/g [32].

This current study selected adsorption with the chelating biomass of *E. crassipes* as the method for removing chromium (VI) from industrial wastewater at a tannery south of Bogota. The primary novelty of this research lies in the use of mathematical models in the adsorption and elution phases of fixed-bed processes. This study aims to create an experimental process for adsorbing and eluting chromium (VI), using fixed-bed columns of *E. crassipes* biomass modified with sodium tripolyphosphate (TPP), and operate it continuously. This study also seeks to develop design tools and determine economic viability by analyzing the adsorption capacity indicators and unit production costs of different biomasses.

2. Materials and Methods

Aquatic plants: The roots and leaves of the aquatic plant *E. crassipes* were taken from the Juan Amarillo wetland in the city of Bogotá D.C. The aim was to obtain particle diameters of 0.212 mm (it should be less than 0.212 mm for better contact between particle and contaminate [1,2]).

Chromium measurement: The samples were analyzed in a flask at each time interval, evaluated for the residual chromium of 20 μm samples, and subsequently centrifuged (KASAI MIKRO 200). Residual chromium was measured using a UV84.

Determination of chromium: This was conducted via the diphenylcarbazide method, which involved preparing a phosphate-buffer solution and adjusting it to a pH equal to 2; the degree of purity was 90% (H_3PO_4). Subsequently, 200 μL of 0.5% diphenylcarbazide was added to an Eppendorf tube, with its purity being 97%, together with acetone w/v; it also had a purity of 97%. An amount of 900 μL of phosphate buffer and 100 μL of the residual sample were obtained, and these were finally taken to an absorption cell, with the absorbance measuring 540 nm. On an Evolution 300 spectrophotometer, changes in absorbance were monitored. All evaluations were performed under the APHA (American Public Health Association Procedure) for standard tests (standard methods for the examination of water and wastewater). All experiments were carried out in triplicate, with final values averaged.

Fixed biomass column experiments: Treatment systems were developed in recycled plastic containers, coupled with interconnection processes between biocapsules (each capsule had 30 g of dry biomass, along with the reagent). The height was 40 cm, and the diameter was 4.5 cm, with an area of 30 cm^2 established and for a total of 1000 cm^3 in volume. The flow rate was 20 mL/min, kept constant by constant flow through the drip.

Preparation of *E. crassipes* beads. Approximately 100 g of *E. crassipes* powder was dissolved in 150 mL of diluted acetic acid (2% (v/v)), and then this solution was gently stirred and mixed with 250 mL and 500 mL of sodium tripolyphosphate (TPP) $Na_5O_{10}P_3$, molecular weight 367.86, solution at pH 8.6 to form the gelled spheres. This procedure had to be repeated five times to obtain sufficient material [32]. Two types of experiments were carried out:

Two similar treatment systems were built, changing only the way they will be distributed in the treatment system.

E. crassipes with 250 mL (TPP) mixed with 100 g EC biomass EC + TPP (1)

E. crassipes with 500 mL (TPP) mixed with 100 g EC biomass EC + TPP (2)

The system is shown in Figure 1.

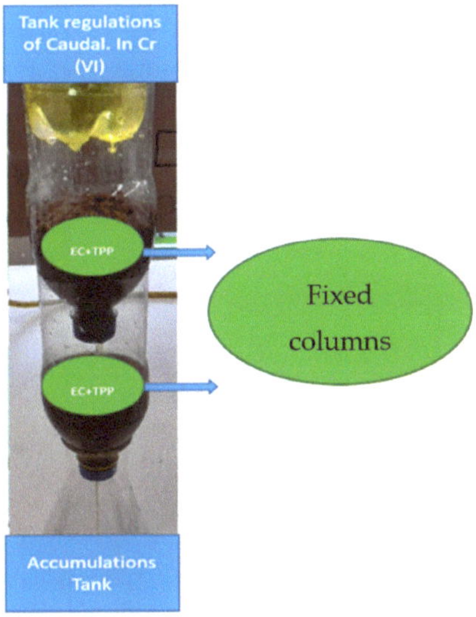

Figure 1. Design of treatment with EC+TPP.

Evaluating initial concentrations of 1000 mg/L of Cr(VI). All the tests were carried out in duplicate, calculating the average between the data obtained and with this the percentage of metal removal.

Adsorption models. Mathematical modelling was used to describe the behavior of the rupture curves, aiding in comprehension and scaling of the system. The biosorption process of Cr(VI) in the fixed-bed configuration was explained by fitting the breakthrough curve data to three different column adsorption models: Yoon-Nelson, Thomas, and Bohart.

The study used the widely accepted Thomas model to estimate the maximum adsorption capacity and predict the rupture curves. The proposed model assumed a second-order kinetic of reversible reactions and the Langmuir isotherm [33]. Additionally, the Thomas model was used to validate the proposed model, demonstrating its effectiveness.

The Yoon-Nelson model assumes that the rate of adsorption decreases proportionally to the curve of adsorbate removal and adsorbent breakdown, without considering factors such as adsorbate properties, adsorbent type, and adsorption column specifications [34].

However, the Bohart equation is commonly used to quantify various types of systems due to its ability to describe the relationship between C/Co and t in a continuous system with clarity and confidence. The model states that the rate of sorption is directly proportional to the remaining capacity of the solid and the concentration of the retained species. This model is specifically used to describe the initial part of the rupture curve. Table 1 presents the various adsorption models [35].

Table 1. Adsorption models.

Thomas model	(1)	$\ln \frac{Co}{C} - 1 = \frac{Kth \times q \times m}{Q} - Kth \times Co \times Tb$
Yoon model	(2)	$\frac{C}{Co} = \frac{1}{1 + \exp(Kyn(y - t))} qN = \frac{TbCoQ}{m}$
Bohart model	(3)	$Tb = \frac{No}{CoU} Z - 1 \ln \frac{1}{KbCo} \ln\left(\frac{co}{c} - 1\right)$
Model Carreño	(4)	$q = \frac{QTbCo}{M} - \frac{QTbCf}{M} - \frac{\varepsilon VCo}{M}$

The Model Carreño contains all the necessary parameters for designing a treatment system, including the relationships between densities, rupture time, and flow. Although this model is not adjusted, it is used to determine the adsorption capacity [36].

Co: initial concentration of Cr(VI); C: final Cr(VI); V: volume; KTh: Thomas constant (mL/mg·min); q: adsorption capacity (mg Cr/g biomass); m: mass of biomass in column (g); Q: flow rate through the column (mL/min); Tb: time of rupture (min); K YN: Yoon and Nelson constant (1/h), q: γ capacity (mg/g), dynamic capacity (mg/dm^{-3}); Z: bed height (cm); U: linear flow rate (cm/min); and Kb: Bohart constant (1/h).

Analysis of reliability. Reliability is a useful tool for establishing efficiency and compliance characteristics in environmental processes [37]. The exponential equation reflects the behavior of data in contaminant removal processes [38].

$$P(X \leq Xs) = 1 - \exp\left\{-\frac{x}{\sigma}\right\},\ x \geq 0 \tag{5}$$

X = Continuous random variable mg/L;

σ = Arithmetic average 0.05 mg/L Cr(VI).

The materials were characterized. This process used a TESCAN FE-MEB LYRA3 scanning electron and focused ion beam microscope. The SEM had an integrated X-ray energy dispersion spectroscopy microanalysis system, EDS (energy dispersive X-ray spectroscopy). EDS is one of the most efficient techniques for the qualitative and quantitative analysis of organic samples and, through the SEM microphotographs, the samples evaluated in the present investigation were observed in detail; the diffraction of an X-ray beam by the atoms of the sample interacts with the X-ray beam, producing regions of diffraction intensity, or peaks, for the diagnosis of each of the elements.

FTIR. The materials were characterized by Fourier transform infrared spectroscopy (79 Jasco FTIR 430) to measure IR spectra in a spectral range of 4000–400 cm^{-1}, a resolution of 4 cm^{-1}, and a scanning speed of 2 mm s^{-1}.

3. Results

FTIR spectral details. When comparing the FTIR spectral details of the cellulose with TPP and the same cellulose after the treatment process, it was observed that the EC-TPP-Cr(VI) showed some new bands compared to that of the untreated biomass in Figure 2.

It can be observed that the biomass without the EC-TPP treatment process exhibits a deep peak due to the presence of hydroxyl groups (OH), in comparison with the biomass of the EC alone, which has a lower wavenumber [1]. This suggests that there is an effective interaction between the sodium tripolyphosphate and the cellulose of *E. crassipes*. Additionally, there is a peak (P=O) at 1210 cm^{-1} and another peak representing (P-OH) at 1038 cm^{-1}; the stretching band is attributed to the presence of phosphorus, while the characteristic band of the amine group in vegetable cellulose at 1650 cm^{-1}, similar to (CH) at 1032 cm^{-1}, is also observable. A graph comparing the two biomasses before and after the treatment processes is presented to identify potential sites for Cr(VI) ion binding.

After conducting the Cr(VI) adsorption process using EC-TPP biomass, a shift towards a lower wavenumber was observed in the peak at 3424.96 cm^{-1} of the (OH) groups, as well as the amine group. This suggests the formation of an interaction between the Cr(VI) ions and the nitrogen atoms, indicating that nitrogen atoms may be the primary adsorption sites for metal ion binding. Furthermore, the intensity of P-O-H was reduced after the adsorption of Cr(VI) ions.

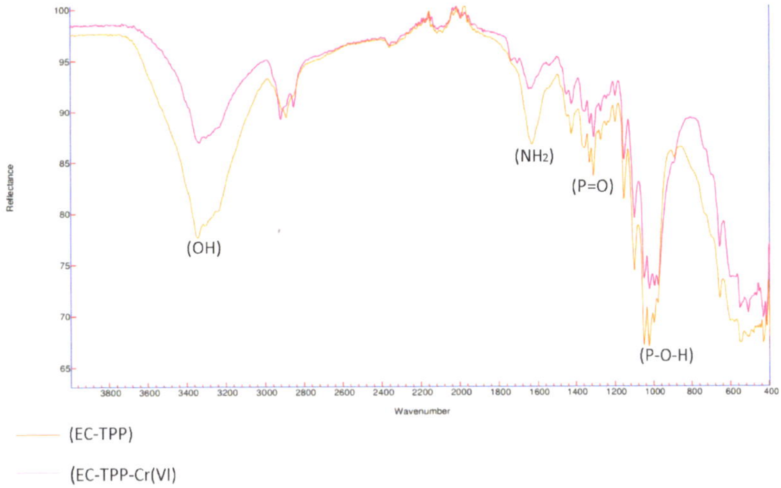

Figure 2. Characterizations with FTIR of EC-TPP.

SEM and EDS analysis. Figure 3 displays images of the EC+TPP biomass prior to the adsorption process of Cr(VI) ions. The surface morphology observed in the image indicates an effective articulation between the sodium tripolyphosphate and the cellulose of *E. crassipes*, with a monolayer also being detectable.

Figure 3. Microphotograph of EC+TPP.

The presence of white particles of Cr(VI) ions on the surface confirms the occurrence of the adsorption process. To better observe the images of each characteristic element, they were identified with colors, as shown in Figure 4a–c.

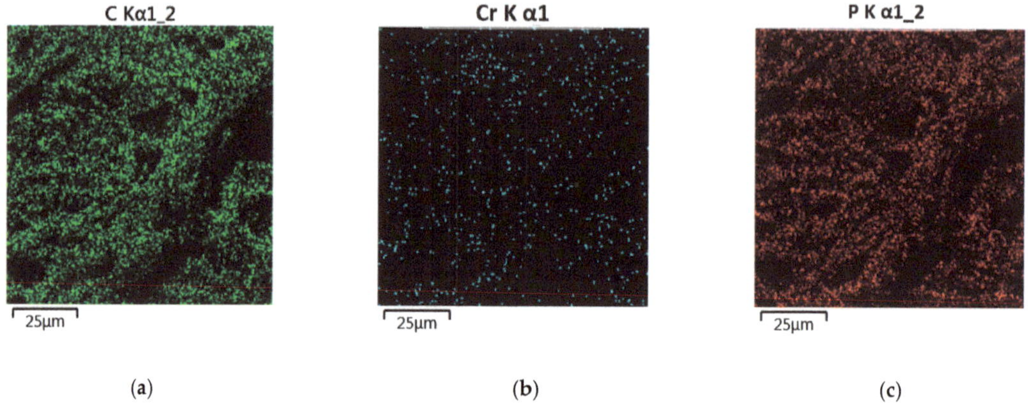

Figure 4. Photomicrographs of EC+TPP with Cr(VI). (**a**) Photograph with the presence of carbon; the presence of Cr(VI) is shown in image (**b**), and the presence of phosphorous in image (**c**).

In the microphotograph displayed in Figure 4, the green dots represent carbon (special components of the *E. crassipes* biomass), red dots represent phosphorus (representative of TPP), and the blue represent the Cr(VI), which remains removed from the EC+TPP biomass through cation exchange processes with the hydroxyl groups (OH) present in the EC+TPP biomass. Table 2 shows the physicochemical characterization of the EC+TPP sample by EDS.

Table 2. Physicochemical characterization of the EC sample.

Element	Weight	Percenteje %
Oxygen	48.64	46.67
Carbon	38.15	36.94
Phosforous	9.13	7.37
Sodium	10.2	9.8

The EDX spectra of the bead suggest that the cross-linking processes between the biomass and the TPP have occurred efficiently, as evidenced by the observations of phosphorus and sodium. The results are summarized in Table 3. Following the Cr(VI) adsorption process, the EDX spectra of the bead were analyzed and the findings are summarized in Table 3.

Table 3. Physicochemical characterization of the EC-TPP+Cr sample.

Element	Weight	Percentage %
Oxygen	43.64	41.67
Carbon	35.15	33.94
Phosphorous	9.13	7.37
Sodium	7.3	7.8
Chromium	9.2	8.3

Table 3 shows the characterization of the elements and it can be observed that at 10% by weight, an adhesion of this heavy metal is evident in the samples. Similar results were presented in [29]. A summary of the physicochemical characteristics evidenced in the FTIR samples and SEM microphotographs are shown in Figure 5, where a cellulose sample transformed with TPP adheres to Cr(VI).

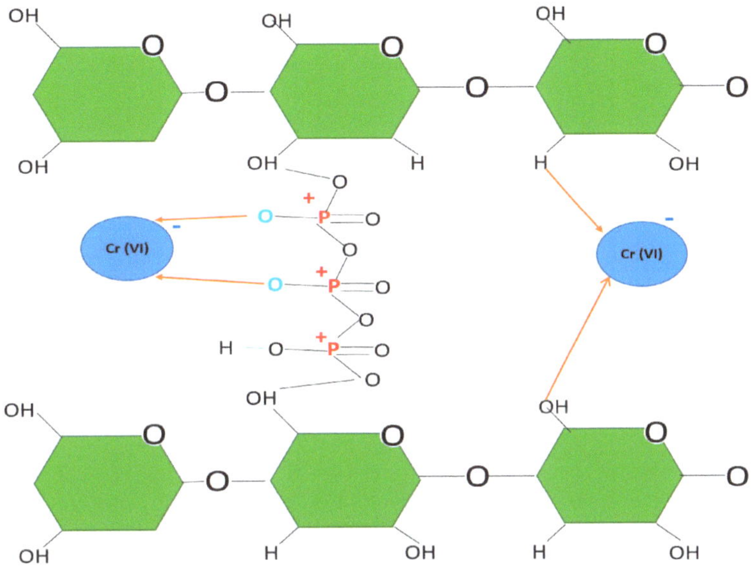

Figure 5. Representations of adsorptions.

Cellulose reacts with TPP to form chelating complexes, increasing the active sites where Cr(VI) will reside. The contact between the Cr_2O_7 dichromate and the biomass loaded with active sites generates reactions between the (H^+) of the biomass and the oxygen of the Cr(VI) structure, reducing it to Cr(III), which is chromium oxide Cr_2O_2.

$$[6(C_6H_{10}O_6)]xPP + Cr_2O_7^{-2} \rightarrow \{[6(C_6H_{10}O_6)]xTPP\}xCr_2 + 7H_2O$$

According to Figure 5, it can be observed that 6 parts of glucose react with dichromate, where ($\{6(C_6H_{10}O_6)\}xTPP$) represents the biomass and $Cr_2O_7^{-2}$ represents Cr(VI). These graphical representations are based on the conclusions drawn from the microphotographs and FTIR analysis.

The results of the process of adsorption of the Cr(VI) column experiments are shown in Figure 6. The configuration of the process used for the removal of Cr(VI) from the actual effluent uses fixed-bed columns in series. The yield is shown with an initial concentration of 1000 mg/L.

The EC+PPT treatment (1) achieved a consistent removal rate of 99% in all cases, as demonstrated in Figure 6a, with a breakpoint occurring at a volume of 4 L. With a flow rate of 20 mL/min, the breaking point is reached after 350 min. In the case of biomass, the EC+PPT treatment (2) resulted in the elimination of 99% of the contaminants after processing approximately 5 L of water. This biomass treated more water and achieved better yields due to its higher PPT content, reaching its breaking point in approximately 420 min. The biomass of *E. crassipes*, without any modification, could treat around 2 L of this same water [1], but with the addition of PPT better yields of up to more than two liters more water could be achieved. The addition of chemical agents to this type of biomass also improved the treatment yields, such is the case of the xanthate biomass of *E. crassipes*, where around 3.5 L of contaminated water was treated under the same concentrations [20]. The addition of PPT achieved better yields due to the expansion of active sites in the EC biomass [29–31].

Analysis of Reliability. The permissible limits for Cr(VI) ions in drinking water prescribed by the World Health Organization (WHO) are 0.05 mg/L. The results shown in Table 4 show the final concentration of Cr(VI) mg/L. Table 4 shows the results of the process of removals.

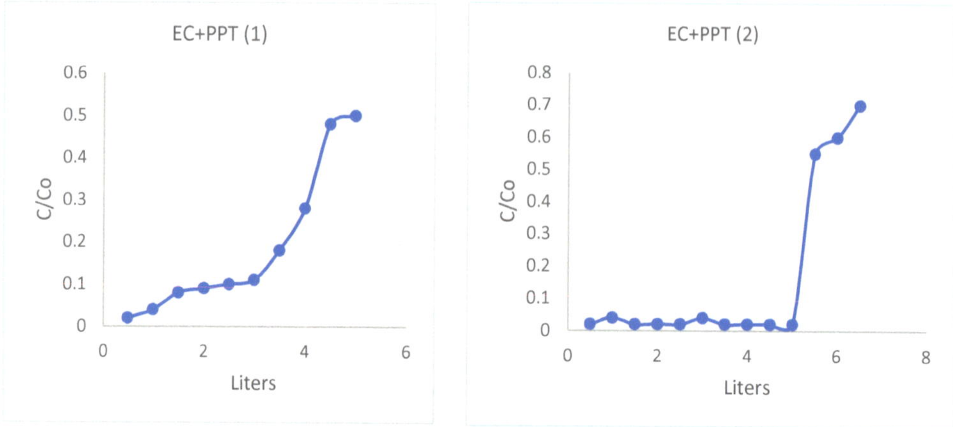

(a) (b)

Figure 6. Treatment with the biomass EC+PPT. Figure (**a**) shows the biomass EC+PPT (1), and the biomass EC+PPT (2) is shown in (**b**).

Table 4. Reliability analysis of the results.

	EC-PPT (1)	**EC-PPT (2)**
Liters	**Concentrations End of Cr(VI) (mg/L)**	
0.5	0.04	0.02
1	0.04	0.01
1.5	0.05	0.01
2	0.04	0.01
2.5	0.05	0.01
3	0.05	0.01
3.5	0.05	0.02
4		0.02
4.5		0.02
5		0.02
5.5		0.02
Average	0.04	0.02
Standard deviation	0.04	0.004

Kolmogorov–Smirnov tests were carried out to check the distribution of the data while the efficiency tests were carried out; the data behave under an exponential distribution in both biomasses. It is ideal to comply with national and international regulations, with Cr(VI) values below 0.05 mg/L of Cr(VI); through Equation (5), the probability of this value was determined.

$$P(X \leq 0.05) = 1 - \exp\left\{-\frac{0.05}{0.02}\right\},$$

The reliability of the system with the EC-PPT biomass (2) is 80%, which exceeds regulatory values due to its high removal efficiency. Meanwhile, the reliability of the EC-PPT biomass (1) is 61% under an exponential distribution, which meets the standard. The most appropriate biomass is selected based on statistical data stabilization. Experiments must comply with regulations and have suitable probability distributions for both simulations and experiment replications [39,40].

Model of adsorptions. This section uses the Thomas model to validate the Carreño Equation and establish the behavior of the treatment process. The Thomas model estimates the maximum adsorption capacity and predicts the breakthrough curves, assuming reversible second-order reaction kinetics and a Langmuir isotherm [41,42]. Tables 5 and 6 present the parameters of the Thomas, Yoon, and Bohart equations.

Table 5. Summary of the experimental results obtained with EC-PPT (1).

	EC-PPT	Bohart	Yoon	Thomas
1000 mg/L	K	0.048 K_b	0.039 K_{YN}	0.055 K_{Th}
	R^2	0.99	0.9444	0.954

Table 6. Summary of the experimental results obtained with EC-PPT (2).

	EC-PPT (2)	Bohart	Yoon	Thomas
1000 mg/L	K	0.061 K_b	0.05 K_{YN}	0.061 K_{Th}
	R^2	0.90	0.9444	0.966

A representative fit of the Thomas equation was observed, although significant adjustments were made to Bohart and Yoon. The Thomas constant was found to be 0.055 mL/mg·min, indicating the chemisorption rate of Cr(VI) into the EC-PPT biomass. The modified biomass has a Cr(VI) adsorption rate of 0.035 mL/mg·min, which is higher than the unmodified biomass but lower than that of the EC-PPT biomass (2), which reached a rate of 0.061 mL/mg·min. The better behavior of the modified biomass in the adsorption of Cr(VI) is evident. Table 6 shows the fitted parameters for EC-PPT biomass (2).

Both the EC-PPT (1) and EC-PPT (2) biomass fit the mathematical Thomas model well, with an R^2 value exceeding 95%. The Langmuir isotherm and second-order kinetics provide evidence of diffusivity in a monolayer for all processes. Equation (4) was validated, and the adsorption capacities were determined for each adsorption and desorption process using the Thomas equation. Table 7 shows a summary of the experimental results obtained.

Table 7. Summary of the experimental results obtained.

	Biomass	q (Thomas)	q (Carreño) mg/g
1000 mg/L	EC-PPT (1)	17.9	17
1000 mg/L	EC-PPT (2)	21.33	22.6

Desorption–Elution and Reuse. Using Equation (4), we proceeded to establish the yield of these biomasses and determine the new adsorption capacity. Figure 7 shows the adsorption capacities in the different elutions.

The elution processes for each of the treated biomasses using EDTA are observable. *E. crassipes* biomass supports different elutions due to the presence of lignin in its plant structure [40–42]. The addition of PPT resulted in up to 7 times more biomass being recycled. The total biomass sum of EC-PPT (2) was 98 mg/g, which is the highest sum reported for *E. crassipes*. The EC-PPT biomass yielded 62 mg/g, a significant parameter as it has half the concentration of PPT and may be more cost-effective for large-scale use.

The costs of the treatment systems. The costs of the treatment systems were determined based on the evaluated biomass, and the unit production costs of 1 kg of material were calculated. The cost of drying, crushing, and logistics to obtain *E. crassipes* biomass is approximately USD 2 per 1 kg [43,44].

The cost of EDTA, which is used in biomasses, is USD 0.5 for about 100 g of the reagent. Sodium tripolyphosphate costs USD 3, and USD 2.5 were spent on 250 mL of it for the EC-PPT biomass, making the total cost USD 5. For the EC-PPT biomass, the cost is USD 8. Acetic acid costs USD 0.5. Table 8 displays the total cost of the two samples.

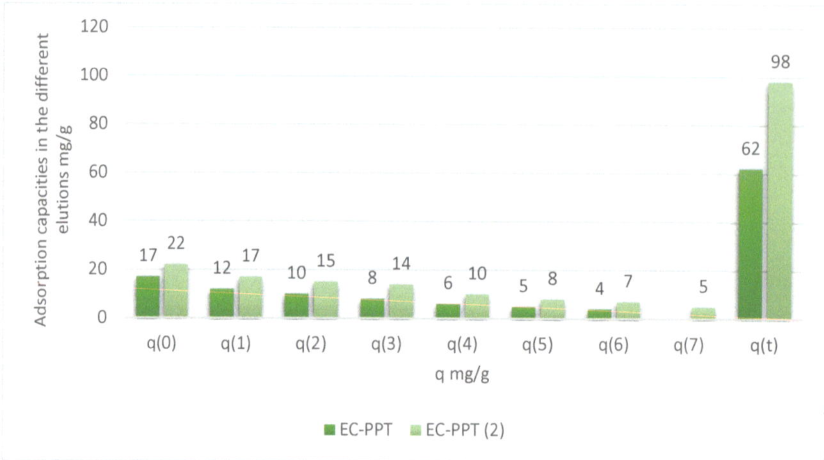

Figure 7. Adsorption capacities in the different elutions.

Table 8. Costs related to treatment systems.

Cost	EC-PPT (1)	EC-PPT (2)
Capacity total g Cr/100 g material	62	98
Cost (USD) 1 kg material	6.2	8.5
g Cr/(USD)	10	11.5

Table 4 demonstrates the project's benefit through the adsorption capacity of Cr(VI) and its elutions. The table displays the cost versus adsorption capacity for each biomass, indicating the cost of producing 1 kg of biomass or composite material. Additionally, the yield of adsorption capacity per dollar spent is included.

The EC-PPT biomass (1) has an adsorption capacity indicator of 10 g Cr(VI) per dollar. The second indicator, EC-PPT biomass (2), has an indicator of 11.5 g/USD due to its effectiveness and low cost. However, the EC biomass, despite being the cheapest, has an indicator of only 1.5 g/USD due to its low adsorption capacity compared to other biomasses.

Representative data were obtained using Equation (4) and various bibliographic references to establish a relationship between adsorption capacities and the production cost of 1 kg. It is important to note that associated costs may vary. Table 9 summarizes several biomass treatment processes. The indicator g HV/(USD) shows the amount of heavy metal removed in relation to the adsorption capacity of the biomass, compared to the cost of production/modification of this biomass in US dollars per kg [45].

Unmodified biomasses [1] have a low unit cost, but their g HV/(USD) indicator does not exceed 7.5. However, chemical modifications can significantly improve the adsorption capacity by increasing active sites, as reflected in this indicator [46].

It is worth noting that the investigated adsorption capacities [47–50] for various heavy metals are exceptionally high, surpassing 150 mg/g. The Humulus biomass with EDTA exhibits an impressive adsorption capacity of 398 mg/g. Processing costs for these biomasses should be taken into account, as indicated by the g HV/(USD) which is comparable to that of poorly processed biomasses. By selecting the appropriate biomass and adjusting it to treat the specific heavy metals present in the water, heavy metal adsorption can be optimized.

Table 9. Research on the process of elutions.

Reference	Biomass	Recycling	Capacity (mg/g) with the Equation (4)	Cost (USD) 1 kg Material	g HV/(USD)
	EC-PPT (2)	EDTA	98	8.5	11.5
	EC-PPT (1)	EDTA	62	6.2	10
[1]	*Crassipes*	-	8	3	2.5
[2]	*Crassipes* + Fe	EDTA	42	3.3	12.7
[2]	*crassipes*	EDTA	23	3.1	7.4
[20]	Xantate of cellulose	EDTA	51	7	7.2
[20]	Cellulose alkaline	EDTA	32	6	5.3
[46]	*Crassipes* + EDTA		45	10	4.5
[47]	*pinecone shells*	HCl	66	10	6.6
[48]	*Citrus maxima peel*		84	10	8.4
[49]	Kraft pulp/carboxyMethyled	HCl	101	40	2.5
[50]	Chitosan biocomposite	EDTA	146	35	5.3
[51]	Humulus lupulus	EDTA	398	50	7.9
[52]	Lignocellulosic biomass	EDTA	156	10	15.6

4. Conclusions

Two treatment processes have been successfully developed on a pilot scale, providing the necessary parameters to design and develop a treatment system on an industrial scale. The resulting system has a unit cost of approximately USD 8, which is significantly lower than conventional treatment systems in the industrial wastewater treatment sector.

The proposed system has an exceptional capacity to adsorb Cr(VI) and can be reused several times by eluting with EDTA. It is cost-effective and promotes sustainability and compliance with landfill regulations in the tannery sector. This helps to preserve the water resources surrounding these production centers.

The elution processes have been optimized to improve the adsorption of heavy metals, resulting in a production cost of only USD 8.5, with an indicator of 11.5 g HV/(USD). Biomass has a remarkable adsorption capacity of 98 mg/g, making it an extremely effective option for the chemisorption of heavy metals, particularly Cr(VI).

The findings suggest that *E. crassipes* biomass, enhanced with PTT and through EDTA elutions, could be a suitable water remediation technology based on chemiadsorption technology. This technology could be used for large-scale decontamination of rivers, wetlands, and other water ecosystems contaminated with various heavy metals.

Author Contributions: Conceptualization, U.F.C.S. and V.A.B.B.; methodology, U.F.C.S.; software, V.A.B.B.; validation, U.F.C.S.; formal analysis, U.F.C.S. and V.A.B.B.; investigation, U.F.C.S.; resources, V.A.B.B.; data curation, U.F.C.S.; writing—original draft preparation, U.F.C.S.; writing—review and editing, U.F.C.S.; visualization, U.F.C.S. and V.A.B.B.; supervision, U.F.C.S.; project administration, U.F.C.S. and V.A.B.B.; funding acquisition, All authors have read and agreed to the published version of the manuscript.

Funding: The "Universidad los Libertadores" funded this research through an internal call.

Data Availability Statement: The datasets used and/or analyzed during the current study are available from the corresponding author upon reasonable request.

Conflicts of Interest: The author declares no conflicts of interest.

References

1. Sayago, U.F.C. Design and development of a biotreatment of E. crassipes for the decontamination of water with Chromium (VI). *Sci. Rep.* **2021**, *11*, 9326. [CrossRef] [PubMed]
2. Carreño Sayago, U.F.; Piñeros Castro, Y.; Conde Rivera, L.R. Design of a Fixed-Bed Column with Vegetal Biomass and Its Recycling for Cr(VI) Treatment. *Recycling* **2022**, *7*, 71. [CrossRef]
3. Afroze, S.; Sen, T.K. A review on heavy metal ions and dye adsorption from water by agricultural solid waste adsorbents. *Water Air Soil Pollut.* **2018**, *229*, 225. [CrossRef]

4. Gradilla-Hernández, M.S.; Díaz-Vázquez, D.; Yebra-Montes, C.; del Castillo, A.F.; Shear, H.; Garcia-Gonzalez, A.; Mazari-Hiriart, M. Assessment of the Potential of Coordinating Two Interacting Monitoring Networks within the Ler-ma-Santiago Hydrologic System in Mexico. *Water* **2022**, *14*, 1687. [CrossRef]
5. Al-Ghouti, M.A.; Al-Absi, R.S. Mechanistic understanding of the adsorption and thermodynamic aspects of cationic methylene blue dye onto cellulosic olive stones biomass from wastewater. *Sci. Rep.* **2022**, *10*, 15928. [CrossRef]
6. Barquilha, C.E.R.; Cossich, E.S.; Tavares, C.R.G.; Silva, E.A. Biosorption of nickel (II) and copper (II) ions in batch and fixed-bed columns by free and immobilized marine algae Sargassum sp. *J. Clean. Prod.* **2017**, *150*, 58–64. [CrossRef]
7. Vinayagam, R.; Dave, N.; Varadavenkatesan, T.; Rajamohan, N.; Sillanpää, M.; Nadda, A.K.; Selvaraj, R. Artificial neural network and statistical modelling of biosorptive removal of hexavalent chromium using macroalgal spent bio-mass. *Chemosphere* **2015**, *296*, 133965. [CrossRef] [PubMed]
8. Ang, T.N.; Young, B.R.; Taylor, M.; Burrell, R.; Aroua, M.K.; Baroutian, S. Breakthrough analysis of continuous fixed-bed adsorption of sevoflurane using activated carbons. *Chemosphere* **2020**, *239*, 124839. [CrossRef] [PubMed]
9. Saade-Cleves, N.; Talero-Munoza, M.D.M.; García-Vargas, D.; Tamayo-Torres, C.S.; Sierra-Pena, J.A.; Torres-Ortiz, M.P.; Palen-cia-Sánchez, F. The Risk Index Used to Assess the Risk of Heavy Metals, Mainly in Water Sediments, on Human Health in Latin America: A Rapid Review of Literature (December 27, 2022). Available online: https://ssrn.com/abstract=4313001 (accessed on 13 March 2024). [CrossRef]
10. Ramírez-Canon, A.; Becerra-Quiroz, A.P.; Herrera-Jacquelin, F. Perfluoroalkyl and polyfluoroalkyl substances (PFAS): First survey in water samples from the Bogotá River, Colombia. *Environ. Adv.* **2022**, *8*, 100223. [CrossRef]
11. Hokkanen, S.; Repo, E.; Suopajärvi, T.; Liimatainen, H.; Niinimaa, J.; Sillanpää, M. Adsorption of Ni (II), Cu (II) and Cd (II) from aqueous solutions by amino modified nanostructured microfibrillated cellulose. *Cellulose* **2014**, *21*, 1471–1487. [CrossRef]
12. Lin, S.; Huang, W.; Yang, H.; Sun, S.; Yu, J. Recycling application of waste long-root Eichhornia crassipes in the heavy metal removal using oxidized biochar derived as adsorbents. *Bioresour. Technol.* **2020**, *314*, 123749. [CrossRef] [PubMed]
13. Ajitha, P.; Vijayalakshmi, K.; Saranya, M.; Gomathi, T.; Rani, K.; Sudha, P.N.; Sukumaran, A. Removal of toxic heavy metal lead (II) using chitosan oligosaccharide-graft-maleic anhydride/polyvinyl alcohol/silk fibroin composite. *Int. J. Biol. Macromol.* **2017**, *104*, 1469–1482.
14. Hokkanen, S.; Repo, E.; Westholm, L.J.; Lou, S.; Sainio, T.; Sillanpää, M. Adsorption of Ni^{2+}, Cd^{2+}, ${PO_4}^{3-}$ and NO^{3-} from aqueous solutions by nanostructured microfibrillated cellulose modified with carbonated hydroxyapatite. *Chem. Eng. J.* **2014**, *252*, 64–74. [CrossRef]
15. Sahota, S.; Pande, K.M.; Suresh, S.; Arisutha, S.; Singh, D.; Shah, G. Biological Pretreatment of Water Hyacinth (Eichhornia crassipes) for Biofuel Production-A Review. *J. Biofuels Bioenergy* **2016**, *2*, 97–101. [CrossRef]
16. Wu, S.; Kan, J.; Dai, X.; Shen, X.; Zhang, K.; Zhu, M. Ternary carboxymethyl chitosan-hemicellulose-nanosized TiO_2 composite as effective adsorbent for removal of heavy metal contaminants from water. *Fibers Polym.* **2017**, *18*, 22–32. [CrossRef]
17. Hokkanen, S.; Bhatnagar, A.; Repo, E.; Lou, S.; Sillanpää, M. Calcium hydroxyapatite microfibrillated cellulose composite as a potential adsorbent for the removal of Cr(VI) from aqueous solution. *Chem. Eng. J.* **2016**, *283*, 445–452. [CrossRef]
18. Lin, S.; Yang, H.; Na, Z.; Lin, K. A novel biodegradable arsenic adsorbent by immobilization of iron oxyhydroxide (FeOOH) on the root powder of long-root Eichhornia crassipes. *Chemosphere* **2018**, *192*, 258–266. [CrossRef] [PubMed]
19. Carreño Sayago, U.F. Design, Scaling, and Development of Biofilters with E crassipes for Treatment of Water Contaminated with Cr(VI). *Water* **2021**, *13*, 1317. [CrossRef]
20. Sayago, U.F.C.; Ballesteros Ballesteros, V. Development of a treith Cr(VI) using cellu-lose xanthogenate from *E. crassipes* on a pilot scale. *Sci. Rep.* **2023**, *13*, 1970. [CrossRef] [PubMed]
21. Tan, L.; Zhu, D.; Zhou, W.; Mi, W.; Ma, L.; He, W. Preferring cellulose of Eichhornia crassipes to prepare xan-thogenate to other plant materials and its adsorption properties on copper. *Bioresour. Technol.* **2008**, *99*, 4460–4466. [CrossRef] [PubMed]
22. Widiyowati, I.I.; Nurhadi, M.; Hatami, M.; Yuan, L.S. Effective TiO_2-Sulfonated Carbon-derived from Eichhornia crassipes in The Removal of Methylene Blue and Congo Red Dyes from Aqueous Solution. *Bull. Chem. React. Eng. Catal.* **2020**, *15*, 476–489. [CrossRef]
23. Sayago, U.F.C. Diseño y evaluación de un biosistema de tratamiento a escala piloto de aguas de curtiembres a tra-vés de la Eichhornia crassipes. *Rev. Colomb. Biotecnol.* **2016**, *18*, 74–81. [CrossRef]
24. Sharma, K.; Singh, N.; Sharma, R. Eichhornia crassipes leaves biomass harvested from heavy metal contaminated water for biodiesel production potential. *Res. Crops* **2021**, *22*, 720–726.
25. Hua, M.; Zhang, S.; Pan, B.; Zhang, W.; Lv, L.; Zhang, Q. Heavy metal removal from water/wastewater by nanosized metal oxides: A review. *J. Hazard. Mater.* **2021**, *211*, 317–331. [CrossRef] [PubMed]
26. Wang, C.; Zhan, Y.; Wu, Y.; Shi, X.; Du, Y.; Luo, Y.; Deng, H. TiO_2/rectorite-trapped cellulose composite nanofibrous mats for multiple heavy metal adsorption. *Int. J. Biol. Macromol.* **2020**, *183*, 245–253. [CrossRef] [PubMed]
27. Ajmani, A.; Patra, C.; Subbiah, S.; Narayanasamy, S. Packed bed column studies of hexavalent chromium adsorption by zinc chloride activated carbon synthesized from Phanera vahlii fruit biomass. *J. Environ. Chem. Eng.* **2020**, *8*, 103825. [CrossRef]
28. Mahmoud, A.E.D.; Fawzy, M.; Hosny, G.; Obaid, A. Equilibrium, kinetic, and diffusion models of chromium (VI) re-moval using Phragmites australis and Ziziphus spina-christi biomass. *Int. J. Environ. Sci. Technol.* **2021**, *18*, 2125–2136. [CrossRef]
29. Vijayalakshmi, K.; Gomathi, T.; Latha, S.; Hajeeth, T.; Sudha, P.N. Removal of copper (II) from aqueous solution using nanochitosan/sodium alginate/microcrystalline cellulose beads. *Int. J. Biol. Macromol.* **2016**, *82*, 440–452. [CrossRef]

30. Carreño-Sayago, U.F. Development of microspheres using water hyacinth (*Eichhornia crassipes*) for treatment of contami-nated water with Cr(VI). *Environ. Dev. Sustain.* **2021**, *23*, 4735–4746. [CrossRef]
31. Wang, J.; Liu, H.; Yang, S.; Zhang, J.; Zhang, C.; Wu, H. Physicochemical characteristics and sorption capacities of heavy metal ions of activated carbons derived by activation with different alkyl phosphate triesters. *Appl. Surf. Sci.* **2014**, *316*, 443–450. [CrossRef]
32. Ammar, N.S.; Elhaes, H.; Ibrahim, H.S.; Ibrahim, M.A. A novel structure for removal of pollutants from wastewater. *Spectrochim. Acta Part A Mol. Biomol. Spectrosc.* **2014**, *121*, 216–223. [CrossRef]
33. da Cunha, L.P.; Casciatori, F.P.; Vicente, I.V.; Garcia, R.L.; Thoméo, J.C. Metarhizium anisopliae conidia produc-tion in packed-bed bioreactor using rice as substrate in successive cultivations. *Process Biochem.* **2020**, *97*, 104–111. [CrossRef]
34. Tikhomirova, T.S.; But, S.Y. Laboratory scale bioreactor designs in the processes of methane bioconversion: Mini-review. *Biotechnol. Adv.* **2021**, *47*, 107709. [CrossRef] [PubMed]
35. Bringas, A.; Bringas, E.; Ibañez, R.; San-Román, M.F. Fixed-bed columns mathematical modeling for selective nickel and copper recovery from industrial spent acids by chelating resins. *Sep. Purif. Technol.* **2023**, *313*, 123457. [CrossRef]
36. Dawodu, F.A.; Akpan, B.M.; Akpomie, K.G. Sequestered capture and desorption of hexavalent chromium from solution and textile wastewater onto low cost Heinsia crinita seed coat biomass. *Appl. Water Sci.* **2024**, *10*, 1–15. [CrossRef]
37. Tallman, P.S.; Collins, S.; Salmon-Mulanovich, G.; Rusyidi, B.; Kothadia, A.; Cole, S. Water insecurity and gender-based violence: A global review of the evidence. *Wiley Interdiscip. Rev. Water* **2023**, *10*, e1619. [CrossRef]
38. Souza Veloso, D.E.; Werderits Silva, D.E.; de Andrade Aguiar, L.G.; Abrãao, R.; de Souza Sampaio, N.A. Application of the exponential distribution to improve environmental quality in a company in the south of Rio de Janeiro State. *GeSec Rev. Gest. E Secr.* **2023**, *14*, 15695–15704. [CrossRef]
39. Lesmana, S.O.; Febriana, N.; Soetaredjo, F.E.; Sunarso, J.; Ismadji, S. Studies on potential applications of biomass for the separation of heavy metals from water and wastewater. *Biochem. Eng. J.* **2009**, *44*, 19–41. [CrossRef]
40. Fan, H.; Chen, S.; Li, Z.; Liu, P.; Xu, C.; Yang, X. Assessment of heavy metals in water, sediment and shellfish organisms in typical areas of the Yangtze River Estuary, China. *Mar. Pollut. Bull.* **2020**, *151*, 110864. [CrossRef] [PubMed]
41. Chatterjee, A.; Abraham, J. Desorption of heavy metals from metal loaded sorbents and e-wastes: A review. *Biotechnol. Lett.* **2019**, *41*, 319–333. [CrossRef]
42. Jiang, X.; An, Q.D.; Xiao, Z.Y.; Zhai, S.R.; Shi, Z. Versatile core/shell-like alginate@ polyethylenimine composites for efficient removal of multiple heavy metal ions (Pb^{2+}, Cu^{2+}, CrO_4^{2-}): Batch and fixed-bed studies. *Mater. Res. Bull.* **2019**, *118*, 110526. [CrossRef]
43. Júnior, W.N.; Silva, M.G.C.; Vieira, M.G.A. Competitive fixed-bed biosorption of Ag (I) and Cu (II) ions on Sargassum filipendula seaweed waste. *J. Water Process Eng.* **2020**, *36*, 101294. [CrossRef]
44. Sayago, U.F.C.; Castro, Y.P.; Rivera, L.R.C.; Mariaca, A.G. Estimation of equilibrium times and maximum capacity of adsorption of heavy metals by *E. crassipes*. *Environ. Monit. Assess.* **2020**, *192*, 141. [CrossRef] [PubMed]
45. Sayago, U.F.C.; Ballesteros Ballesteros, V. Recent Advances in the Treatment of Industrial Wastewater from Different Celluloses in Continuous Systems. *Polymers* **2023**, *15*, 3996. [CrossRef] [PubMed]
46. Sayago, U.F.C. Design and Development of a Pilot-Scale Industrial Wastewater Treatment System with Plant Biomass and EDTA. *Water* **2023**, *15*, 3484. [CrossRef]
47. Amar, M.B.; Mallek, M.; Valverde, A.; Monclús, H.; Myers, T.G.; Salvadó, V.; Cabrera-Codony, A. Competitive heavy metal adsorption on pinecone shells: Mathematical modelling of fixed-bed column and surface interaction insights. *Sci. Total Environ.* **2024**, *917*, 170398. [CrossRef] [PubMed]
48. Chao, H.P.; Chang, C.C.; Nieva, A. Biosorption of heavy metals on Citrus maxima peel, passion fruit shell, and sugarcane bagasse in a fixed-bed column. *J. Ind. Eng. Chem.* **2014**, *20*, 3408–3414. [CrossRef]
49. Islam, M.S.; Rahaman, M.S.; Barbeau, B. Removal of Pb (II), Zn (II), Cu (II), and As (III) ions from water using kraft pulp-based carboxymethylated cellulose in a fixed-bed column adsorption process. *J. Environ. Chem. Eng.* **2023**, *11*, 111181. [CrossRef]
50. Solgi, M.; Mohamed, M.H.; Udoetok, I.A.; Steiger, B.G.; Wilson, L.D. Evaluation of a granular Cu-modified chitosan biocomposite for sustainable sulfate removal from aqueous media: A batch and fixed-bed column study. *Int. J. Biol. Macromol.* **2024**, 129275.
51. Perendija, J.; Ljubić, V.; Popović, M.; Milošević, D.; Arsenijević, Z.; Đuriš, M.; Cvetković, S. Assessment of waste hop (*Humulus Lupulus*) stems as a biosorbent for the removal of malachite green, methylene blue, and crystal violet from aqueous solution in batch and fixed-bed column systems: Biosorption process and mechanism. *J. Mol. Liq.* **2024**, *394*, 123770. [CrossRef]
52. Safardastgerdi, M.; Ardejani, F.D.; Mahmoodi, N.M. Lignocellulosic biomass functionalized with EDTA dianhydride for removing Cu (II) and dye from wastewater: Batch and fixed-bed column adsorption. *Miner. Eng.* **2023**, *204*, 108423. [CrossRef]

water

MDPI

Article

Immobilization of Horseradish Peroxidase and Myoglobin Using Sodium Alginate for Treating Organic Pollutants

Xinyu Wang [1,†], Hossein Ghanizadeh [2,†], Shoaib Khan [3], Xiaodan Wu [1], Haowei Li [4], Samreen Sadiq [5], Jiayin Liu [6,*], Huimin Liu [3] and Qunfeng Yue [1,*]

1 College of Chemistry and Chemical Engineering, Harbin Normal University, Harbin 150030, China; xinyu4533@gmail.com (X.W.); smile_200325@163.com (X.W.)
2 School of Agriculture and Environment, Massey University, Palmerston North 4410, New Zealand; h.ghanizadeh@massey.ac.nz
3 College of Horticulture and Landscape Architecture, Northeast Agricultural University, Harbin 150030, China; shoaibkhanagriculture@gmail.com (S.K.); liuhm0423@163.com (H.L.)
4 College of Food Science, Northeast Agricultural University, Harbin 150030, China; lhw1661800871@163.com
5 School of Biotechnology, Jiangsu University of Science and Technology, Zhenjiang 212100, China; samreensadiq2022@gmail.com
6 College of Arts and Sciences, Northeast Agricultural University, Harbin 150030, China
* Correspondence: liujiayin@neau.edu.cn (J.L.); qunfengyue@126.com (Q.Y.)
† These authors contributed equally to this work.

Abstract: Removing organic pollutants from wastewater is crucial to prevent environmental contamination and protect human health. Immobilized enzymes are increasingly being explored for wastewater treatment due to their specific catalytic activities, reusability, and stability under various environmental conditions. Peroxidases, such as horseradish peroxidase (HRP) and myoglobin (Mb), are promising candidates for immobilized enzymes utilized in wastewater treatment due to their ability to facilitate the oxidation process of a wide range of organic molecules. However, the properties of the carrier and support materials greatly influence the stability and activity of immobilized HRP and Mb. In this research, we developed immobilized HRP and Mb using support material composed of sodium alginate and $CaCl_2$ as carriers and glutaraldehyde as a crosslinking agent. Following this, the efficacy of immobilized HRP and Mb in removing aniline, phenol, and p-nitrophenol was assessed. Both immobilized enzymes removed all three organic pollutants from an aqueous solution, but Mb was more effective than HRP. After being immobilized, both enzymes became more resilient to changes in temperature and pH. Both immobilized enzymes retained their ability to eliminate organic pollutants through eight treatment cycles. Our study uncovered novel immobilized enzyme microspheres and demonstrated their successful application in wastewater treatment, paving the way for future research.

Keywords: immobilization; peroxidase; sodium alginate; reusability; organic pollutants

Citation: Wang, X.; Ghanizadeh, H.; Khan, S.; Wu, X.; Li, H.; Sadiq, S.; Liu, J.; Liu, H.; Yue, Q. Immobilization of Horseradish Peroxidase and Myoglobin Using Sodium Alginate for Treating Organic Pollutants. *Water* **2024**, *16*, 848. https://doi.org/10.3390/w16060848

Academic Editor: Yung-Tse Hung

Received: 7 January 2024
Revised: 7 March 2024
Accepted: 8 March 2024
Published: 15 March 2024

1. Introduction

Enzymes are biochemical catalysts that are very selective for the substrates on which they act [1]. Peroxidase is a class of enzyme capable of catalyzing the oxidation of aromatic compounds [2]. The roots of horseradish plants (*Armoracia rusticana*, syn. *Cochlearia armoracia*) contain horseradish peroxidase (HRP), which is utilized extensively in environmental protection, wastewater treatment, and biotechnology [3,4]. Myoglobin (Mb) is an abundant iron-containing heme protein in vertebrate cardiac and skeletal muscle [5] which has peroxidase activity [6]. Both HRP and Mb catalyze organic compounds, including phenolic compounds, aniline, and dyes [7–10]. HRP is a multifunctional peroxidase that oxidizes substrates in the presence of hydrogen peroxide [11]. The efficiency of HRP in breaking down a wide range of contaminants has been demonstrated, thereby making a valuable contribution to environmental remediation [12]. Mb can bind and transport oxygen, suggesting it has potential applications in enzymatic degradation [13]. Due to its strong affinity for contaminant binding, Mb can be exploited to efficiently remove pollutants [14].

Both HRP and Mb are highly substrate-specific, environmentally friendly, and require mild catalytic conditions and low energy [15], making them ideal for enzymatic wastewater treatment and organic pollutant removal [9]. Nevertheless, the application of unbound HRP and Mb for wastewater treatment is not cost-effective [16]. In addition, unbound HRP and Mb are unstable and sensitive to denaturants, such as temperature, pH, metal ions, and surfactants [16–18]. Additionally, free enzymes, particularly those that are more costly and difficult to purify, have a restricted capacity for a single application [19]. Enzyme immobilization is a potent method for overcoming these drawbacks and providing superior features for practical applications of peroxidases [20–23]. Immobilization improves the stability, storage, recovery, and recycling of peroxidases [22].

The stability and activity of immobilized enzymes are largely determined by the characteristics of the carrier, reaction conditions, the nature of binding (e.g., the number of bonds formed), the microenvironment of the enzyme molecule, and the characteristics of the spacer connecting enzyme molecules to the carrier [24]. An ideal immobilization carrier needs to be cost-effective, biodegradable and non-toxic [25]. Various carriers, including silica nanoparticles [26], mesoporous carbon [27], carbon nanotubes [28], and alginate microspheres [29], have been employed for enzyme immobilization. Alginate is a naturally derived biocompatible and non-toxic biopolymer which provides a suitable environment for enzyme immobilization [30]. Alginate is frequently combined with calcium chloride ($CaCl_2$) and glutaraldehyde for enzyme immobilization. Calcium chloride is employed in the synthesis of alginate gels or microspheres due to its low toxicity [31]. Glutaraldehyde is a cross-linking agent that forms covalent bonds between amino groups of enzymes and functional groups in alginate [32,33] to efficiently impeded enzyme leaching, hence prolonging the functionality of immobilized enzymes [34–38].

Wastewater, particularly industrial wastewater, often contains organic contaminants such as aniline, phenol, and p-nitrophenol, as these compounds are frequently used in manufacturing and can be released into water supply through industrial discharges [39]. Aniline is an aromatic amine which is used in the production of various chemicals, dyes, and pharmaceuticals [40]. Phenol, chemically referred to as carbolic acid, is a crystalline solid that possesses a white coloration and is predominantly employed in the manufacturing processes of pharmaceuticals, polymers, and resins [41]. p-Nitrophenol is a substituted phenol and is primarily utilized in the synthesis of dyes, pesticides, and pharmaceuticals [42]. When remain untreated, these three organic substances are capable of disturbing the balance of microbial communities and harming aquatic organisms, both of which are detrimental to aquatic ecosystems [39]. In addition, because of their potential environmental and public health risks, it is essential to properly manage and treat wastewater containing these organic contaminants before they are released into the environment.

While HRP has been extensively studied and applied in wastewater treatment, the potential of Mb in wastewater treatment is an emerging area of research, and limited investigation has been devoted to the synthesis of immobilized Mb. In addition, the efficacy of Mb in removing aniline, phenol, and p-nitrophenol compared to HRP has not been well-understood. Here, we developed immobilized HRP and Mb using a support material composed of sodium alginate and $CaCl_2$ as carriers and glutaraldehyde as a crosslinking agent to remove these organic contaminants from wastewater. The stability and efficacy of the synthesized immobilized enzymes in removing all three organic contaminants in wastewater were also evaluated under various storage periods, as well as pH and temperature conditions. To the best of our knowledge, this is the first report on the removal of aniline, phenol, and p-nitrophenol from wastewater using HRP and Mb immobilized with sodium alginate, $CaCl_2$, and glutaraldehyde.

2. Materials and Methods

2.1. Materials

Sodium alginate (SA), calcium chloride ($CaCl_2$), sodium dihydrogen phosphate (NaH_2PO_3), disodium hydrogen phosphate (Na_2HPO_3), citric acid, sodium citrate, sodium chloride (NaCl), phenol (C_6H_5OH), aniline ($C_6H_5NH_2$), and p-nitrophenol ($C_6H_4(NO_2)OH$) were purchased from Beijing chemical Co., Ltd., Beijing, China (analytical grades). Horseradish peroxidase (HRP), myoglobin (Mb), and 2,2′-azinobis-(3-ethylbenzthiazoline-6-sulphonate) (ABTS) were purchased from Aladdin Industrial Co., Ltd., Shanghai, China. All chemicals used in this research were of an analytical grade.

2.2. Preparation of Immobilized HRP and Mb

The procedure for immobilizing HRP and Mb in calcium alginate microspheres is shown in Figure 1. The HRP and Mb microspheres were prepared using sodium alginate and glutaraldehyde as the carrier and crosslinking agent, respectively. For this, 5 mL of each enzyme (1 mg/mL) were mixed with 10 mL sodium alginate (3.5% w/v), and the mixture was added dropwise to 50 mL of calcium chloride (3.0% w/v) and allowed to harden for 6 h. Subsequently, the hardened material was neutralized and crosslinked in 100 mL of 0.2% glutaraldehyde solution for 1 h. The immobilized HRP and Mb microspheres were stored at 4 °C for later use.

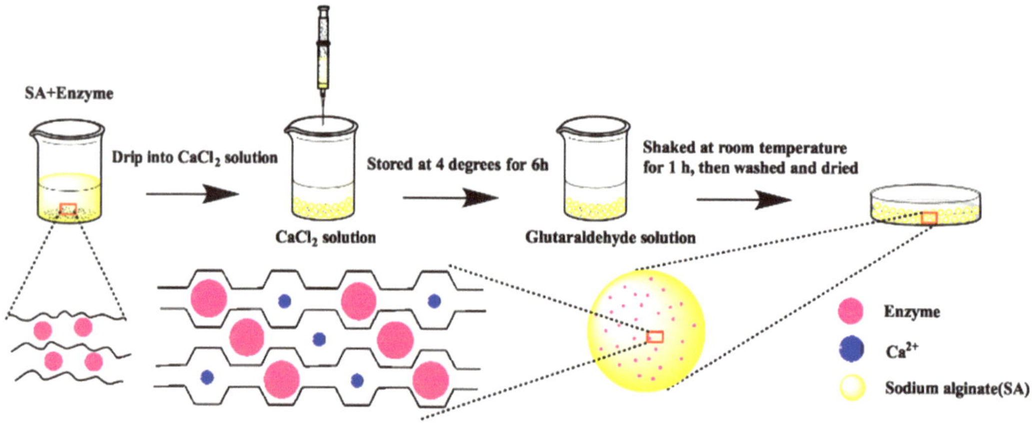

Figure 1. The process involved in the immobilization of myoglobin (Mb) and immobilized horseradish peroxidase (HRP) in calcium alginate microspheres.

2.3. Characterization of Immobilized HRP and Mb

A field emission scanning electron microscope (SEM) (Hitachi, Tokyo, Japan) equipped with an energy dispersive X-ray spectrometer su8010 (FESEM-EDS) was used to examine the structure and appearance of immobilized Mb and HRP microspheres (FESEM-EDS). The element distribution of the microspheres of both immobilized enzymes was analyzed using FESEM-EDS with the following parameters: electron beam energy = 20 kV, and electron beam current = 80.0 μA

2.4. Enzymatic Activity and Stability of Immobilized HRP and Mb

The activity of immobilized HRP and Mb was evaluated in the presence of hydrogen peroxide with ABTS as a substrate using the method described by Andrade et al. [43]. Briefly, 20 immobilized HRP and Mb microspheres were added to a mixture of 0.12 mmol·L^{-1} of ABTS, 25 mmol·L^{-1} of H_2O_2, and 0.03 mol·L^{-1} phosphate buffer solution (pH 6.0) and reacted at 25 °C for 3 min. Subsequently, the oxidation of ABTS was evaluated at 420 nm using a spectrophotometer (UV2550, Shimadzu, Kyoto, Japan). The amount of immobilized

HRP and Mb needed to generate 1 nmol of ABTS+ per minute was defined as one unit of activity. To assess the stability, the activity of both immobilized HRP and Mb was assessed under a wide range of temperatures (i.e., 15 to 55 °C), pH (i.e., 3.0 to 8.0) and storage periods (i.e., 0 days to 32 days at 25 °C), using the same ABTS assay outlined above.

2.5. Removal Efficiency and Reusability of Immobilized HRP and Mb

The efficiency of the immobilized HRP and Mb microspheres to remove organic pollutants from wastewater was assessed and compared with enzyme-free sodium alginate microspheres (control), in the presence of 15 mg/L H_2O_2. For this, we established single-factor assays to investigate the effect of treatment time (0.5 to 6 h), the initial concentration of organic compounds (5 to 50 mg/L), and amounts of microspheres (5 to 25) on the removal efficiency of phenol, aniline, and p-nitrophenol. To prepare the aqueous solution of organic contaminants, a stock solution of each contaminant was prepared by adding 50 mg of each compound to 1 L of water. Subsequently, by employing serial dilution, different concentrations of each organic compound were produced (i.e., 5 mg/L, 10 mg/L, 20 mg/L, 30 mg/L, 40 mg/L, 50 mg/L). To determine the removal efficacy, the concentration of the organic compounds in treated aqueous solutions was evaluated with a UV–Vis spectrophotometer (UV2550, Shimadzu, Kyoto, Japan) at 270, 251, and 371 nm for phenol, aniline, and p-nitrophenol [44].

To assess the reusability of the immobilized microspheres, the microspheres were applied to aqueous solutions containing 5 mM of each phenol, aniline, and p-nitrophenol. The microspheres were then removed from aqueous solutions using filters, and washed with purified water before using in the next cycle. The reusability experiment consisted of eight cycles, and the residue of each organic compound was assessed using the spectrophotometric method outlined above to examine the reusability of the immobilized enzymes. To ensure for the maximum removal of organic compounds, the recovery of microspheres and the assessment of residuals were performed at 6 h intervals.

2.6. Statistical Analyses

All enzymatic assays were performed using a randomized design with three replicates for each treatment. All experiments were repeated in time, and as the variance of the data from separate experiments were equal according to a Levene's test, the data were pooled. Experiments measuring enzymatic activity and stability were subjected to a *t*-test at a significance level of 5%. For studies testing removal efficiency and reusability, we used one-way ANOVA, and the means were compared using Tukey's test at a 5% level of probability. All statistical analyses were performed with SPSS v.21.

3. Results and Discussion

3.1. Characterization of Immobilized HRP and Mb

Myoglobin and HRP solutions (0.1 mmol·L^{-1}) were mixed with calcium alginate (3.5% *w*/*v*) in order to synthesize the microspheres, as shown in Figure 1. The capsulation of HRP and Mb enzymes yielded 3-mm microspheres with a uniform size and a smooth surface (Figure 2).

The SEM analysis of the microspheres revealed that their inner structure was collapsed and damaged, likely as a result of water loss during freeze-drying, which led to the formation of a lamellar structure (Figure 3a,c). It was noted that the inner structure of immobilized microspheres contains evenly distributed small particles (Figure 3b,d). These small particles likely represent the crystals of immobilized enzymes formed after the freeze-drying treatment [23], as no particles were observed in the empty microspheres of sodium alginate carries (Figure 3e,f). This result indicates that HRP and Mb were successfully immobilized in the microspheres.

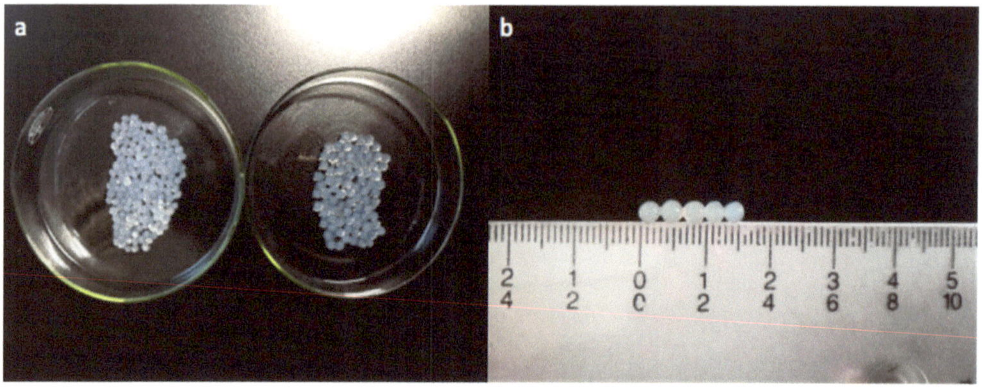

Figure 2. (**a**) Synthesized sodium alginate microspheres, (**b**) size of sodium alginate microspheres.

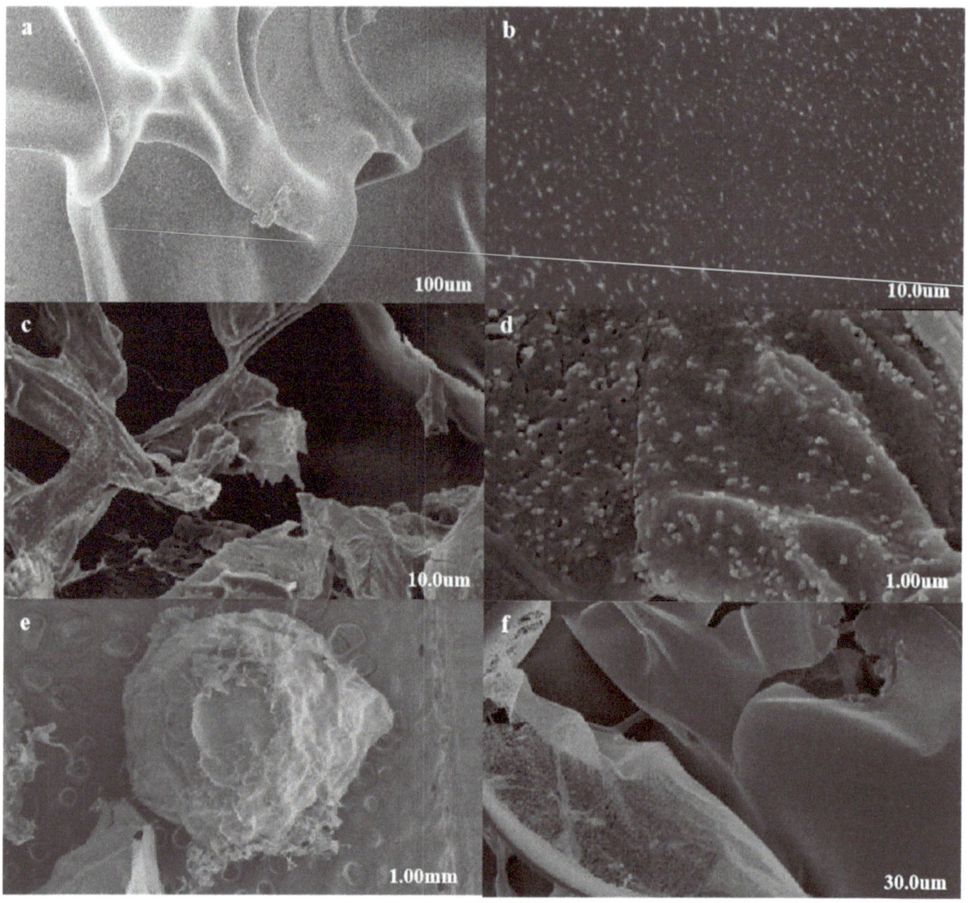

Figure 3. SEM images of morphology of microspheres (**a**,**b**), SEM images of immobilized myoglobin (**c**,**d**), SEM images of immobilized horseradish peroxidase (**c**,**d**), and SEM images of empty morphology (**e**,**f**) of sodium alginate microspheres.

To further prove that HRP and Mb were successfully entrapped in the carries, the content of nitrogen was compared between the immobilized sodium alginate carrier and the empty ones (Table 1). The results revealed that the nitrogen content of immobilized microspheres was much higher than that of empty microspheres. The nitrogen content of immobilized Mb and HRP was 14.1% and 11.9%, respectively.

Table 1. Elemental analysis of empty sodium alginate and immobilized myoglobin (Mb), and horseradish peroxidase (HRP).

Sample	Nitrogen Content (%)	Carbon Content (%)
Empty sodium alginate	0.1	29.0
Immobilized Mb	14.1	32.8
Immobilized HRP	11.9	36.2

As nitrogen is a crucial component of enzymes [22], this result provides further evidence for the immobilization of HRP and Mb in sodium alginate carriers. The peroxidase activity of immobilized HRP and Mb was investigated using an ABTS assay, and the results revealed that the enzyme activity of the immobilized Mb and HRP was 123.78 and 85.20 U/mg, respectively (Figure 4). This indicates that an immobilized HRP concentration of nearly 1.5 times that of Mb is required to achieve similar absolute enzymatic activity. These results are in agreement with a previously published study [45].

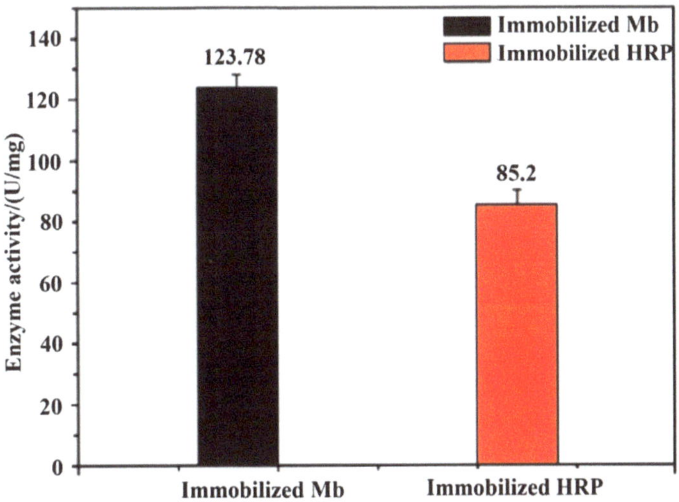

Figure 4. Absolute enzyme activity of immobilized myoglobin (Mb) and immobilized horseradish peroxidase (HRP).

3.2. Stability of Immobilized Enzyme

The effect of temperature, pH, and storage periods on the stability of immobilized HRP and Mb was assessed (Figure 5). According to the results, the activity of immobilized Mb and its corresponding free enzymes showed a generally increasing trend with increasing the temperature, with their highest activity peaking at 35 °C (Figure 5a). However, a downward trend was recorded for the activity of both immobilized Mb and its native enzyme at temperatures greater than 35 °C. The immobilized HRP activity likewise increased with increasing the temperature up to 35 °C, whereas the free HRP activity peaked at 45 °C (Figure 5b). Overall, the results showed that immobilization of both enzymes improve their thermal stability when compared to their corresponding native enzymes, suggesting the immobilization increased the rigidity of the HRP and Mb structures, hence limiting conformational changes in both enzymes at elevated temperatures [46]. However, the activity of immobilized Mb appeared to be less affected at low temperatures compared to the immobilized HRP. It is probable that the immobilization of Mb utilizing the methodology implemented in this study provided Mb with greater resistance to low temperatures than HRP. Immobilization offers a level of protection for enzymes, enhancing their stability and minimizing their susceptibility to environmental influences, ensuring a more sustained enzyme activity and improving processing effects [47].

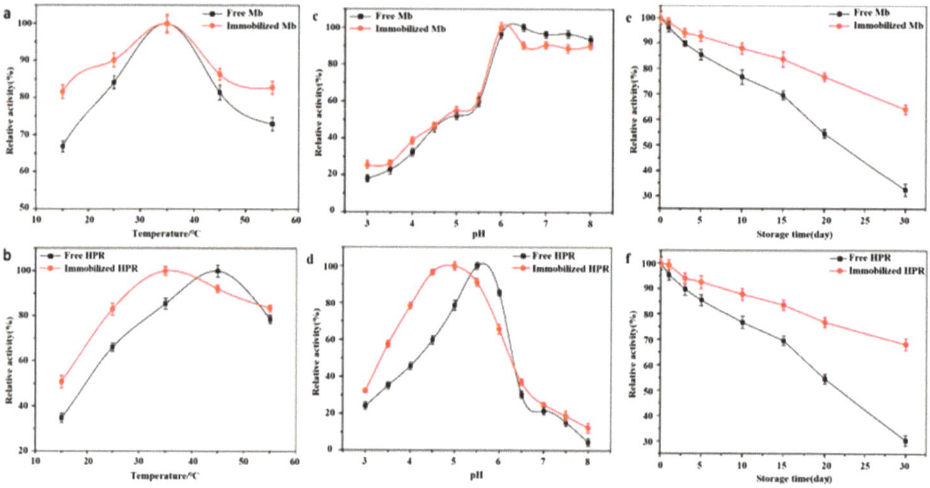

Figure 5. The stability (**a**,**b**), various temperatures (**c**,**d**), and pH levels (**e**,**f**) of free and immobilized myoglobin (Mb) and horseradish peroxidase (HRP) storage periods.

The effect of pH on the activity of both immobilized Mb and HRP was investigated and compared to their corresponding free enzymes. Generally, a different trend was recorded between the response of both immobilized enzymes with increasing the level of pH. The results showed that increasing pH progressively increased the activity of free and immobilized Mb, and both responded similarly when pH increased from 3 to 6 (Figure 5c). However, the greatest activity of the immobilized Mb was recorded at pH of 6, while the free Mb was the most active at pH of 6.5. Despite this small difference, the activity of the immobilized Mb and its native form remained at above 80% at a pH range of 6 to 8. On the other hand, a different pattern in response to increasing the level of pH was recorded for the immobilized HRP and its native enzyme. The results showed that the activity of the immobilized HRP and its native form peaked at pH of 5 and 5.5, respectively. However, there was a general downward trend in their activity at higher pH levels (Figure 5d). In addition, the activity of immobilized HRP appeared to be greater at a pH range between 3 to 5, suggesting that immobilization improved acidic pH tolerance of HRP, possibly owing to the hydroxyl and carboxyl groups on the surface of the support acting as buffer to create a favorable environment for the enzyme. Overall, it appears that the immobilization method employed in this research conferred broader pH tolerance to Mb than HRP, making Mb preserve its activity at a wider range of pH [29,48].

The effect of storage on the stability of immobilized HRP and Mb was evaluated, and the results showed that the activity of both immobilized enzymes and their corresponding native forms progressively reduced with increasing the storage period (Figure 5e,f). However, the activity of the immobilized enzymes was less affected during storage compared to their native forms. For instance, both immobilized enzymes retained over 85% of their initial activity when they were stored for 15 days at 4 °C, while less than 70% activity was recorded for the free enzymes over the same storage period. Additionally, after 30 days of storage, the activity of free HRP and Mb dropped to 30% of their initial activity, while the activity of immobilized HRP and Mb was greater than 65% after 30 days. These results indicated that immobilization provided a relatively stable microenvironment for HRP and Mb molecules, making them less prone to inactivation during storage [31,49]. This finding is consistent with prior research indicating that calcium alginate does not react with the by-product and can effectively prolong the enzyme's durability, hence resulting in enhanced removal efficiency [50]. It has been suggested that the "egg-box" structure formed by sodium alginate and calcium chloride can protect the secondary and tertiary structures of embedded enzymes from damage during long-term storage periods [51].

3.3. Treatment of Organic Wastewater by Immobilized HRP and Mb

3.3.1. Single Factor Exploration

The effect of reaction time, initial concentration of organic pollutants, and the number of microspheres on the efficiency of immobilized HRP and Mb in eliminating aniline, phenol, and p-nitrophenol was evaluated. According to the results, the immobilized HRP and Mb showed improved aniline, phenol, and p-nitrophenol removal efficacy with longer reaction times (Figure 6). Immobilized Mb, however, consistently showed higher removal efficiency of all organic pollutants than immobilized HRP. For instance, at 4 h after treatment, almost 65%, 85%, and 32% of aniline, phenol, and p-nitrophenol, respectively, was removed by the immobilized Mb, while the immobilized HRP only removed 40%, 75%, and 8% of aniline, phenol, and p-nitrophenol, respectively, over the same reaction period. The greater removal efficacy recorded for the immobilized Mb under prolonged reaction time is probably due to a greater synergistic effect between the Mb and the support material compared to the immobilized HRP [52].

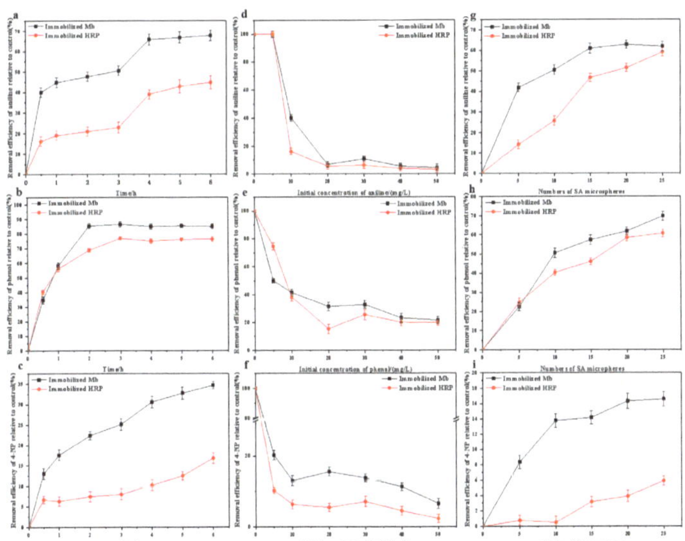

Figure 6. The effect of different reaction periods (**a**–**c**), concentration of pollutants (**d**–**f**), and number of microspheres (**g**–**i**) on the removal efficacy of aniline, phenol, and p-nitrophenol by immobilized myoglobin (Mb) and horseradish peroxidase (HRP) relative to control (enzyme-free sodium alginate microspheres).

Increasing the initial concentration of organic pollutants negatively influenced the removal efficiency of both immobilized enzymes. According to the results, almost 100% of aniline was removed by both immobilized enzymes when the concentration of this organic pollutant was 5 mg/L (Figure 6d). However, increasing the concentration of aniline to 10 mg/L dramatically reduced the removal efficiency of the immobilized Mb and HRP to 40% and 18%, respectively. The removal efficiency of the immobilized HRP and Mb was less than 10% when the concentration of aniline was greater than 20 mg/L. The greatest removal efficiency of both immobilized enzymes was achieved at 5 mg/L of phenol, and it was 74% and 50% for the immobilized HRP and Mb, respectively (Figure 6e). The removal efficiency of the immobilized HRP was reduced dramatically with increasing the concentration of phenol, and it was 40% and 12% at 10 and 20 mg/L of phenol, respectively. Increasing the concentration of phenol also reduced the removal efficiency of the immobilized Mb, though not as prominent as that of the immobilized HRP. According to the results, the removal efficiency of the immobilized Mb at 10 and 20 mg/L of phenol was 41% and 30%, respectively, which was greater than that recorded for the immobilized HRP. The immobilized Mb was also found to be less affected by increasing the concentration of p-nitrophenol compared to the immobilized HRP (Figure 6f). The results showed that at 5 mg/L of p-nitrophenol, the immobilized Mb and HRP removed almost 20% and 10% of this organic pollutant, respectively. At a concentration range of 10 to 50 mg/L of p-nitrophenol, the removal efficiency of the immobilized HRP notably reduced to less than 10%, in contrast to the immobilized Mb, which had a greater removal efficiency at this concentration range. Overall, these results indicate that the removal efficacy of both immobilized enzymes is highly dependent on the concentration of pollutants, which is in agreement with prior research [53,54].

A positive relationship was recoded between increasing the number of microspheres and removal efficiency of the immobilized HRP and Mb, though the removal efficiency of the immobilized Mb was always greater than that of the immobilized HRP, regardless of the type of organic pollutants (Figure 6g–i). For instance, when the solutions containing organic pollutants were treated with 15 microspheres, 60%, 55%, and 14% of aniline, phenol, and p-nitrophenol, respectively, were removed by the immobilized Mb. While only 45%, 42%, and 3% of aniline, phenol, and p-nitrophenol, respectively, were removed by the immobilized HRP, using the same number of microspheres.

The removal efficiency of both immobilized enzymes was assessed when the initial concentration of organic pollutants was 10 mg/L, the number of microspheres was 10 and 15 for the immobilized Mb and HRP, respectively, and the reaction time was 2 h. The results showed that the immobilized Mb had greater removal efficiency than the immobilized HRP, regardless of the type of organic pollutants, as shown in (Figure 7). It was also noted that organic pollutant removal efficiency followed the order of p-nitrophenol < aniline < phenol for both immobilized Mb and HRP.

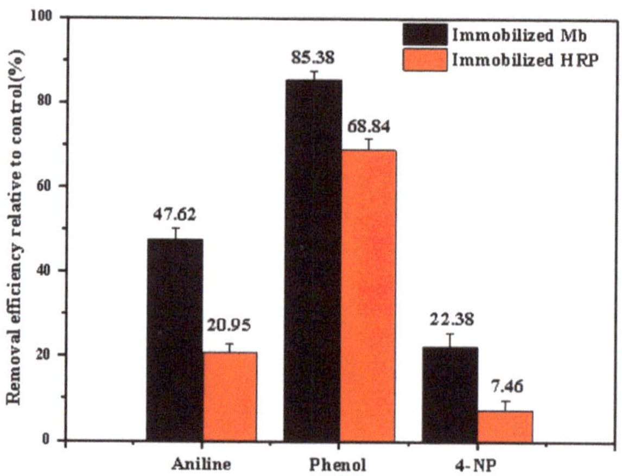

Figure 7. The removal efficiency of aniline, phenol, and p-nitrophenol (4-NP) by myoglobin (Mb) and horseradish peroxidase (HRP) relative to control (enzyme-free sodium alginate microspheres) when the initial concentration of organic pollutants was 10 mg/L, the number of microspheres was 10 and 15 for immobilized Mb and HRP, respectively, and the reaction time was 2 h.

3.3.2. Reusability

The reusability of the immobilized Mb and HRP was assessed when the initial concentration of organic pollutants was 10 mg/L, the number of microspheres was 10 and 15 for the immobilized Mb and HRP, respectively, and the reaction time was 1 h. The results showed that both immobilized enzymes retained their removal efficiency of all three organic pollutants up to eight cycles of treatment (Figure 8). According to the results, the removal efficiency of the immobilized Mb after eight cycles of treatments remained at 48%, 55%, and 14% for aniline, phenol, and p-nitrophenol, respectively. The removal efficiency of the immobilized HRP after eight cycles of treatment was 22%, 49%, and 2% for aniline, phenol, and p-nitrophenol, respectively. Although the activity of Mb may not be as high as that of HRP, the enhanced stability it achieved through immobilization resulted in more efficient catalytic performance. Taken together, these results indicate that both immobilized enzymes can be recycled up to eight times with a negligible loss in their activity during the recycling process. The recyclability and stability of immobilized enzymes play an important role in their applicability. Immobilized enzymes with improved stability and reusability can make wastewater treatment procedures cost-effective [23,48,55].

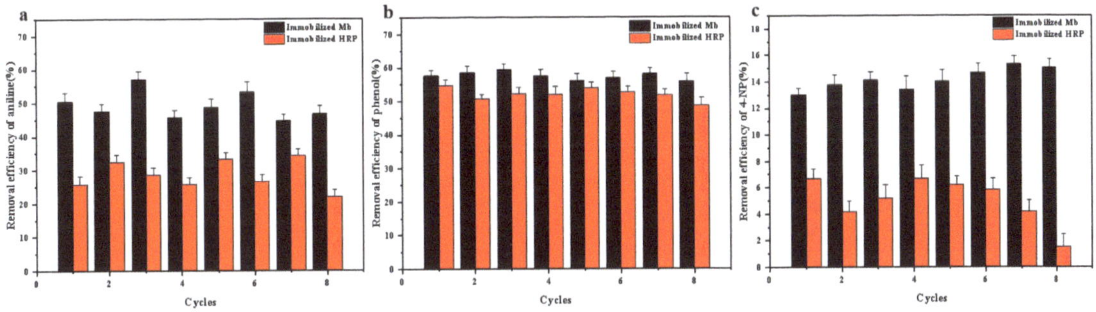

Figure 8. Reusability of myoglobin (Mb) and horseradish peroxidase (HRP) for treating solutions containing (**a**) aniline, (**b**) phenol, and (**c**) p-nitrophenol.

4. Conclusions

In this study, we immobilized HRP and Mb on microspheres made of sodium alginate and calcium chloride, with glutaraldehyde as a crosslinking agent. The efficiency of the immobilized HRP and Mb in eliminating aniline, phenol, and p-nitrophenol, three commonly found organic pollutants in industrial wastewater, was subsequently evaluated. Both immobilized enzymes could remove all three organic pollutants from aqueous solutions, but the immobilized Mb was more effective than the immobilized HRP. Both enzymes were more stable and rigid after immobilization against temperature and pH changes. After 30 days of storage at 4 °C, the immobilized enzymes retained higher activity than free enzymes, and their ability to remove organic pollutants was maintained through eight treatment cycles. Our research has led to the discovery of novel immobilized enzyme microspheres and their application in wastewater treatment, which opens up new avenues for further investigation. The findings of this research revealed that the immobilized Mb performed better as the concentration of organic compounds and reaction period increased, however, the underlying mechanism for this improvement is still unknown and needs additional research. The limitation of our study is that it did not assess the efficacy of both immobilized enzymes using industrial wastewater containing aniline, phenol, and p-nitrophenol. In the future, a large-scale treatment system for organic wastewater will be used to test the removal efficiency of immobilized HRP and Mb. In addition, further research is required to facilitate the retrieval of microspheres from treated water by the implementation of magnetic material techniques. Important to consider in biotechnological applications is the environmental impact of immobilized enzymes. Although immobilization frequently improves the stability and reusability of enzymes, the environmental repercussions of the materials and techniques utilized can differ. Therefore, in order to obtain a thorough understanding of the trade-off between the effectiveness of the immobilized Mb and HRP, and sustainable environmental practices, a holistic environmental impact assessment should be employed.

Author Contributions: Writing—original draft, writing—review and editing, X.W. (Xinyu Wang); writing—review and editing, resources, H.G.; writing—review and editing, S.K.; writing—review and editing, S.S. and H.L. (Haowei Li); investigation, conceptualization, resources, supervision, J.L.; methodology, conceptualization, resources, H.L. (Huimin Liu) and X.W. (Xiaodan Wu); conceptualization, resources, supervision, Q.Y. All authors have read and agreed to the published version of the manuscript.

Funding: This research received no external funding.

Data Availability Statement: The data presented in this study are available on request from the corresponding author. The data are not publicly available due to privacy reasons.

Acknowledgments: The authors are grateful and would like to thank Prince Sultan University Riyadh Saudi Arabia for financial support.

Conflicts of Interest: The authors declare that they have no known competing financial interests or personal relationships that could have appeared to influence the work reported in this paper.

References

1. Contesini, F.; Figueira, J.; Kawaguti, H.; Fernandes, P.; Carvalho, P.; Nascimento, M.; Sato, H. Potential applications of carbohydrases immobilization in the food industry. *Int. J. Mol. Sci.* **2013**, *14*, 1335–1369. [CrossRef]
2. Nicell, J.A.; Bewtra, J.K.; Taylor, K.E.; Biswas, N.; St. Pierre, C. Enzyme catalyzed polymerization and precipitation of aromatic compounds from wastewater. *Water Sci. Technol.* **1992**, *25*, 157–164. [CrossRef]
3. Koeller, K.M.; Wong, C.H. Enzymes for chemical synthesis. *Nature* **2001**, *409*, 232. [CrossRef]
4. Ge, J.; Lu, D.; Liu, Z.; Zheng, L. Recent advances in nanostructured biocatalysts. *Biochem. Eng. J.* **2009**, *44*, 53–59. [CrossRef]
5. Ordway, G.A.; Garry, D.J. Myoglobin: An essential hemoprotein in striated muscle. *J. Exp. Biol.* **2004**, *207*, 3441–3446. [CrossRef] [PubMed]
6. Oohora, K.; Hayashi, T. Chapter Nineteen—Reconstitution of heme enzymes with artificial metalloporphyrinoids. In *Methods in Enzymology*; Pecoraro, V.L., Ed.; Academic Press: Cambridge, MA, USA, 2016; Volume 580, pp. 439–454.
7. Preethi, S.; Anumary, A.; Ashokkumar, M.; Thanikaivelan, P. Probing horseradish peroxidase catalyzed degradation of azo dye from tannery wastewater. *SpringerPlus* **2013**, *2*, 341. [CrossRef] [PubMed]
8. Van Haandel, M.J.H.; Claassens, M.M.J.; Van der Hout, N.; Boersma, M.G.; Vervoort, J.; Rietjens, I.M.C.M. Differential substrate behaviour of phenol and aniline derivatives during conversion by horseradish peroxidase. *Biochim. Biophys. Acta Protein Struct. Mol. Enzymol.* **1999**, *1435*, 22–29. [CrossRef]
9. Xiang, H.-F.; Xu, J.-K.; Liu, J.; Yang, X.-Z.; Gao, S.-Q.; Wen, G.-B.; Lin, Y.-W. Efficient biodegradation of malachite green by an artificial enzyme designed in myoglobin. *RSC Adv.* **2021**, *11*, 16090–16095. [CrossRef]
10. Carlsen, C.U.; Skovgaard, I.M.; Skibsted, L.H. Pseudoperoxidase activity of myoglobin: Kinetics and mechanism of the peroxidase cycle of myoglobin with H_2O_2 and 2,2-azino-bis(3-ethylbenzthiazoline-6-sulfonate) as substrates. *J. Agric. Food Chem.* **2003**, *51*, 5815–5823. [CrossRef] [PubMed]
11. Mao, L.; Luo, S.; Huang, Q.; Lu, J. Horseradish peroxidase inactivation: Heme destruction and influence of polyethylene glycol. *Sci. Rep.* **2013**, *3*, 3126. [CrossRef]
12. Zhu, C.; Wei, F.; Jiang, H.; Lin, Z.; Zhong, L.; Wu, Y.; Sun, X.; Song, L. Exploration of the structural mechanism of hydrogen (H2)-promoted horseradish peroxidase (HRP) activity via multiple spectroscopic and molecular dynamics simulation techniques. *Int. J. Biol. Macromol.* **2024**, *258*, 128901. [CrossRef]
13. Guo, W.-J.; Xu, J.-K.; Liu, J.-J.; Lang, J.-J.; Gao, S.-Q.; Wen, G.-B.; Lin, Y.-W. Biotransformation of lignin by an artificial heme enzyme designed in myoglobin with a covalently linked heme group. *Front. Bioeng. Biotechnol.* **2021**, *9*, 664388. [CrossRef]
14. Liu, C.; Yuan, H.; Liao, F.; Wei, C.-W.; Du, K.-J.; Gao, S.-Q.; Tan, X.; Lin, Y.-W. Unique Tyr-heme double cross-links in F43Y/T67R myoglobin: An artificial enzyme with a peroxidase activity comparable to that of native peroxidases. *Chem. Commun.* **2019**, *55*, 6610–6613. [CrossRef]
15. Xu, L.; Zhang, N.; Wang, Q.; Yuan, J.; Yu, Y.; Wang, P.; Fan, X. Eco-friendly grafting of chitosan as a biopolymer onto wool fabrics using horseradish peroxidase. *Fibers Polym.* **2019**, *20*, 261–270. [CrossRef]
16. Torres-Salas, P.; Monte-Martinez, A.D.; Cutio-Avila, B.; Rodriguez-Colinas, B.; Alcalde, M.; Ballesteros, A.O.; Plou, F.J. *Immobilized Biocatalysts: Novel Approaches and Tools for Binding Enzymes to Supports*; Wiley Online Library: Hoboken, NJ, USA, 2011.
17. Jun, L.Y.; Yon, L.S.; Mubarak, N.M.; Bing, C.H.; Pan, S.; Danquah, M.K.; Abdullah, E.C.; Khalid, M. An overview of immobilized enzyme technologies for dye, phenolic removal from wastewater. *J. Environ. Chem. Eng.* **2019**, *7*, 102961. [CrossRef]

18. Carlsson, N.; Gustafsson, H.; Th?Rn, C.; Olsson, L.; Holmberg, K.; Åkerman, B. Enzymes immobilized in mesoporous silica: A physical–chemical perspective. *Adv. Colloid Interface Sci.* **2014**, *205*, 339–360. [CrossRef]
19. Yang, X.; Yan, X.H.; Guo, Q.; Ghanizadeh, H.; Li, M.H.; Tuo, H.H.; Wen, Z.M.; Li, W. Effects of different management practices on plant community and soil properties in a restored grassland. *J. Soil Sci. Plant Nutr.* **2022**, *22*, 3811–3821. [CrossRef]
20. Liu, D.M.; Dong, C. Recent advances in nano-carrier immobilized enzymes and their applications. *Process Biochem.* **2020**, *92*, 464–475. [CrossRef]
21. Chen, C.; Sun, W.; Lv, H.; Li, H.; Wang, Y.; Wang, P. Spacer arm-facilitated tethering of laccase on magnetic polydopamine nanoparticles for efficient biocatalytic water treatment. *Chem. Eng. J.* **2018**, *350*, 949–959. [CrossRef]
22. Cui, J.; Ren, S.; Sun, B.; Jia, S. Optimization protocols and improved strategies for metal-organic frameworks for immobilizing enzymes: Current development and future challenges. *Coord. Chem. Rev.* **2018**, *370*, 22–41. [CrossRef]
23. Liu, J.; Ghanizadeh, H.; Li, X.; An, L.; Qiu, Y.; Zhang, Y.; Chen, X.; Wang, A. Facile synthesis of core\shell Fe_3O_4@mSiO_2(Hb) and its application for organic wastewater treatment. *Environ. Res.* **2022**, *203*, 111796. [CrossRef] [PubMed]
24. Zhang, D.-H.; Yuwen, L.-X.; Peng, L.-J. Parameters affecting the performance of immobilized enzyme. *J. Chem.* **2013**, *2013*, 946248. [CrossRef]
25. Xin, Y.; Wang, G.; Han, W.; Shen, Y.; Uyama, H. An ideal enzyme immobilization carrier: A hierarchically porous cellulose monolith fabricated by phase separation method. *Pure Appl. Chem.* **2018**, *90*, 1055–1062. [CrossRef]
26. Wang, F.; Guo, C.; Yang, L.R.; Liu, C.Z. Magnetic mesoporous silica nanoparticles: Fabrication and their laccase immobilization performance. *Bioresour. Technol.* **2010**, *101*, 8931–8935. [CrossRef] [PubMed]
27. Liu, Y.; Zeng, Z.; Zeng, G.; Tang, L.; Pang, Y.; Li, Z.; Liu, C.; Lei, X.; Wu, M.; Ren, P. Immobilization of laccase on magnetic bimodal mesoporous carbon and the application in the removal of phenolic compounds. *Bioresour. Technol.* **2012**, *115*, 21–26. [CrossRef] [PubMed]
28. Tavares, A.P.; Silva, C.G.; Dražić, G.; Silva, A.M.; Loureiro, J.M.; Faria, J.L. Laccase immobilization over multi-walled carbon nanotubes: Kinetic, thermodynamic and stability studies. *J. Colloid Interface Sci.* **2015**, *454*, 52–60. [CrossRef]
29. Jiang, D.S.; Long, S.Y.; Huang, J.; Xiao, H.Y.; Zhou, J.Y. Immobilization of *Pycnoporus sanguineus* laccase on magnetic chitosan microspheres. *Biochem. Eng. J.* **2005**, *25*, 15–23. [CrossRef]
30. Sánchez-Machado, D.I.; López-Cervantes, J.; Correa-Murrieta, M.A.; Sánchez-Duarte, R.G.; Cruz-Flores, P.; de la Mora-López, G.S. Chapter 4.2—Chitosan. In *Nonvitamin and Nonmineral Nutritional Supplements*; Nabavi, S.M., Silva, A.S., Eds.; Academic Press: Cambridge, MA, USA, 2019; pp. 485–493.
31. Liu, J.; Han, Z.; An, L.; Ghanizadeh, H.; Wang, A. Evaluation of immobilized microspheres of *Clonostachys rosea* on *Botrytis cinerea* and tomato seedlings. *Biomaterials* **2023**, *301*, 122217. [CrossRef]
32. Manrich, A.; Galv?o, C.M.A.; Jesus, C.D.F.; Giordano, R.C.; Giordano, R.L.C. Immobilization of trypsin on chitosan gels: Use of different activation protocols and comparison with other supports. *Int. J. Biol. Macromol.* **2008**, *43*, 54–6110. [CrossRef]
33. Adriano, W.S.; Filho, E.; Silva, J.A.; Giordano, R.; GonçAlves, L. Stabilization of penicillin G acylase by immobilization on glutaraldehyde-activated chitosan. *Braz. J. Chem. Eng.* **2005**, *22*, 529–538. [CrossRef]
34. Silva, J.A.; Macedo, G.P.; Rodrigues, D.S.; Giordano, R.; GonçAlves, L. Immobilization of *Candida antarctica* lipase B by covalent attachment on chitosan-based hydrogels using different support activation strategies. *Biochem. Eng. J.* **2012**, *60*, 16–24. [CrossRef]
35. Urrutia, P.; Bernal, C.; Escobar, S.; Santa, C.; Mesa, M.; Wilson, L.; Illanes, A. Influence of chitosan derivatization on its physicochemical characteristics and its use as enzyme support. *J. Appl. Polym. Sci.* **2014**, *131*, 631–644. [CrossRef]
36. Trizna, E.Y.; Baydamshina, D.R.; Kholyavka, M.G.; Sharafutdinov, I.S.; Kayumov, A.R. Soluble and immobilized papain and trypsin as destroyers of bacterial biofilms. *Genes Cells* **2015**, *10*, 106–112.
37. Li, H.; Du, Y.; Xu, Y. Adsorption and complexation of chitosan wet-end additives in papermaking systems. *J. Appl. Polym. Sci.* **2004**, *91*, 2642–2648. [CrossRef]
38. Kumar, M.; Muzzarelli, R.; Muzzarelli, C.; Sashiwa, H.; Domb, A.J. Chitosan chemistry and pharmaceutical perspectives. *Chem. Rev.* **2005**, *104*, 6017–6084. [CrossRef] [PubMed]
39. Zhu, L.; Chen, B.; Shen, X. Sorption of phenol, p-nitrophenol, and aniline to dual-cation organobentonites from water. *Environ. Sci. Technol.* **2000**, *34*, 468–475. [CrossRef]
40. Gheni, S.A.; Ali, M.M.; Ta, G.C.; Harbin, H.J.; Awad, S.A. Toxicity, hazards, and safe handling of primary aromatic amines. *ACS Chem. Health Saf.* **2024**, *31*, 8–21. [CrossRef]
41. Saputera, W.H.; Putrie, A.S.; Esmailpour, A.A.; Sasongko, D.; Suendo, V.; Mukti, R.R. Technology advances in phenol removals: Current progress and future perspectives. *Catalysts* **2021**, *11*, 998. [CrossRef]
42. Xu, J.; Wang, B.; Zhang, W.-h.; Zhang, F.-J.; Deng, Y.-d.; Wang, Y.; Gao, J.-J.; Tian, Y.-S.; Peng, R.-H.; Yao, Q.-H. Biodegradation of p-nitrophenol by engineered strain. *AMB Express* **2021**, *11*, 124. [CrossRef]
43. Andrade, C.T.; Barros, L.A.M.; Lima, M.C.P.; Azero, E.G. Purification and characterization of human hemoglobin: Effect of the hemolysis conditions. *Int. J. Biol. Macromol.* **2004**, *34*, 233–240. [CrossRef]
44. Wang, Y.; Sun, H.; Duan, X.; Ang, H.M.; Tadé, M.O.; Wang, S. A new magnetic nano zero-valent iron encapsulated in carbon spheres for oxidative degradation of phenol. *Appl. Catal. B Environ.* **2015**, *172*, 73–81. [CrossRef]
45. Rong, J.; Zhou, Z.; Wang, Y.; Han, J.; Li, C.; Zhang, W.; Ni, L. Immobilization of horseradish peroxidase on multi-armed magnetic graphene oxide composite: Improvement of loading amount and catalytic activity. *Food Technol. Biotechnol.* **2019**, *57*, 260–271. [CrossRef] [PubMed]

46. Liu, C.; Tan, L.; Zhang, K.; Wang, W.; Ma, L. Immobilization of horseradish peroxidase for phenol degradation. *ACS Omega* **2023**, *8*, 26906–26915. [CrossRef] [PubMed]
47. Datta, S.; Christena, L.R.; Rajaram, Y.R.S. Enzyme immobilization: An overview on techniques and support materials. *3 Biotech* **2013**, *3*, 1–9. [CrossRef]
48. Qian, G.; Yang, C.; Zhang, J.; Pu, W. Immobilization of hemoglobin on platinum nanoparticles-modified glassy carbon electrode for H_2O_2 sensing. *Wuhan Univ. J. Nat. Sci.* **2010**, *15*, 160–164. [CrossRef]
49. Gu, Y.; Yuan, L.; Li, M.; Wang, X.; Rao, D.; Bai, X.; Shi, K.; Xu, H.; Hou, S.; Yao, H. Co-immobilized bienzyme of horseradish peroxidase and glucose oxidase on dopamine-modified cellulose-chitosan composite beads as a high-efficiency biocatalyst for degradation of acridine. *RSC Adv.* **2022**, *12*, 23006–23016. [CrossRef] [PubMed]
50. Nawaz, M.A.; Rehman, H.U.; Bibi, Z.; Aman, A.; Ul Qader, S.A. Continuous degradation of maltose by enzyme entrapment technology using calcium alginate beads as a matrix. *Biochem. Biophys. Rep.* **2015**, *4*, 250–256. [CrossRef]
51. Wang, B.; Wan, Y.; Zheng, Y.; Lee, X.; Liu, T.; Yu, Z.; Huang, J.; Ok, Y.S.; Chen, J.; Gao, B. Alginate-based composites for environmental applications: A critical review. *Crit. Rev. Environ. Sci. Technol.* **2019**, *49*, 318–356. [CrossRef]
52. Huang, J.; Chang, Q.; Ding, Y.; Han, X.; Tang, H. Catalytic oxidative removal of 2,4-dichlorophenol by simultaneous use of horseradish peroxidase and graphene oxide/Fe_3O_4 as catalyst. *Chem. Eng. J.* **2014**, *254*, 434–442. [CrossRef]
53. Xiao, J.; Lu, Q.; Cong, H.; Shen, Y.; Yu, B. Microporous poly(glycidyl methacrylate-co-ethylene glycol dimethyl acrylate) microspheres: Synthesis, functionalization and applications. *Polym. Chem.* **2021**, *12*, 6050–6070. [CrossRef]
54. Liu, J.; Guan, J.; Lu, M.; Kan, Q.; Li, Z. Hemoglobin immobilized with modified "fish-in-net" approach for the catalytic removal of aniline. *J. Hazard. Mater.* **2012**, *217*, 156–163. [CrossRef] [PubMed]
55. Zhang, C.; Cai, X. Immobilization of horseradish peroxidase on Fe_3O_4/nanotubes composites for biocatalysis-degradation of phenol. *Compos. Interfaces* **2019**, *26*, 379–396. [CrossRef]

water

MDPI

Article

Xenobiotic Removal by *Trametes hirsuta* LE-BIN 072 Activated Carbon-Based Mycelial Pellets: Remazol Brilliant Blue R Case Study

Olga. A. Glazunova *, Konstantin V. Moiseenko and Tatyana V. Fedorova

A. N. Bach Institute of Biochemistry, Research Center of Biotechnology, Russian Academy of Sciences, Leninsky Ave. 33/2, Moscow 119071, Russia; mr.moiseenko@gmail.com (K.V.M.); fedorova_tv@mail.ru (T.V.F.)
* Correspondence: olga.a.glas@gmail.com

Abstract: As a toxic xenobiotic compound, the anthraquinone dye Remazol Brilliant Blue R (RBBR) poses a serious threat to aquatic ecosystems. In the present study, the ability of *Trametes hirsuta* to remove RBBR from the medium was investigated, and the role of adsorption by fungal mycelium and biodegradation by fungal enzymes was evaluated. It was shown that the whole fungal culture was able to remove up to 97% of the dye within the first four hours of incubation. Based on enzymatic activities in the culture broth, laccases were proposed to be the main enzymes contributing to RBBR degradation, and RT-qPCR measurements demonstrated an increase in transcription for the two laccase genes—*lacA* and *lacB*. Composite mycelial pellets of *T. hirsuta* with improved adsorption ability were prepared by adding activated carbon to the growth medium, and the induction of laccase activity by carbon was shown. For composite pellets, the RBBR decolorization degree was about 1.9 times higher at 1 h of incubation compared to carbon-free pellets. Hence, it was shown that using fungal mycelium pellets containing activated carbon can be an effective and economical method of dye removal.

Keywords: white-rot fungi; *Trametes hirsuta*; laccases; ligninolytic peroxidases; RBBR; dye biodegradation; biosorption; mycelial pellets; activated carbon; RT-qPCR

Citation: Glazunova, O.A.; Moiseenko, K.V.; Fedorova, T.V. Xenobiotic Removal by *Trametes hirsuta* LE-BIN 072 Activated Carbon-Based Mycelial Pellets: Remazol Brilliant Blue R Case Study. *Water* **2024**, *16*, 133. https://doi.org/10.3390/w16010133

Academic Editors: Yung-Tse Hung, Issam A. Al-Khatib, Rehab O. Abdel Rahman and Tsuyoshi Imai

Received: 26 November 2023
Revised: 15 December 2023
Accepted: 25 December 2023
Published: 29 December 2023

1. Introduction

The textile industry is one of the most important sectors in the modern economy. At the same time, it produces an enormous volume of dye-containing wastewater, causing a serious impact on the environment. For example, about 10–50% of reactive dyes remain unbound after the dyeing process [1] and can be discharged into industrial effluents. After being released into water bodies, textile dyes reduce the penetration of solar light, which, in turn, inhibits the primary productivity of aquatic ecosystems and increases biological and chemical oxygen demand [2]. In addition, almost all dyes are typical xenobiotics that cannot be naturally formed in the environment and, therefore, can have a toxic, mutagenic, or carcinogenic effect on its inhabitants [3–5]. Therefore, the appropriate treatment of dye-containing wastewater is an issue of utmost importance.

Currently, many physical or chemical methods for dye effluent treatment are proposed. These methods include coagulation, ozonation, photocatalysis, absorption, etc. [6]. However, all these methods have several disadvantages, such as the use of additional chemicals and high operating costs for large remediation volumes. Moreover, commercial dyes are generally recalcitrant and light-resistant aromatic compounds, and their removal by the methods mentioned above is usually inefficient [6,7]. On the contrary, many studies have demonstrated the effectiveness of biological methods for dye removal. In addition, biological aquaremediation does not cause secondary pollution from the chemicals used and, as a result, is more environmentally friendly [8–10].

One of the promising environmentally friendly, effective, and inexpensive biological alternatives to physical and chemical methods of treating wastewater from synthetic dyes is the removal of pollutants using the mycelial biomass of various filamentous fungi [7]. Growing in liquid media, filamentous fungi can form pellets—dense spherical mycelial structures that can be regarded as ready-to-use granular biosorbents. This advantageously distinguishes fungi from bacteria, the immobilization of which requires additional materials. The fungal cell wall contains various functional groups that provide effective biosorption for a wide range of compounds [11].

Among all filamentous fungi, the fungi that cause white rot of wood (i.e., white-rot fungi) are of particular interest due to their unique ability to effectively metabolize many types of highly recalcitrant phenolic compounds, including aromatic dyes [12–14]. The first stages of the metabolic assimilation of aromatic compounds by white-rot fungi are their extracellular and intracellular oxidative degradation. Currently, the main extracellular oxidative enzymes of white rot fungi are considered to be laccases and class II (ligninolytic) peroxidases [15], both of which are capable of degrading many types of aromatic dyes through direct and indirect free radical oxidation processes [16–18]. The intracellular oxidative enzyme system of white-rot fungi is primarily composed of cytochrome P450 mixed-function oxidases that are capable of catalyzing a wide range of reactions, including hydroxylation, epoxidation, dealkylation, sulfoxidation, deamination, desulfurization, dehalogenation, and nitro reduction [19].

Trametes hirsuta (Wulfen) Lloyd 1924 is a white-rot fungus belonging to the order Polyporales. Its ligninolytic system has been extensively studied for the degradation of lignin, xenobiotics, and monolignol-related compounds [20–23]. It was shown that, depending on the structure of the compound being degraded, *T. hirsuta* may prefer either laccases or peroxidases for its degradation. In the present study, the process of xenobiotic removal by mycelial pellets of *T. hirsuta* was investigated using Remazol Brilliant Blue R (RBBR) synthetic dye as a model compound. The aim of this study was to evaluate the contribution of biosorption and enzymatic biodegradation to the overall dye removal process and to enhance the overall removal capacity of the fungal pellets by incorporating activated carbon into their structure. An additional goal of this study was to determine which oxidative enzymes are primarily responsible for RBBR degradation by the mycelial pellets of *T. hirsuta*.

2. Materials and Methods

2.1. Strain and Culture Conditions

The strain *T. hirsuta* LE-BIN 072 was obtained from the Komarov Botanical Institute Basidiomycetes Culture Collection (LE-BIN; St. Petersburg, Russia). It was stored on slant worth agar until further use. For all experiments, *T. hirsuta* was pre-cultured on glucose-peptone (GP) medium without agitation for 7 days at 28 °C. The composition of the GP medium was as follows: ($g \cdot L^{-1}$): glucose—10.0; peptone—3.0; KH_2PO_4—0.6; $ZnSO_4 \times 7H_2O$—0.001; $K_2HPO_4 \times 3H_2O$—0.4; $FeSO_4 \times 7H_2O$—0.0005; $MnSO_4 \times 5H_2O$—0.05; $MgSO_4 \times$ —0.5; $CaCl_2$—0.25.

For all decolorization experiments, *T. hirsuta* was grown on a liquid-agitated GP medium at 25 °C and 180 rpm. The dry weight of biomass was measured after reaching a constant weight at 70 °C.

2.2. RBBR Decolorization by the Whole Fungal Culture

For the whole culture (i.e., mycelial pellets and culture broth) decolorization experiment, *T. hirsuta* was grown for 5 days, and then RBBR was added to the final concentration of 240 $\mu M \cdot L^{-1}$. For residual RBBR content analysis and RT-qPCR analysis, samples of culture broth were taken at 1, 2, 4, and 24 h of cultivation. Heat-inactivated (65 °C for 3 days) samples were used as a control. For cytochrome P450 inhibition experiments, 500 μL of piperonyl butoxide (PBO) per 200 mL flask was added. All experiments were performed in three biological replicates.

2.3. RBBR Decolorization by Fungal Pellets

For RBBR decolorization by fungal pellets (FP), *T. hirsuta* was grown for 5, 7, and 10 days; pellets were collected using filtration through nylon mesh and washed twice with distilled water. Composite fungal pellets supplemented with activated carbon (FP-AC) were grown as previously described in Section 2.1. The only exception was the addition of activated carbon (1 g per 200 mL flask) ground into powder (<0.25 mm). Activated carbon was purchased from Sigma (Sigma, St. Louis, MO, USA), product number C-2889.

RBBR decolorization by fungal pellets was performed at 25 °C and 70 rpm. About 1.0 g (wet weight) of mycelial pellets was added to RBBR dissolved in 10 mL of citrate-phosphate buffer (pH 4.5). The RBBR concentrations were 50, 100, 200, or 300 µM. Heat-inactivated samples (65 °C for 3 days) were used as a control. All experiments were performed in three biological replicates.

2.4. RBBR Decolorization Assay

Prior to measurements, samples were centrifuged at 13,000 rpm for 3 min. Residual RBBR content in the samples was measured spectrophotometrically at 592 nm (absorbance maximum) using a Lambda 35 UV/VIS spectrophotometer (Perkin Elmer, Waltham, MA, USA). The decolorization degree was calculated as follows: $(OD_{592}{}^{initial} - OD_{592}{}^{final})/OD_{592}{}^{initial} \times 100$.

2.5. RT-qPCR Analysis

The RNA extraction, reverse transcription, and qPCR analysis for laccase genes were performed as described in Moiseenko et al. [21]. Briefly, collected fungal mycelium was ground in liquid nitrogen; RNA was extracted using Trizol reagent (Invitrogen, Waltham, MA, USA) and treated with DNase I (Thermo Scientific, Waltham, MA, USA) according to the manufacturer's protocol. The reverse transcription was performed with the MMLV RT kit (Evrogen, Moscow, Russia) using 100 ng of RNA and polyA-specific primers. All qPCR reactions were performed on the StepOnePlus Real-Time PCR System (Thermo Fisher Scientific, Waltham, MA, USA) using qPCRmix-HS SYBR+ROX (Evrogen, Moscow, Russia) according to the manufacturer's protocol. All primers are available in Moiseenko et al. [21].

The transcription levels of target genes relative to the transcription of the control gene, β-tubulin, were calculated as follows: $RQ = 2^{Ct(laccase)}/2^{Ct(\beta\text{-tubulin})}$.

2.6. Enzyme Activity Assays

Laccase activity was measured using 2,2′-azino-bis(3-ethylbenzothiazoline-6-sulfonic acid) (ABTS) as a substrate [24]. The reaction mixture contained 50 µL of sample and 2 mL of 1 mM ABTS in sodium acetate buffer (0.1 M; pH 4.5). Absorbance increase at 436 nm (ε_{436} = 29,500 $M^{-1}cm^{-1}$) was monitored for 3 min using a Lambda 35 UV/Vis spectrophotometer (Perkin Elmer, Waltham, MA, USA).

Manganese-dependent peroxidase activity was measured using Mn^{2+} as a substrate. The method was adopted from work [25] and slightly modified. The reaction mixture contained 1760 µL of sodium tartrate buffer (0.1 M; pH 3.0), 200 µL of 1 mM $MnSO_4$, 40 µL of 5 mM H_2O_2, and 50 µL of sample. The formation of the Mn^{3+}-tartrate complex was monitored at 238 nm (ε_{238} = 6500 $M^{-1}cm^{-1}$) for 3 min using a Lambda 35 UV/Vis spectrophotometer (Perkin Elmer, Waltham, MA, USA).

Manganese-independent peroxidase activity was measured using veratryl alcohol as a substrate [26]. The reaction mixture contained 1760 µL of sodium tartrate buffer (0.1 M; pH 3.0 or 5.0), 200 µL of 100 mM veratryl alcohol, 40 µL of 5 mM H_2O_2, and 50 µL of sample. The formation of veratryl aldehyde was monitored at 310 nm (ε_{310} = 9300 $M^{-1}cm^{-1}$) for 3 min using a Lambda 35 UV/Vis spectrophotometer (Perkin Elmer, Waltham, MA, USA).

All enzymatic activities were expressed as units (U) per mL, where 1 U corresponds to the formation of 1 µmol of product per minute.

3. Results

3.1. Decolorization of RBBR by the Whole Fungal Culture

The role of sorption and enzymatic degradation in the process of RBBR (Figure 1B) decolorization by the whole culture of *T. hirsuta* (i.e., mycelial pellets and culture broth) was assessed during short-term incubation in the presence of the dye. To evaluate the role of extracellular enzymes, a culture broth without mycelium was used for RBBR decolorization. To evaluate the role of intracellular enzymes from the cytochrome P450 family, the inhibitor of cytochrome P450, PBO, was added to the whole culture prior to RBBR decolorization. To evaluate the role of sorption by fungal mycelium, the whole fungal culture was heat inactivated until extracellular enzymatic activities were totally inhibited.

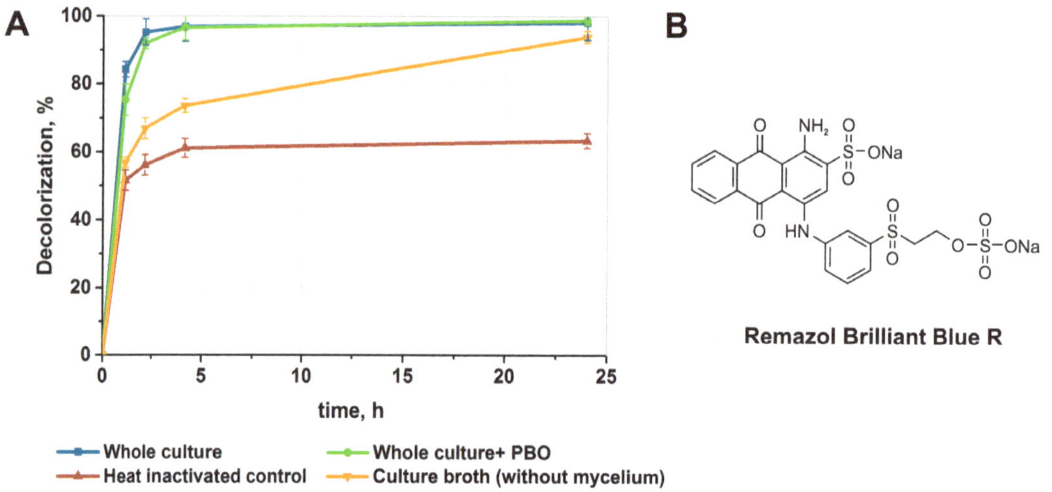

Figure 1. (**A**) The degree of RBBR decolorization by the whole culture, PBO-inhibited culture, culture broth, and heat-inactivated whole culture of *T. hirsuta*. (**B**) RBBR structure.

The whole culture of *T. hirsuta* decolorizes RBBR up to 97% within the first four hours of incubation (Figure 1A), and the addition of PBO to the whole culture did not significantly affect the decolorization degree. The latter suggests that cytochrome P450 enzymes were not largely responsible for RBBR decolorization in the chosen conditions. At the same time, extracellular enzymes contained in the culture broth were able to degrade up to 94% of RBBR in 24 h, but the decolorization rate was much lower than that of the whole culture. Heat-inactivated culture decolorizes about 60% of RBBR within the first 4 h with no significant changes afterwards. Thus, we can conclude that adsorption occurs mainly in the first 4 h of incubation.

Enzymatic activities of laccase, manganese-dependent peroxidase, and manganese-independent peroxidase were measured before the addition of RBBR and after 24 h of incubation with the dye. Laccase activity did not change significantly after the addition of RBBR; only a weak tendency ($0.05 < p < 0.1$, Student's *t*-test) for its increase was observed (26 ± 5 before RBBR addition vs. 31 ± 7 U·mL^{-1} after 24 h of incubation). The manganese-dependent peroxidase activity was 11 ± 3 U·mL^{-1} before the addition of RBBR and almost depleted after 24 h of incubation with the dye. Manganese-independent peroxidase activity was absent before and after RBBR addition.

3.2. Transcriptional Response of Laccases to RBBR

As demonstrated in Section 3.1, laccase activity was predominant in the culture broth during RBBR decolorization. It was previously demonstrated that the genome of *T. hirsuta* contains seven nonallelic copies of laccase genes (*lacA-lacG*) [21]. To find out which laccases were induced by the presence of RBBR, a classical before-and-after experimental design with three biological replicates (i.e., three flasks) was adopted. From each flask, mycelium was collected before (0 h) the addition of RBBR and at certain time points (2, 4, and 24 h) after the introduction of the dye. The obtained values of gain scores (i.e., differences in $\log_2(RQ)$ and $\log_2(RQ_{0h})$) were averaged for all biological replicates and represented as a clustered heat-map in Figure 2.

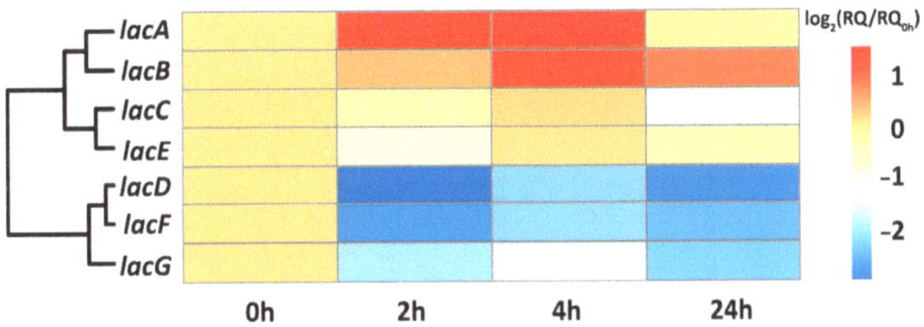

Figure 2. Clustered heat-map representing fold-changes (with a factor of two) in the transcriptional levels of laccase genes after RBBR introduction. The time point immediately before the RBBR introduction is designated as 0 h.

With respect to their transcriptional patterns, all laccase genes formed three clusters. The first cluster comprised *lacA* and *lacB*, whose transcription was induced by the presence of RBBR or its oxidation products in the culture broth. Transcription of *lacA* demonstrated almost three-fold induction at 2 and 4 h of cultivation, after which it returned to the same level as at the beginning of the experiment. In comparison with *lacA*, transcription of *lacB* was induced with a delay; it was unchanged at 2 h of cultivation and two-fold induced at 4 and 24 h. This suggests that the transcription of *lacB* may be more sensitive to the oxidation products of RBBR than to the dye itself. The second cluster comprised *lacC* and *lacE*, whose transcription levels were unaltered by the introduction of RBBR. The third cluster comprised *lacD*, *lacF*, and *lacG*, whose transcription levels were almost four-fold suppressed at all studied time points.

3.3. Formation of Composite Fungal Pellets Supplemented with Activated Carbon

In order to improve the adsorption properties of fungal pellets, activated carbon was added to the GP medium, and *T. hirsuta* was cultivated for 5, 7, and 10 days. The content of activated carbon in the composite FP+AC pellets was calculated based on the amount of carbon remaining after the cultivation. *T. hirsuta* pellets grown without the addition of activated carbon were used as a control.

The addition of activated carbon inhibited the growth of *T. hirsuta*. While FP achieved its maximal dry biomass weight on the 7th day of cultivation, FP+AC achieved the same dry biomass weight on the 10th day (Table 1). The activated carbon content was the highest on the 5th day of cultivation (64%), and it significantly decreased on the following days. Interestingly, activated carbon was mostly entrapped in fungal mycelium on the first days of cultivation. Fungal morphology also changed when activated carbon was added to the growth medium. While FP had developed hairy region, the hairy region was almost absent in FP+AC. Instead, FP+AC had a smooth black surface after 5 and 7 days of cultivation. However, on the 10th day of cultivation, a black core of FP+AC was totally covered by light and hairy fungal mycelium (Supplementary Figure S1), despite the fact that more than half of the initial amount of activated carbon remained in the cultural media at that moment.

Along with growth inhibition, strong laccase induction was observed during the cultivation of FP+AC. Compared to FP culture, laccase activity was induced more than 60 times in FP+AC, reaching 1518 ± 399 U·mL^{-1} on the 10th day of cultivation. In addition, during the cultivation of FP, laccase activity decreased from 5 to 10 days, but the opposite situation was observed during the cultivation of FP+AC. Manganese-dependent peroxidase activity was slightly increased in FP+AC; in both FP and FP+AC, its growth was observed from day 5 to day 10. Manganese-independent peroxidase activity was observed at the trace level only in FP+AC on the 10th day of cultivation.

Table 1. Properties of fungal pellets (FP) and fungal pellets supplemented with activated carbon (FP+AC).

	FP			FP+AC		
Day	5	7	10	5	7	10
Dry weight, g	0.65 ± 0.11	1.23 ± 0.28	0.91 ± 0.20	0.34 ± 0.12	0.80 ± 0.19	1.52 ± 0.15
Activated carbon content, %	-	-	-	64	32	23
Laccase activity, U·mL^{-1}	25 ± 3	16 ± 4	7 ± 1	936 ± 196	1254 ± 395	1518 ± 399
Mn-dependent peroxidase activity, U·mL^{-1}	9 ± 2	12 ± 4	19 ± 3	0	15 ± 1	25 ± 3
Mn-independent peroxidase activity, U·mL^{-1}	0	0	0	0	0	0.4 ± 0.1

3.4. Decolorization of RBBR by Composite Fungal Pellets Supplemented with Activated Carbon

The decolorization of RBBR by composite fungal pellets supplemented with activated carbon was studied in comparison with carbon-free fungal pellets. For both pellet types, heat-inactivated controls were used to estimate their adsorption properties in the absence of enzymatic activity. To avoid interference with the activity of already secreted enzymes, pellets were thoroughly washed prior to the RBBR decolorization measurements.

Both FP and FP+AC were tested for RBBR decolorization ability on different days of cultivation (Figure 3). The FP collected on the 5th and 7th days of cultivation were the most effective and demonstrated 84% and 87% of RBBR decolorization after 4 h of incubation, respectively. Both FP collected on the 5th and 7th days of cultivation were able to decolorize 93–95% of RBBR after 24 h of incubation; also, for these FP, the contribution of sorption and biodegradation to the overall RBBR removal was almost the same. The FP collected on the 10th day of cultivation were much less effective in RBBR decolorization. However, according to the data on heat-inactivated FP, FP collected on the 10th day of cultivation had more pronounced adsorption properties than FP collected on the 5th and 7th days. This could be explained by the increase in the adsorbing area of the mycelium due to its erosion, since at the late stages of cultivation (from 7 to 10 days of cultivation), the degradation of biomass was observed. Also, during the first 2 h of RBBR decolorization by FP collected on the 10th day of cultivation, only adsorption was observed, suggesting that oxidative enzymes were secreted by these FP only at the late stages of incubation with the dye. Thus, it was shown that the pellets collected somewhere between 5 and 8 days of cultivation are more effective for dye decolorization since, besides demonstrating good adsorption properties, they also actively secrete oxidative enzymes.

As with the decolorization of RBBR by the whole fungal culture, the dye was absorbed by FP within the first few hours of incubation, and the remaining dye was further degraded by the fungal extracellular enzymes. It is worth noting that the absorbed dye that was visible inside the pellets was degraded within the following 24 h (Supplementary Figure S1). Thus, after being absorbed by FP, RBBR was subjected to biodegradation, making the desorption of the dye from the pellets back into the environment impossible.

The overall tendency of RBBR decolorization by FP+AC was similar to that observed for FP (Figure 3); however, their adsorption efficiency was significantly increased. The most effective were FP+AC collected on the 7th day of cultivation, followed by FP+AC collected on the 5th and 10th days. Interestingly, the content of activated carbon in FP+AC collected on the 7th day of cultivation (32%) was lower than that in FP+AC collected on the 5th day (64%). This may be due to the different porosities and accessibilities of activated carbon entrapped in the pellets collected at different growth stages. The low adsorption efficiency of FP+AC collected on the 10th day of cultivation can be explained by the different pellet morphology since access to activated carbon in the core of the pellets was hindered by the covering mycelium (see Section 3.2).

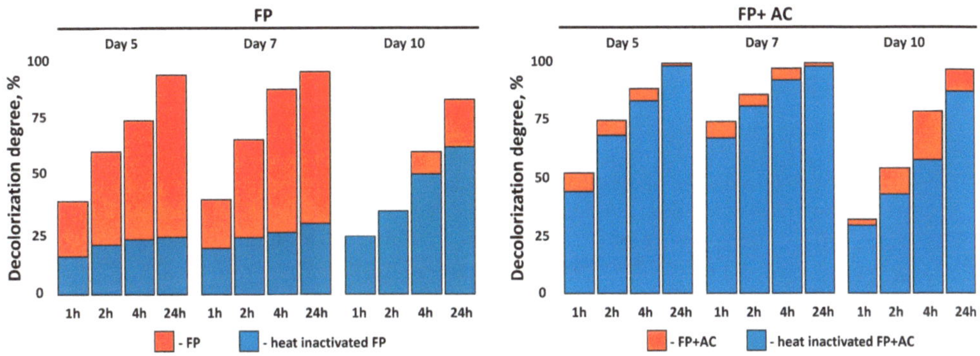

Figure 3. The decolorization of RBBR (with a starting concentration of 50 μM) by fungal pellets (FP) and fungal pellets supplemented with activated carbon (FP+AC) that were collected on the 5th, 7th, and 10th day of *T. hirsuta* cultivation.

The role of biodegradation by enzymes in FP+AC was not as significant as in the case of carbon-free pellets, since RBBR was readily adsorbed by pellets at the first hours of incubation with the dye. It should also be noted that the UV/Vis spectra of RBBR solutions decolorized by FP+AC showed significantly lower absorbance at ~500 nm (Supplementary Figure S2). This can be explained by the adsorption of not only RBBR but also its degradation products. Since some products of RBBR degradation can also be toxic to the environment [27], their adsorption is an additional advantage of FP+AC use.

Since at the initially tested concentration (50 μM) adsorption of RBBR to FP+AC prevailed over biodegradation in the overall dye decolorization process, the best-performing FP+AC, collected on the 5th and 7th day of cultivation, were tested at elevated RBBR concentrations. As can be seen from the presented data (Figure 4), with increasing concentrations of RBBR, the contribution of adsorption to the overall decolorization process was decreasing, and the contribution of biodegradation through the enzymatic process was increasing. At the same time, it can be noted that the efficiency of FP+AC collected on the 5th day of cultivation drops significantly, while FP+AC collected on the 7th day still decolorized about 90% of RBBR in 4 h of incubation, even at the highest concentration of RBBR (300 μM).

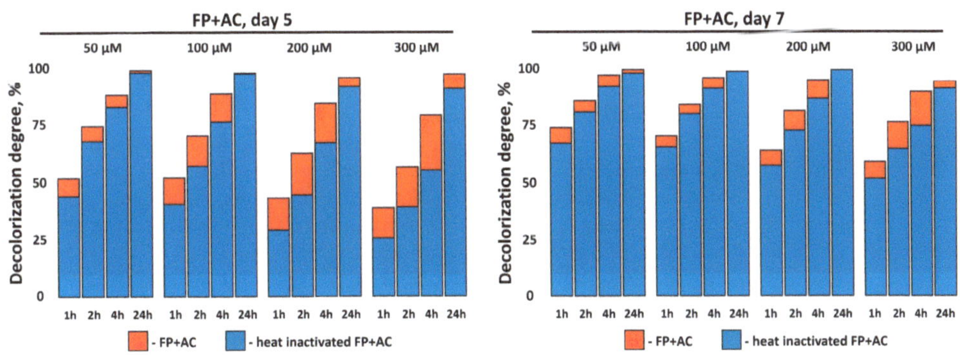

Figure 4. The decolorization of RBBR by FP+AC collected on the 5th and 7th days of cultivation at various RBBR concentrations.

4. Discussion

As a toxic xenobiotic compound, RBBR poses a serious threat to aquatic ecosystems. Belonging to anthraquinone dyes, RBBR has a complex aromatic structure and, therefore, is extremely recalcitrant to degradation [28]. Currently, there are two promising bioremediation methods suitable for removing complex, recalcitrant aromatic compounds from wastewater: biosorption and enzymatic biodegradation. It is noteworthy that mycelial pellets, which can be formed by white-rot fungi under certain cultivation conditions, can be both an efficient adsorbent of aromatic compounds and a source of oxidative enzymes that are able to degrade different xenobiotics of aromatic nature [7,9,29]. Moreover, the formation of mycelium-based composites by the inclusion of various materials into the fungal pellets can significantly improve their bioremediation potential. Fungal immobilization could be performed with lignocellulose agricultural residues (wheat straw, rice hull, corn cob, etc.), wood chips, polyurethane foam, alginate beads, biochar, activated carbon, and others [2,9,10,30]. Immobilized fungal cultures could have enhanced stability and enzymatic activity. Also, the use of fungal pellets and mycelium-based composites solves the very important problem in xenophobic adsorption—regeneration of adsorbents; since xenobiotics concentrated by fungal pellets or mycelium-based composites can be further locally degraded by fungal enzymes and even totally metabolized by fungal mycelium [2].

Despite their huge potential in bioremediation, there is a limited amount of information regarding the use of fungal pellets, especially fungal mycelium-based composites, for the removal of textile dyes. Most of the published studies are still concentrated on pure adsorption or pure enzymatic degradation processes. The current work showed that self-immobilized on activated carbon mycelial pellets of *T. hirsuta* can effectively decolorize RBBR in the concentration range of 50–300 μM in less than 24 h and showed the substantial contribution of both biosorption and biodegradation to this process.

From the perspective of enzymatic degradation, several previous works have investigated which enzymes may be involved in the degradation of RBBR by cultures of white-rot fungi. Based on enzymatic activities in a culture broth, it was shown that for the cultures of *Irpex lacteus* [31–33], *Ischnoderma resinosum*, and *Pleurotus calyptratus* [34], the main degrading enzymes were manganese-dependent peroxidases. At the same time, fungi such as *Pleurotus ostreatus* [35,36], *Trametes pubescens* [35], *Funalia trogii* [37], and *Trametes versicolor* [38] mainly relied on laccases in the process of dye degradation. Also, for *Dichomitus squalens*, the work of Šušla et al. [39] demonstrated the main role of laccases in the RBBR decolorization process, while the work of Eichlerová et al. reported cooperation between laccases and manganese-dependent peroxidases [34]. In our work, on the basis of enzymatic activities in the culture broth, it was shown that for *T. hirsuta*, the main RBBR-degrading enzymes were laccases. It should be noted that in the above-mentioned works, the maximum laccase activities, measured with ABTS as a substrate, varied from 0.05 to 2 $U{\cdot}mL^{-1}$ [34–36,39], while in our work, *T. hirsuta* demonstrated laccase activity of 31.7 $U{\cdot}mL^{-1}$. This drastic difference may be due to different fungal cultivation conditions in those studies, specific features of *T. hirsuta* as a good laccase-producing strain, and individual characteristics of the major *T. hirsuta* laccase isozyme, LacA, which possesses high catalytic efficiency towards ABTS [40].

Since, on the basis of enzymatic activity in the culture broth, it is almost impossible to determine which of the seven laccase isozymes presented in the *T. hirsuta* genome was induced by RBBR, we used RT-qPCR to clarify this point. As a result, it was shown that for *T. hirsuta*, the presence of RBBR induced the transcription of *lacA* and *lacB*, while the transcription of the other laccase genes was either suppressed or unaffected. Previously, we have already shown that, as a universal "expressed by default" oxidative exoenzyme, LacA plays an important role in the detoxification of small aromatic compounds by *T. hirsuta* [21,23]. With respect to LacB, most probably its delayed induction was related to the accumulation of specific degradation products formed by the action of LacA. The sequential induction of several laccase genes by RBBR with simultaneous suppression of others was already reported for white-rot fungi. For *Ganoderma lucidum*, transcription of five laccase genes increased immediately after RBBR introduction, and transcription of all laccase genes except one was suppressed after 10 h of cultivation [41]. For *P. ostreatus*, RBBR induced the transcription of four laccases [42]; however, it is worth noting that in this work, samples were taken within 120–552 h, and data on the depletion of the dye in the medium were not provided, so it is difficult to conclude whether the induction of laccases occurs as a result of RBBR addition or the appearance of its oxidation products.

From the perspective of biosorption, although some industrial dyes have already been tested for biosorption by the mycelia of white-rot fungi [43–46], there is very little information available on RBBR. In the studies of Erkurt et al., microscopic examinations of pellets formed by *P. ostreatus*, *T. versicolor*, and *F. trogii* did not show adsorption of RBBR to mycelium [47]. Similarly, several works in which adsorption was investigated via the extraction of residual dye from fungal mycelium did not report this mechanism as relevant to the decolorization process [34,48,49]. However, in later works, the action of intracellular or cell-wall-associated enzymes could be underestimated as they were not inactivated during the experiments. In our work, using the heat inactivation procedure, it was demonstrated that for *T. hirsuta*, the first rapid step in RBBR decolorization was the adsorption of the dye by fungal mycelium, and biodegradation followed thereafter. According to our data, adsorption of mycelium contributes a lot (up to 60%) to the RBBR decolorization by *T. hirsuta*. Importantly, similar conclusions were made for *F. trogii* when a heat-inactivated control was used [50]. Hence, heat-inactivated control seems to be extremely relevant for the evaluation of dye adsorption by fungal mycelium.

Although activated carbon is an effective adsorbent for many industrial dyes [51], the necessity of its regeneration seriously hinders its applications. Typically, pollutants adsorbed by activated carbon remain entrapped in its porous structure but are not degraded. This problem can be overcome by including activated carbon in mycelium-based composites, since these composites combine the advantages of physical adsorption on activated carbon, biosorption on fungal mycelium, and biodegradation by fungal enzymes. Previously, composite fungal pellets containing activated carbon or biochar were studied for the degradation of different dyes [44,52,53], phenanthrene [54], and heavy metal ions [55,56]. Remarkably, in our work, activated carbon not only significantly improved the adsorption capacity of fungal pellets but also drastically increased the laccase activity during the cultivation of *T. hirsuta*. To the best of our knowledge, this is the first report on the induction of laccase in white-rot fungus by activated carbon. In previous reports on the growth of *F. trogii* [57] and *Phlebia radiata* [58] in the presence of carbon, laccase induction was absent. There is only one report on laccase induction by carbon in *Physiporiopsis rivulosus* [59]; however, in addition to carbon, the cultivation medium was supplemented with sawdust.

A possible explanation for the induction of *T. hirsuta* laccases by activated carbon is the presence of polycyclic aromatic hydrocarbons or other aromatic compounds formed during pyrolysis in its composition [59]. As it was previously shown, these compounds could be laccase inducers [23,60–64]. Thus, at least in the case of *T. hirsuta*, the inclusion of activated carbon not only improves the adsorption properties of fungal pellets but also increases the production of laccase. However, to elucidate the exact mechanism of laccase induction by activated carbon, further investigations are needed.

5. Conclusions

Currently, bioremediation of synthetic dyes using pellets formed by filamentous fungi is a hot topic of research, and basidiomycete fungi causing white rot of the wood are extremely promising candidates for developing this technology. In this article, it was demonstrated that pellets of *T. hirsuta* were able to efficiently decolorize RBBR. The overall process of dye decolorization included two main mechanisms—biosorption onto fungal mycelium and biodegradation by the extracellularly secreted oxidative enzymes. While in the first few hours, biosorption was the predominant process, in the later stages of incubation, biodegradation became more and more pronounced. It was determined that the main enzyme used by *T. hirsuta* for RBBR degradation was laccase, and it was proposed that the fungus mainly used two out of seven laccase isoenzymes (i.e., products of different nonallelic genes) encoded in its genome—LacA and LacB. The presence of activated carbon as a part of the mycelium-based composite not only increased the adsorption capability of the fungal pellets but also stimulated laccase secretion by the fungus. As a result of this synergistic effect, self-immobilized on activated carbon mycelial pellets of *T. hirsuta* were much more efficient than carbon-free mycelial pellets for RBBR decolorization. Therefore, fungal pellets of *T. hirsuta* supplemented with activated carbon could be a promising RBBR-removing agent that combines both adsorption by activated carbon, biosorption by fungal mycelium, and biodegradation by fungal enzymes.

Supplementary Materials: The following supporting information can be downloaded at: https://www.mdpi.com/article/10.3390/w16010133/s1. Figure S1: Mycelium appearance at different time points of RBBR degradation. Figure S2: UV/Vis spectra of RBBR before degradation (black line) and after degradation by fungal pellets (FP, red line) and fungal pellets supplemented with activated carbon (FP+AC, blue line) for 24 h.

Author Contributions: Conceptualization, O.A.G., K.V.M. and T.V.F.; methodology, O.A.G. and K.V.M.; validation, O.A.G. and K.V.M.; formal analysis, O.A.G. and K.V.M.; investigation, O.A.G. and K.V.M.; data curation, O.A.G. and K.V.M.; writing—original draft preparation, O.A.G. and K.V.M.; writing—review and editing, O.A.G., K.V.M. and T.V.F.; visualization, O.A.G. and K.V.M.; supervision, O.A.G. and T.V.F.; project administration, T.V.F.; funding acquisition, T.V.F. All authors have read and agreed to the published version of the manuscript.

Funding: This research was funded by the Russian Science Foundation, grant number 23-46-00018.

Data Availability Statement: The data presented in this study are available within this article and supplementary materials.

Conflicts of Interest: The authors declare no conflict of interest.

References

1. Benkhaya, S.; M'rabet, S.; Lgaz, H.; El Bachiri, A.; El Harfi, A. Dyes: Classification, Pollution, and Environmental Effects. In *Dye Biodegradation, Mechanisms and Techniques*; Springer: Berlin/Heidelberg, Germany, 2022; pp. 1–50.
2. Rodríguez Couto, S. Dye removal by immobilised fungi. *Biotechnol. Adv.* **2009**, *27*, 227–235. [CrossRef] [PubMed]
3. Qadir, A.; Hashmi, M.Z.; Mahmood, A. Xenobiotics, Types, and Mode of Action. In *Xenobiotics in the Soil Environment*; Springer: Berlin/Heidelberg, Germany, 2017; pp. 1–7.
4. Manzoor, J.; Sharma, M. *Impact of Textile Dyes on Human Health and Environment*; IGI Global: Hershey, PA, USA, 2020; pp. 162–169.
5. Lellis, B.; Fávaro-Polonio, C.Z.; Pamphile, J.A.; Polonio, J.C. Effects of textile dyes on health and the environment and bioremediation potential of living organisms. *Biotechnol. Res. Innov.* **2019**, *3*, 275–290. [CrossRef]
6. Katheresan, V.; Kansedo, J.; Lau, S.Y. Efficiency of various recent wastewater dye removal methods: A review. *J. Environ. Chem. Eng.* **2018**, *6*, 4676–4697. [CrossRef]
7. Legorreta-Castañeda, A.J.; Lucho-Constantino, C.A.; Beltrán-Hernández, R.I.; Coronel-Olivares, C.; Vázquez-Rodríguez, G.A. Biosorption of water pollutants by fungal pellets. *Water* **2020**, *12*, 1155. [CrossRef]
8. Harms, H.; Schlosser, D.; Wick, L.Y. Untapped potential: Exploiting fungi in bioremediation of hazardous chemicals. *Nat. Rev. Microbiol.* **2011**, *9*, 177–192. [CrossRef] [PubMed]
9. Wang, L.; Yu, T.; Ma, F.; Vitus, T.; Bai, S.; Yang, J. Novel self-immobilized biomass mixture based on mycelium pellets for wastewater treatment: A review. *Water Environ. Res.* **2019**, *91*, 93–100. [CrossRef] [PubMed]
10. Mir-Tutusaus, J.A.; Baccar, R.; Caminal, G.; Sarrà, M. Can white-rot fungi be a real wastewater treatment alternative for organic micropollutants removal? A review. *Water Res.* **2018**, *138*, 137–151. [CrossRef]
11. Sintakindi, A.; Ankamwar, B. Fungal biosorption as an alternative for the treatment of dyes in waste waters: A review. *Environ. Technol. Rev.* **2021**, *10*, 26–43. [CrossRef]
12. Herath, I.S.; Udayanga, D.; Jayasanka, D.J.; Hewawasam, C. Textile dye decolorization by white rot fungi—A review. *Bioresour. Technol. Reports* **2024**, *25*, 101687. [CrossRef]
13. Latif, W.; Ciniglia, C.; Iovinella, M.; Shafiq, M.; Papa, S. Role of White Rot Fungi in Industrial Wastewater Treatment: A Review. *Appl. Sci.* **2023**, *13*, 8318. [CrossRef]
14. Kathiravan, A.; Gnanadoss, J.J. White-rot fungi-mediated bioremediation as a sustainable method for xenobiotic degradation. *Environ. Exp. Biol.* **2021**, *19*, 103–119. [CrossRef]
15. Dashtban, M.; Schraft, H.; Syed, T.A.; Qin, W. Fungal biodegradation and enzymatic modification of lignin. *Int. J. Biochem. Mol. Biol.* **2010**, *1*, 36–50. [PubMed]

16. Rodríguez-Couto, S. Industrial and environmental applications of white-rot fungi. *Mycosphere* **2017**, *8*, 456–466. [CrossRef]
17. Hammel, K.E.; Cullen, D. Role of fungal peroxidases in biological ligninolysis. *Curr. Opin. Plant Biol.* **2008**, *11*, 349–355. [CrossRef] [PubMed]
18. Loi, M.; Glazunova, O.; Fedorova, T.; Logrieco, A.F.; Mulè, G. Fungal laccases: The forefront of enzymes for sustainability. *J. Fungi* **2021**, *7*, 1048. [CrossRef] [PubMed]
19. Lin, S.; Wei, J.; Yang, B.; Zhang, M.; Zhuo, R. Bioremediation of organic pollutants by white rot fungal cytochrome P450: The role and mechanism of CYP450 in biodegradation. *Chemosphere* **2022**, *301*, 134776. [CrossRef] [PubMed]
20. Vasina, D.V.; Moiseenko, K.V.; Fedorova, T.V.; Tyazhelova, T.V. Lignin-degrading peroxidases in white-rot fungus *Trametes hirsuta* 072. Absolute expression quantification of full multigene family. *PLoS ONE* **2017**, *12*, e0173813. [CrossRef]
21. Moiseenko, K.V.; Vasina, D.V.; Farukshina, K.T.; Savinova, O.S.; Glazunova, O.A.; Fedorova, T.V.; Tyazhelova, T.V. Orchestration of the expression of the laccase multigene family in white-rot basidiomycete *Trametes hirsuta* 072: Evidences of transcription level subfunctionalization. *Fungal Biol.* **2018**, *122*, 353–362. [CrossRef]
22. Moiseenko, K.V.; Glazunova, O.A.; Savinova, O.S.; Vasina, D.V.; Zherebker, A.Y.; Kulikova, N.A.; Nikolaev, E.N.; Fedorova, T.V. Relation between lignin molecular profile and fungal exo-proteome during kraft lignin modification by *Trametes hirsuta* LE-BIN 072. *Bioresour. Technol.* **2021**, *335*, 125229. [CrossRef]
23. Moiseenko, K.V.; Glazunova, O.A.; Savinova, O.S.; Fedorova, T.V. Exoproteomic Study and Transcriptional Responses of Laccase and Ligninolytic Peroxidase Genes of White-Rot Fungus *Trametes hirsuta* LE-BIN 072 Grown in the Presence of Monolignol-Related Phenolic Compounds. *Int. J. Mol. Sci.* **2023**, *24*, 13115. [CrossRef]
24. Niku-Paavola, M.L.; Karhunen, E.; Salola, P.; Raunio, V. Ligninolytic enzymes of the white-rot fungus *Phlebia radiata*. *Biochem. J.* **1988**, *254*, 877–883. [CrossRef] [PubMed]
25. Martinez, M.J.; Ruiz-Duenas, F.J.; Guillen, F.; Martinez, A.T. Purification and catalytic properties of two manganese peroxidase isoenzymes from *Pleurotus eryngii*. *Eur. J. Biochem.* **1996**, *237*, 424–432. [CrossRef] [PubMed]
26. Tien, M.; Kirk, T.K. Lignin peroxidase of *Phanerochaete cyrysosporium*. *Methods Enzymol.* **1988**, *161*, 238–249. [CrossRef]
27. Eichlerová, I.; Homolka, L.; Benada, O.; Kofroňová, O.; Hubálek, T.; Nerud, F. Decolorization of Orange G and Remazol Brilliant Blue R by the white rot fungus *Dichomitus squalens*: Toxicological evaluation and morphological study. *Chemosphere* **2007**, *69*, 795–802. [CrossRef] [PubMed]
28. Routoula, E.; Patwardhan, S.V. Degradation of Anthraquinone Dyes from Effluents: A Review Focusing on Enzymatic Dye Degradation with Industrial Potential. *Environ. Sci. Technol.* **2020**, *54*, 647–664. [CrossRef] [PubMed]
29. Li, L.; Liang, T.; Liu, W.; Liu, Y.; Ma, F. A Comprehensive Review of the Mycelial Pellet: Research Status, Applications, and Future Prospects. *Ind. Eng. Chem. Res.* **2020**, *59*, 16911–16922. [CrossRef]
30. Madadi, R.; Bester, K. Fungi and biochar applications in bioremediation of organic micropollutants from aquatic media. *Mar. Pollut. Bull.* **2021**, *166*, 112247. [CrossRef]
31. Novotný, Č.; Svobodová, K.; Kasinath, A.; Erbanová, P. Biodegradation of synthetic dyes by *Irpex lacteus* under various growth conditions. *Int. Biodeterior. Biodegrad.* **2004**, *54*, 215–223. [CrossRef]
32. Kasinath, A.; Novotný, Č.; Svobodová, K.; Patel, K.C.; Sasek, V. Decolorization of synthetic dyes by *Irpex lacteus* in liquid cultures and packed-bed bioreactor. *Enzyme Microb. Technol.* **2003**, *23*, 167–173. [CrossRef]
33. Šušla, M.; Svobodová, K. Effect of various synthetic dyes on the production of manganese-dependent peroxidase isoenzymes by immobilized Irpex lacteus. *World J. Microbiol. Biotechnol.* **2008**, *24*, 225–230. [CrossRef]
34. Eichlerová, I.; Homolka, L.; Lisá, L.; Nerud, F. Orange G and Remazol Brilliant Blue R decolorization by white rot fungi *Dichomitus squalens*, *Ischnoderma resinosum* and *Pleurotus calyptratus*. *Chemosphere* **2005**, *60*, 398–404. [CrossRef] [PubMed]
35. Casieri, L.; Varese, G.C.; Anastasi, A.; Prigione, V.; Svobodová, K.; Filippelo Marchisio, V.; Novotný, Č. Decolorization and detoxication of reactive industrial dyes by immobilized fungi *Trametes pubescens* and *Pleurotus ostreatus*. *Folia Microbiol.* **2008**, *53*, 44–52. [CrossRef] [PubMed]
36. Palmieri, G.; Cennamo, G.; Sannia, G. Remazol Brilliant Blue R decolourisation by the fungus *Pleurotus ostreatus* and its oxidative enzymatic system. *Enzyme Microb. Technol.* **2005**, *36*, 17–24. [CrossRef]
37. Deveci, T.; Unyayar, A.; Mazmanci, M.A. Production of Remazol Brilliant Blue R decolourising oxygenase from the culture filtrate of *Funalia trogii* ATCC 200800. *J. Mol. Catal. B Enzym.* **2004**, *30*, 25–32. [CrossRef]
38. Sari, A.A.; Tachibana, S. Muryanto Correlation of ligninolytic enzymes from the newly-found species *Trametes versicolor* U97 with RBBR decolorization and DDT degradation. *Water. Air. Soil Pollut.* **2012**, *223*, 5781–5792. [CrossRef]
39. Šušla, M.; Novotný, Č.; Svobodová, K. The implication of *Dichomitus squalens* laccase isoenzymes in dye decolorization by immobilized fungal cultures. *Bioresour. Technol.* **2007**, *98*, 2109–2115. [CrossRef] [PubMed]
40. Glazunova, O.A.; Shakhova, N.V.; Psurtseva, N.V.; Moiseenko, K.V.; Kleimenov, S.Y.; Fedorova, T.V. White-rot basidiomycetes *Junghuhnia nitida* and *Steccherinum bourdotii*: Oxidative potential and laccase properties in comparison with *Trametes hirsuta* and *Coriolopsis caperata*. *PLoS ONE* **2018**, *13*, e0197667. [CrossRef]
41. Qin, P.; Wu, Y.; Adil, B.; Wang, J.; Gu, Y.; Yu, X.; Zhao, K.; Zhang, X.; Ma, M.; Chen, Q.; et al. Optimization of laccase from *Ganoderma lucidum* decolorizing remazol brilliant blue R and Glac1 as main laccase-contributing gene. *Molecules* **2019**, *24*, 3914. [CrossRef]

42. Garrido-Bazán, V.; Téllez-Téllez, M.; Herrera-Estrella, A.; Díaz-Godínez, G.; Nava-Galicia, S.; Villalobos-López, M.Á.; Arroyo-Becerra, A.; Bibbins-Martínez, M. Effect of textile dyes on activity and differential regulation of laccase genes from *Pleurotus ostreatus* grown in submerged fermentation. *AMB Express* **2016**, *6*, 93. [CrossRef]
43. Binupriya, A.R.; Sathishkumar, M.; Swaminathan, K.; Kuz, C.S.; Yun, S.E. Comparative studies on removal of Congo red by native and modified mycelial pellets of *Trametes versicolor* in various reactor modes. *Bioresour. Technol.* **2008**, *99*, 1080–1088. [CrossRef]
44. Zhang, F.; Yu, J. Decolourisation of Acid Violet 7 with complex pellets of white rot fungus and activated carbon. *Bioprocess Eng.* **2000**, *23*, 295–301. [CrossRef]
45. Bonugli-Santos, R.C.; Vieira, G.A.L.; Collins, C.; Fernandes, T.C.C.; Marin-Morales, M.A.; Murray, P.; Sette, L.D. Enhanced textile dye decolorization by marine-derived basidiomycete *Peniophora* sp. CBMAI 1063 using integrated statistical design. *Environ. Sci. Pollut. Res.* **2016**, *23*, 8659–8668. [CrossRef] [PubMed]
46. Cing, S.; Yesilada, O. Astrazon Red dye decolorization by growing cells and pellets of *Funalia trogii*. *J. Basic Microbiol.* **2004**, *44*, 263–269. [CrossRef] [PubMed]
47. Erkurt, E.A.; Ünyayar, A.; Kumbur, H. Decolorization of synthetic dyes by white rot fungi, involving laccase enzyme in the process. *Process Biochem.* **2007**, *42*, 1429–1435. [CrossRef]
48. Rigas, F.; Dritsa, V. Decolourisation of a polymeric dye by selected fungal strains in liquid cultures. *Enzyme Microb. Technol.* **2006**, *39*, 120–124. [CrossRef]
49. Moreira Neto, S.L.; Esteves, P.J.; Santos, V.T.O.; Paranhos, A.P.; Cescato, F.; Vitali, V.M.; Machado, K.M.G. Novel salt and alkali tolerant neotropical basidiomycetes for dye decolorisation in simulated textile effluent. *World J. Microbiol. Biotechnol.* **2011**, *27*, 2665–2673. [CrossRef]
50. Trupkin, S.; Levin, L.; Forchiassin, F.; Viale, A. Optimization of a culture medium for ligninolytic enzyme production and synthetic dye decolorization using response surface methodology. *J. Ind. Microbiol. Biotechnol.* **2003**, *30*, 682–690. [CrossRef]
51. Dutta, S.; Gupta, B.; Srivastava, S.K.; Gupta, A.K. Recent advances on the removal of dyes from wastewater using various adsorbents: A critical review. *Mater. Adv.* **2021**, *2*, 4497–4531. [CrossRef]
52. Yildirim, S.C.; Yesilada, O. A comparative study on decolorization of reactive azo and indigoid dyes by free/immobilized pellets of *Trametes versicolor* and *Funalia trogi*. *J. Environ. Biol.* **2015**, *37*, 1393–1400.
53. Alam, M.Z.; Khan, M.J.H.; Kabbashi, N.A.; Sayem, S.M.A. Development of an Effective Biosorbent by Fungal Immobilization Technique for Removal of Dyes. *Waste Biomass Valorization* **2018**, *9*, 681–690. [CrossRef]
54. Zhang, Y.; Xiao, X.; Zhu, X.; Chen, B. Self-assembled fungus-biochar composite pellets (FBPs) for enhanced co-sorption-biodegradation towards phenanthrene. *Chemosphere* **2022**, *286*, 131887. [CrossRef] [PubMed]
55. Bai, S.; Wang, L.; Ma, F.; Zhu, S.; Xiao, T.; Yu, T.; Wang, Y. Self-assembly biochar colloids mycelial pellet for heavy metal removal from aqueous solution. *Chemosphere* **2020**, *242*, 125182. [CrossRef] [PubMed]
56. Zheng, Z.; Ali, A.; Su, J.; Zhang, S.; Su, L.; Qi, Z. Biochar fungal pellet based biological immobilization reactor efficiently removed nitrate and cadmium. *Chemosphere* **2022**, *296*, 134011. [CrossRef] [PubMed]
57. Birhanli, E.; Erdogan, S.; Yesilada, O.; Onal, Y. Laccase production by newly isolated white rot fungus *Funalia trogii*: Effect of immobilization matrix on laccase production. *Biochem. Eng. J.* **2013**, *71*, 134–139. [CrossRef]
58. Mäkelä, M.R.; Lundell, T.; Hatakka, A.; Hildén, K. Effect of copper, nutrient nitrogen, and wood-supplement on the production of lignin-modifying enzymes by the white-rot fungus *Phlebia radiata*. *Fungal Biol.* **2013**, *117*, 62–70. [CrossRef] [PubMed]
59. Hilber, I.; Blum, F.; Schmidt, H.P.; Bucheli, T.D. Current analytical methods to quantify PAHs in activated carbon and vegetable carbon (E153) are not fit for purpose. *Environ. Pollut.* **2022**, *309*, 119599. [CrossRef] [PubMed]
60. Piscitelli, A.; Giardina, P.; Lettera, V.; Pezzella, C.; Sannia, G.; Faraco, V. Induction and transcriptional regulation of laccases in fungi. *Curr. Genomics* **2011**, *12*, 104–112. [CrossRef]
61. Pozdnyakova, N.; Dubrovskaya, E.; Chernyshova, M.; Makarov, O.; Golubev, S.; Balandina, S.; Turkovskaya, O. The degradation of three-ringed polycyclic aromatic hydrocarbons by wood-inhabiting fungus *Pleurotus ostreatus* and soil-inhabiting fungus *Agaricus bisporus*. *Fungal Biol.* **2018**, *122*, 363–372. [CrossRef]
62. Park, H.; Choi, I.G. Genomic and transcriptomic perspectives on mycoremediation of polycyclic aromatic hydrocarbons. *Appl. Microbiol. Biotechnol.* **2020**, *104*, 6919–6928. [CrossRef]
63. Imam, A.; Suman, S.K.; Vempatapu, B.P.; Tripathi, D.; Ray, A.; Kanaujia, P.K. Pyrene remediation by *Trametes maxima*: An insight into secretome response and degradation pathway. *Environ. Sci. Pollut. Res.* **2022**, *29*, 44135–44147. [CrossRef]
64. Scheel, T.; Hofer, M.; Ludwig, S.; Holker, U. Differential expression of manganese peroxidase and laccase in white-rot fungi in the presence of manganese or aromatic compounds. *Appl. Microbiol. Biotechnol.* **2000**, *54*, 686–691. [CrossRef] [PubMed]

water

MDPI

Article

Degradation of Phenol by Immobilized *Alcaligenes faecalis* Strain JH1 in Fe_3O_4-Modified Biochar from Pharmaceutical Residues

Zhi Zeng, Jiahui Xiao, Manzhi Li, Jiahui Wu and Taiping Zhang *

School of Environment and Energy, South China University of Technology, Guangzhou 510006, China; 202121048364@mail.scut.edu.cn (Z.Z.); oldpinker@163.com (J.X.); lmz15626153672@163.com (M.L.); 15007919777@163.com (J.W.)

* Correspondence: lckzhang@scut.edu.cn; Tel.: +86-15918769501

Abstract: The effect and mechanism of phenol removal by immobilized microorganisms in Fe_3O_4 nanomaterial-modified pharmaceutical residue biochar was investigated to develop efficient biochar immobilizing microbial technology. Plant residue biochar (Y3, Y5, and Y7) was prepared from *Andrographis paniculata* plant residues as the raw material at 300 °C, 500 °C, and 700 °C, respectively. Y7 was modified with Fe_3O_4 nanomaterial (Fe-Y7). These four kinds of biochars were used as carriers to adsorb immobilized *Alcaligenes faecalis* strain JH1, JY3, JY5, JY7, and Fe-Y7 to investigate the mechanism of phenol removal, and eight cycles were performed to analyze their immobilization performance. Compared with suspended bacteria, biochar-immobilized bacteria could improve their tolerance in different environments. At temperatures of 25 °C to 40 °C, pH = 5~9, initial phenol concentration of 300–500 mg/L, and salinity of 3%, the bacteria could still grow and maintain strong activity within 48 h. The water-extractable organic carbon of biochar was also tested for the degradation of phenol by bacteria, which was found to have different stimulating effects on bacteria. In the batch experiments, as the number of cycles increased, the bacteria grew and adhered rapidly to the biochar, eventually forming a thick and sticky biofilm. After the sixth cycle, all the biochar-immobilized bacteria could remove 300 mg/L phenol solution within 12 h. The removal rate of phenol by JFe-Y7 was relatively fast in the eighth cycle. The results indicated that biochar-immobilized bacteria have good durability, stability, and reproducibility and that Fe_3O_4 nanoparticle modification could improve the removal of phenol by increasing the phenol adsorption amount, the adsorption capacity of bacteria, and the enzymatic activity of bacteria.

Keywords: biochar; immobilization; bacteria; phenol; mechanism

Citation: Zeng, Z.; Xiao, J.; Li, M.; Wu, J.; Zhang, T. Degradation of Phenol by Immobilized *Alcaligenes faecalis* Strain JH1 in Fe_3O_4-Modified Biochar from Pharmaceutical Residues. *Water* **2023**, *15*, 4084. https://doi.org/10.3390/w15234084

Academic Editors: Issam A. Al-Khatib, Rehab O. Abdel Rahman, Tsuyoshi Imai and Yung-Tse Hung

Received: 26 October 2023
Revised: 14 November 2023
Accepted: 20 November 2023
Published: 24 November 2023

1. Introduction

Phenol is widely utilized in various industries such as pharmaceuticals, textiles, coal gasification, leather production, resin synthesis, pulp and paper manufacturing, paint production, and wood processing [1]. It is important to note that phenol has also been classified as a priority pollutant by the United States Environmental Protection Agency (US EPA) [2]. Phenol has a wide range of toxic effects for both acute and chronic exposures, is carcinogenic, and is one of the endocrine-disrupting compounds that cause serious hazardous damage to human health and the environment [3,4].

At present, the primary methods employed for treating phenol-containing wastewater encompass adsorption, biodegradation, solvent extraction, and chemical oxidation [5]. Among these methods, biodegradation has emerged as the most promising, practical, and cost-effective approach for effectively removing phenol from chemical production processes [6]. However, the application of bioremediation at contaminated sites is greatly limited because microorganisms usually exhibit unsatisfactory survivability and activity in the face of load shock and highly heterogeneous environmental media [7]. To tackle these concerns, one potential solution is the utilization of cell immobilization in microbial treatment, where microorganisms are confined to a specific spatial region. This confinement ensures the sustained biological activity and rapid proliferation rates, thereby enhancing the overall efficiency [8].

Biochar, a carbon-rich and porous substance, is created through the controlled thermal conversion of biomass in a low-oxygen environment, typically below temperatures of 600 °C [9]. Due to its large specific surface area and the presence of negatively charged organic functional groups, it possesses strong adsorption capacity, which makes it an effective adsorbent for the removal of various pollutants from wastewater, including metals and organic compounds [10]. It is highly efficient, low cost, easy to apply, and produces few byproducts compared to conventional alternatives [11]. However, biochar obtained by the high-temperature cracking of biomass has limited functional groups and porosity, so it is modified before and after biomass cracking by oxidation, sulfonation, amination, and compounding to adjust the surface functionality and introduce abundant functional groups or nanomaterials into the surface of biochar to change its porosity and surface functionality [12,13].

Immobilized microbial technology is a bioengineering technique that uses physical or chemical means to confine free microorganisms to a specific area so that they can maintain a relatively high level of biological activity in a highly dense state and can be used repeatedly. The most common microorganisms immobilized by carriers are fungi as well as bacteria [14]. Research has shown that the biomass of calcium alginate-immobilized *Bacillus cereus* strain, capable of degrading phenolic compounds in petroleum wastewater, exhibits excellent biodegradation efficiency. Microbial treatment has been found to reduce Chemical Oxygen Demand (COD) levels and phenolic compound concentrations by a remarkable 95% [15]. Although traditional immobilization techniques have their own advantages, they all have shortcomings. Therefore, to improve the immobilization effect and treatment efficiency, compound immobilization methods have been derived from basic immobilization methods, which can enhance immobilization techniques by co-fixing mixed strains of bacteria, or bacteria and algae, or by improving the carriers of immobilized microorganisms.

Biochar is an ideal immobilization carrier, characterized by its large surface area, porous structure, and strong stability [16]. The immobilization of microorganisms with biochar occurs through the coordinated impregnation of pollutant-degrading microorganisms and biochar. This synergistic approach enhances the mass transfer of pollutants from the contaminated environment to the degrading microbial community, while promoting the enrichment of degrading bacteria to form a biofilm [17]. Zhuang [18] investigated Fe_3O_4 nanoparticles loaded with bamboo charcoal (Fe_3O_4/BC) and immobilized *Streptomyces* sp. N01 for quinoline removal from water. The results showed that bamboo charcoal enhanced the enzymatic activity as a barrier for bacteria, while Fe_3O_4 nanoparticles improved the cell permeability; thus, the quinoline degradation efficiency was significantly improved. However, there is still a scarcity of systematic studies on the removal of phenol by microorganisms immobilized on modified biochar. Singh and Balomajumder [19] investigated the effect of the simultaneous removal of *S. odorifera* (MTCC 5700) immobilized on the surface of coconut shell-activated carbon (CSAC) from single- and two-component aqueous solutions in an intermittent reactor on phenol. The simultaneous removal of phenol using *S. odorifera* (MTCC 5700) immobilized on the CSAC biosorbent surface exhibited superior performance in the binary component system compared to the single component biosorbent system.

In this paper, Andrographis herb residue, which is not easily disposed, was selected as the raw material for biochar. Biochar pyrolyzed under different temperatures and modified with Fe_3O_4 was selected for the adsorption and immobilization of bacteria to determine the differences caused by these factors and their effects on bacterial immobilization. Simultaneous degradation of phenol by dioxygenases requires a metal cofactor, most commonly Fe(II) or Fe(III) [20]. The nanomaterial Fe_3O_4 was chosen for modification to form loaded biochar. The biochar was used as a carrier to adsorb the immobilized strain JH1 to investigate the mechanisms of phenol removal and to analyze the performance of multiple recycling.

2. Materials and Methods

2.1. Chemicals and Culture Medium

Phenol (>99% purity) was acquired from Aladdin (Shanghai, China). All other chemicals used in this study were of analytical grade and were obtained from local suppliers. The Luria-Bertani (LB) medium consisted of 10.0 g/L tryptone, 5.0 g/L yeast extract, and 10.0 g/L NaCl (pH = 7.0). The mineral salt medium (MSM) was comprised of 1.0 g/L Na_2HPO_4, 0.5 g/L KH_2PO_4, 0.03 g/L $MgSO_4{\cdot}7H_2O$, and 1 mL of trace element solution (pH = 7.0). The trace element solution contained 0.2 g/L $MnCl_2{\cdot}4H_2O$, 0.24 g/L $CoCl_2{\cdot}6H_2O$, 0.15 g/L $NaMoO_4{\cdot}2H_2O$, 0.15 g/L $FeCl_3{\cdot}6H_2O$, 0.16 g/L $ZnSO_4{\cdot}7H_2O$, and 0.2 g/L $NiCl_2{\cdot}6H_2O$. Addition of AGAR (15 g/L) to form solid medium. The LB medium or MSM medium were incubated in an orbital shaker at 30 °C and 150 rpm.

A dominant phenol-degrading strain was domesticated from coking wastewater in Nanchang, Jiangxi Province, China, and named *Alcaligenes faecalis* strain JH1. In brief, 30 mL of coking wastewater sludge supernatant was added to a conical flask containing 70 mL of MSM medium with 10 mg phenol. The samples were then incubated in an orbital shaker at 30 °C and 150 rpm. After 2 to 7 days of culturing, a 30 mL portion of the culture solution was transferred and underwent subculturing under the same conditions. After subculturing for 6 generations, strain JH1 could metabolize and remove phenol effectively with phenol as the only carbon source.

2.2. Fe_3O_4 Nanoparticle-Modified Biochar Preparation

Andrographis paniculata slag residue was obtained from the Guangdong Pharmaceutical University laboratory. The biomass materials were prepared by the slow pyrolysis method. First, the dried biomass materials were processed by crushing them using a high-speed multifunctional crusher, sifted through a 2 mm sieve, and then transferred to a quartz-tube furnace (SK-G06123K, Tianjin, China). Then, under a constant supply of nitrogen, biomass materials were pyrolyzed separately at 300 °C, 500 °C and 700 °C for 2 h at a heating rate of 5°/min. After cooling to room temperature, they were ground through a 100-mesh sieve, washed with deionized water, and dried for 12 h in an oven at 80 °C. Finally, the biochars of medicinal residue (Y3, Y5, Y7) were packed in Ziplock bags and stored in a desiccator.

In this study, Y7 was modified by Fe_3O_4 nanoparticles [1]: 5 g biochars were suspended in a mixture of 100 mL $FeCl_3{\cdot}6H_2O$ (2.9 g, 10.8 mmol) and $FeSO_4{\cdot}7H_2O$ (6.0 g, 21.5 mmol) at 70 °C, and then 10 mL NaOH solution (5 mol/L) was added to precipitate Fe_3O_4 nanoparticles under an N_2 atmosphere. The obtained materials were dried for 24 h in an oven at 70 °C and expressed as Fe-Y7. The chemical formula of the reaction process is $Fe^{2+} + Fe^{3+} + 8OH^- = Fe_3O_4{\downarrow} + 4H_2O$.

2.3. Material Characterization

The percentage of C, H, N, and S in the elemental composition of biochar was directly determined by a UNICUBE elemental analyzer, and the content of O could be calculated by difference subtraction. The specific surface area and pore size of biochar were analyzed using an automatic specific surface and porosity analyzer by measuring the adsorption/desorption isotherm of N_2 at 77 K. Zeta (ζ) potential measurements were carried out using a NanoBrook Omni instrument (Bruker, Billerica, MA, USA). The surface of the sample was analyzed using the Brunauer–Emmett–Teller (BET) method with N_2 gas adsorption at 77.3 K, utilizing the high-speed surface analyzer ASAP 2460 (Micromeritics, Mack Instruments Ltd., Norcross, GA, USA). X-ray diffraction (XRD) patterns were acquired using an X-ray diffractometer (Empyrean, PANalytical, Almelo, The Netherlands) at a scanning range of 2θ from 10° to 80°, and a scanning rate of 10°/min. These measurements were performed to analyze the crystal structure of the biochar. X-ray photoelectron spectroscopy (XPS) using the K-Alpha instrument (Thermo Scientific, Waltham, MA, USA) was employed to analyze the surface chemical composition of the modified biochar as well as the chemical status of iron. The obtained data were analyzed and processed using Advantage 5.5 software. The morphology was observed by transmission electron microscopy (JEM

2100, JEOL, Tokyo, Japan) at various magnifications with an accelerating voltage of 5.0 kV. The obtained images were then used to determine the formation of Fe_3O_4 nanoparticles on the modified biochar by Digital Micrograph software(Digital Micrograph 2021).

2.4. Immobilization of Strain JH1 to Biochar

Strain JH1 was cultured in LB medium and incubated on a shaking table for 24 h at 30 °C and 150 rpm. The bacteria were centrifuged at 5000 r/min, cleaned with PBS, collected, and suspended again to achieve the logarithmic growth phase (1.2×10^9 CFU/mL). The biochar and cell suspension of strain JH1 were mixed at a ratio of 5:100 (w/v) and shaken on a shaking table for 24 h at 30 °C and 150 rpm to make the bacteria adsorb and adhere on the biochar, and then centrifuged at 1000 rpm for 10 min. The supernatant was filtered and dried at 30 °C. These four biochar-immobilized bacteria were named JY3, JY5, JY7, and JFe-Y7.

2.5. Determination of the Microbial Biomass

The amount of immobilized strain JH1 on the biochar was expressed by measuring biological phosphorus content [21]. Biochar (0.5 g) attached to microorganisms was used to extract phospholipids from cell membranes in a separation funnel containing a mixture of chloroform, ethanol, and water with a volume ratio of 1:2:0.8. The extraction time was 2–24 h. Then, by adding more chloroform and water, the mixture was partitioned into a chloroform phase containing lipids and a methanol–water phase, resulting in a final ratio of chloroform–methanol–water of 1:1:0.9. After standing for an additional 12 h, the lipid-containing chloroform was divided into 50 mL colorimetric tubes, and the chloroform was evaporated in 70 °C water baths. Ammonium molybdate spectrophotometry (TU-1901, Beijing General Instrument Co., Ltd., Beijing, China) was used to determine the content of biological phosphorus. The final result was expressed as nmol P/g biochar.

2.6. Biosorption and Reusability Studies: Batch Experiments

The biosorption experiments were conducted in 100 mL Erlenmeyer flasks, where 0.2 g of either biochars or biochar-immobilized cells were added to 20 mL of phenol solutions. Additionally, 5 mL of cell suspension (1.2×10^9 CFU/mL, Dilution plate counting was used to obtain) was incorporated as the background. The mixture was incubated on a shaking table with a constant agitation rate of 150 rpm at 30 °C for 24 h, unless otherwise stated. The impact of several factors, such as initial phenol concentration, pH, and temperature, on the degradation of phenol by immobilized cells was investigated. Briefly, the experimental conditions were as follows: initial phenol concentrations of 300, 400, 500, and 600 mg/L; incubation temperatures of 25, 30, 35, and 40 °C; and initial pH levels of 5.0, 6.0, 7.0, 8.0, and 9.0. Samples were periodically collected and filtered using a 0.22 μm organic filter membrane (PES) for subsequent analysis of residual phenol concentration.

Biochar-immobilized cells were repeatedly used in phenol degradation for 8 cycles to test their reusability. In brief, 0.5 g of biochar-immobilized cells were incubated in 20 mL MSM medium containing 300 mg/L phenol at pH 7.0 on a shaking table at 150 rpm at 30 °C in 100 mL Erlenmeyer flasks. Samples were collected periodically through a 0.22 μm organic filter membrane for residual phenol concentration analysis. The first cycle ended when phenol reached 99% in solution within 0 to 4 days. The Erlenmeyer flask was allowed to stand for 1–2 min to precipitate the biochar, after which the supernatant was withdrawn with a syringe. Then, sterilized distilled water and phenol were added to adjust the phenol concentration in the solution to 300 mg/L again. Then, the second cycle was started, and the experimental operation was the same as the first cycle, repeated for eight cycles. The eight cycles were named 1C, 2C, 3C, 4C, 5C, 6C, 7C, and 8C. In particular, during the experiment at 8 °C, samples were taken every 2 h to determine the concentration of phenol in the solution, and the sampling was stopped when the phenol concentration of all solutions dropped to 0. At the same time, different cycles were selected for scanning electron microscope (SEM) imaging and to determine the biomass of biochar-immobilized cells.

2.7. Effect of Biochar Water-Extractable Organic Carbon on Strain JH1

To study whether the water-extractable organic carbon (WEOC) of biochar affects the degradation of bacteria, the WEOC of biochar was investigated according to the study of Graber [22]. Biochar was added to the MSM medium at a ratio of 0.1% (w/v), and then the medium was oscillated at 150 rpm at 30 °C for 24 h. The biochar was then removed with a 0.22 μm organic filter membrane to obtain a filtrate containing biochar WEOC and MSM medium. A 6×10^7 CFU/mL cell suspension was added to 50 mL of the above filtrate containing 300 mg/L phenol, and MSM medium without cells was used for the control experiment. All media were incubated in the dark for 24 h at 30 °C, 150 rpm, pH 7 on a shaking table. Samples were collected periodically through a 0.22 μm organic filter membrane for residual phenol concentration analysis. Meanwhile, after 24 h of degradation, the enzyme activities of bacteria in the solution were determined. Catechol 2,3-dioxygenase and catechol 1,2-dioxygenase activities were determined by measuring the production of either 2-hydroxymuconic semialdehyde at 375 nm or muconic acid at 260 nm [23] using a UV–VIS spectrophotometer (TU-1901, PERSEE, China).

2.8. Fourier Transform Infrared Spectroscopy (FTIR) Analysis

Fourier transform infrared spectroscopy (FTIR, Nicolet is 50, Thermo Scientific, Waltham, MA, USA) was used to detect whether extracellular polymeric substances (EPS) were generated during the interaction between bacteria and biochar. Alternatively, it can be used to determine the group types of biochar before and after the adsorption of phenol to measure the mechanism of biochar adsorption of phenol. All FTIR spectra were collected within the wavenumber range of 400~4000 cm^{-1}. Each spectrum was obtained by averaging 32 scans, and the spectral resolution was set at 4 cm^{-1}. The spectra were displayed in terms of absorbance.

2.9. Scanning Electron Microscopy (SEM) Analysis

The morphology of biochar-immobilized bacteria was observed by scanning electron microscopy (Merlin, Zeiss, Jena, Germany). The biochar-immobilized bacteria were fixed in 2.5% (w/v) glutaraldehyde phosphate buffer at 4 °C for 12 h, and then the residual glutaraldehyde was washed twice with PBS. Then, the dehydrated samples were dehydrated with a multiconcentration gradient (30%, 50%, 70%, 90%, and 100%) of ethanol and incubated for 10~15 min at each stage. The dehydrated samples were freeze-dried in a vacuum. The distribution was observed by a scanning electron microscope after spraying gold.

2.10. Analytical Methods of Phenol

The concentration of phenol was determined using the 4-AAP spectrophotometric method with a UV–VIS spectrophotometer (TU-1901, PERSEE, China) at 510 nm [24]. Samples were centrifuged at 6000 rpm for 10 min, and the supernatant was collected to analyze the phenol concentration. Biochar-containing samples were filtered through a 0.22 μm organic filter membrane, and the filtrate was collected to analyze the phenol concentration. Microsoft Excel 2019 and Origin 2017 were used for data processing and charting, respectively. pH was measured using a pH meter (ST300, OHAUS, USA).

3. Results and Discussion

3.1. Characterization of Biochar and the Modified Biochar

Table 1 shows the yield, elemental composition, and elemental content ratio of the *Andrographis paniculata* medicinal residue biochar prepared at different pyrolysis temperatures. The biochar yield of the medicinal residue decreases with increasing charring temperature, mainly because during the pyrolysis process, the biomass decomposes rapidly into volatiles and biochar [12], resulting in the content of volatiles decreasing, while after the charring temperature is >700 °C, the content of volatiles remains approximately the same, while the biochar yield also remains the same [25,26]. The order of strength for stability (O/C) was Fe-Y7 > Y3 > Y7 > Y5, for aromaticity (H/C) was Y7 > Fe-Y7 > Y5 > Y3, and for polarity ((O + N)/C) was Fe-Y7 > Y3 > Y7 > Y5. The aromaticity of *Andrographis paniculata* medicinal residue biochar decreased with increasing pyrolysis temperature. After biochar modification, Fe-Y7 had stronger stability and polarity, but the aromaticity decreased, which indicated that the decrease in polar functional groups on the surface led to a decrease in polarity.

Table 1. Chemical composition and pore structure of biochar samples.

Samples	Yield (%)	Elemental Composition (%)					Atomic Ratio		
		C	H	O	N	S	O/C	H/C	(O + N)/C
Y3	44	59.44	5.19	32.70	2.54	0.13	0.550	0.087	0.593
Y5	29.3	67.47	3.08	27.49	1.79	0.18	0.407	0.046	0.434
Y7	25.1	67.51	1.71	29.18	1.43	0.17	0.432	0.025	0.453
Fe-Y7	—	61.2	1.70	35.51	1.34	0.26	0.580	0.028	0.602

Analysis of the data presented in Table 2 reveals a clear trend: as the pyrolysis temperature of the Andrographis paniculata medicinal residue biochar increased, both the specific surface area and pore capacity exhibited a notable increase. Additionally, it was observed that the average pore size decreased as the pyrolysis temperature increased. The average pore sizes of Y3 and Y5 ranged from 2 to 50 nm, indicating that their pore sizes were mainly mesopores, while the average pore size of Y7 was close to that of micropores, which was 2.0441 nm. The results indicated that as the pyrolysis temperature increased, the biochar formed more micropores, which effectively increased the specific surface area and contact with contaminants, as well as providing more attachment and proliferation sites for immobilized bacteria [25]. Fe-Y7 has a larger specific surface area and higher pore capacity than Y7, suggesting that the loading of Fe_3O_4 nanoparticles on the biochar surface increases the contact area with contaminants, resulting in an increase in the overall specific surface area. It also resulted in a larger average pore size and the formation of more mesopores and macropores [27]. The relatively high specific surface area values for Y7 and Fe-Y7, compared to Y3 and Y5, illustrate the effect of temperature on carbonization. In general, the increase in specific surface area values at high-pyrolysis temperatures is mainly due to the removal of volatiles. During pyrolysis, the rapid release of volatiles opens and connects blind and closed pores, as well as forms new cracks, micropores, and mesopores, leading to a significant increase in specific surface area [28].

Table 2. The pore structure parameters of biochars.

Samples	Specific Surface Area (m^2/g)	Pore Capacity (cm^3/g)	Average Aperture (nm)
Y3	1.8796	0.003046	6.4819
Y5	3.2917	0.005034	6.1169
Y7	138.8126	0.070936	2.0441
Fe-Y7	164.0975	0.097943	2.3874

Figure 1 shows the N_2 adsorption–desorption isotherm curves of the four biochars. The N_2 adsorption–desorption isotherms of the four biochars are all type IV, and the adsorption hysteresis loop appears in the middle section. According to the IUPAC classification, the hysteresis loops of Y7 and Fe-Y7 are of type H4. The hysteresis loops of type H4 have no obvious saturation adsorption plateau, and the N_2 adsorption rises rapidly in the low-pressure region (relative pressure $P/P_0 < 0.1$), which is caused by the filling adsorption of micropores, indicating that the biochar contains abundant micropores, narrow fissure pores or laminar structure, and this type of hysteresis loop indicates that the pore structure of biochar is very irregular, mainly consisting of mesopores and macropores.

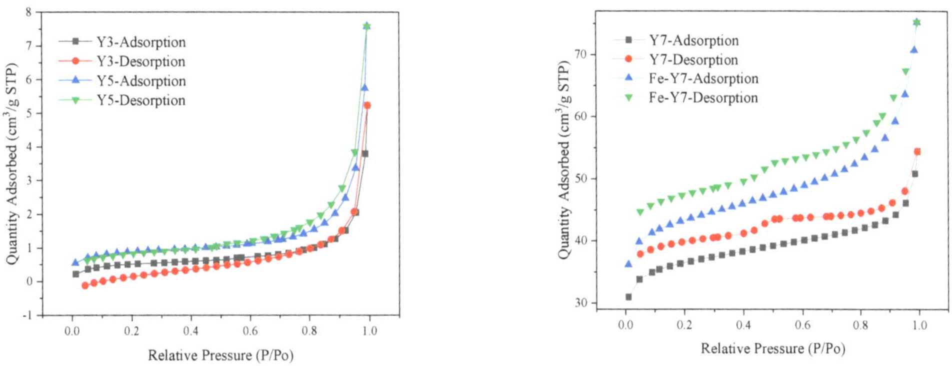

Figure 1. N_2 adsorption−desorption isotherms of biochars.

The zeta potential of the four biochars was measured under different pH conditions, and the pH value corresponding to a zeta potential of 0 was the zero point charge (pH_{PZC}) of the biochar. Figure 2a shows that the isoelectric point of the biochar is lower ($pH_{PZC} < 4$) [29]. The pH_{PZC} of all four biochars is less than 7.0 and negatively charged, so the surfaces of both phenol and biochar are negatively charged, and electrostatic repulsion is easily generated between them, resulting in biochar maintaining a low adsorption capacity for phenol. However, the good adsorption of phenol by biochar prepared under high-temperature conditions implies that electrostatic adsorption does not play a major role in phenol adsorption by biochar and may also be influenced by strong hydrogen bonds and π–π interactions [30]. For the modified biochar, the pH_{PZC} of Fe-Y7 was higher than that of Y7 by 2.53 and 2.19, respectively.

The crystalline minerals of biochar are mainly SiO_2 and $CaCO_3$ at different pyrolysis temperatures Figure 2b. Biochar prepared under low-temperature pyrolysis (Y3) has no obvious crystalline peaks and shows only a small amount of SiO_2 crystalline minerals, among which Y3 shows several significant peaks at 2θ values of 15°~25°, which are mainly amorphous C peaks related to the change in crystalline structure of cellulose in biomass [25]. As the pyrolysis temperature increased, the crystalline structure of the drug residue biochar improved, and the content of crystalline minerals increased. The most abundant crystalline mineral in the residue biochar is $CaCO_3$, and there is a most obvious $CaCO_3$ signal peak at a 2θ value of 29.4° (d = 3.03 Å), and the signal peak becomes increasingly intense with increasing temperature. The diffraction peak of SiO_2 was enhanced and then weakened with increasing temperature. Compared with the unmodified biochar, the basic crystal structure of the modified biochar is the same, except that the diffraction peaks of Fe_3O_4 with different crystallographic planes appear, and the 2θ values of Fe-Y7 are 35.5° (d = 2.52 Å) and 47.3° (d = 1.92 Å), which indicates that the biochar is successfully loaded with Fe_3O_4 nanomaterials.

To further verify the successful loading of Fe_3O_4 nanomaterials, TEM and XPS analyses were performed. The XPS spectrum of Fe-Y7 is shown in Figure 2c, which shows the presence of Fe 2p (711.74 eV), O 1s (532.88 eV), C ls (285.27 eV) and Si (102.71 eV) on the surface of Fe-Y7. The peaks at 711.06 and 720.95 eV belong to Fe 2p3/2 and Fe 2p1/2, respectively, and are decomposed into four peaks at 711.1, 714.1, 723.7 and 725.3 eV [31], where 714.1 and 723.7 eV are associated with Fe(II) of Fe_3O_4 and 711.1 and 725.3 eV are associated with Fe(III). It was demonstrated that Fe_3O_4 nanoparticles containing both Fe(II) and Fe(III) were successfully loaded on the modified biochar surface. TEM (Figure 2d–f) revealed (110) lattice planes of SiO_2 with a characteristic spacing of 0.2465 nm, (012) and (104) lattice planes of $CaCO_3$ with a characteristic spacing of 0.39 and 0.30 nm, and (311) and (222) lattice planes of Fe_3O_4 with characteristic spacings of 0.25 and 0.24 nm. These data correspond to the diffraction peaks of SiO_2, $CaCO_3$, and Fe_3O_4 in the XRD spectra, which also confirm the formation of Fe_3O_4.

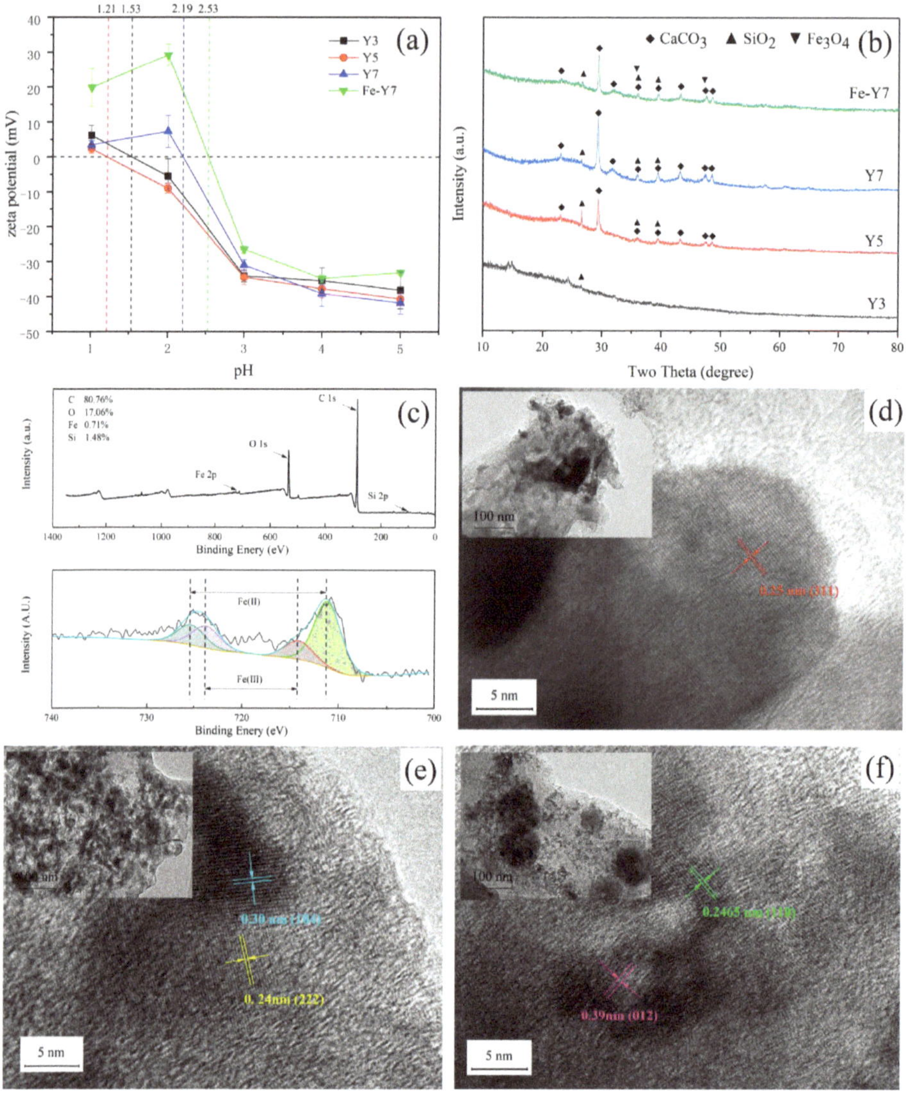

Figure 2. (**a**) Zeta potential of different biochars, (**b**) XRD patterns of biochars, (**c**) XPS spectrum of Fe-Y7 of full scan and Fe 2p, and (**d**–**f**) TEM and HR-TEM images.

3.2. Fourier Infrared Spectroscopy (FTIR) Analysis of Biochar Adsorption Strain JH1

Once attached to the carrier, bacteria begin to secrete a surrounding matrix called the extracellular polymer (EPS), which is a protective matrix of some polymers, such as polysaccharides, proteins, and nucleic acids, among other polymers. By the FTIR technique, it is possible to detect whether bacteria adhere to the surface of biochar through EPS substances [32]. Cells composed of a large number of proteins contribute to hydrophobicity, while polysaccharides on the cell surface confer hydrophilicity [33]. Figure 3 shows the FTIR plots of biochar before and after the adsorption of A. faecalis JH1. For the mixed region of proteins and fatty acids from 1500~1200 cm^{-1}, there is a significant difference between the residue biochar before and after the adsorption of A. faecalis JH1. The region 1200~900 cm^{-1} belong to the polysaccharides within the cell wall, and there are significant stretching vibrations of both the residue biochar and the modified Fe-Y7. Thus, FTIR analysis showed that A. faecalis JH1 could adhere to the surface of biochar through substances such as secreted polysaccharides or proteins. Proteins and polysaccharides in EPS play a key role in promoting initial adhesion and biofilm development [34].

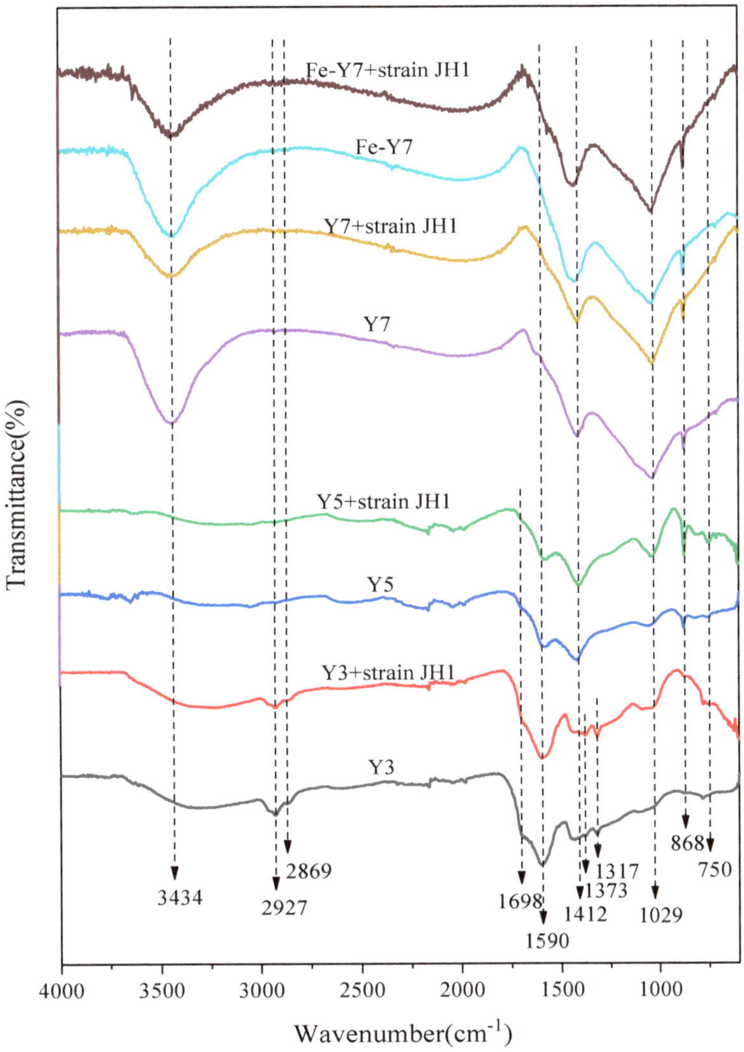

Figure 3. FTIR spectra of biochars before and after adsorption of A. faecalis JH1.

The degree of hydrophobicity acquired by bacteria on the substrate affects the number of cells attached to the biochar, while the hydrophilicity/hydrophobicity of the biochar surface also affects the ability of bacteria to adhere; for example, the presence of -CH_2 fatty acids on the bacterial surface causes bacteria to selectively adhere to the more hydrophobic biochar [32]. From the (O + N)/C in Table 1, it is clear that Fe-Y7 is more hydrophilic than Y7, which also corresponds to the functional groups on FTIR. Thus, the surface of the pharmaceutical residue biochar has many hydrophobic and hydrophilic groups, and their adsorption of bacteria is a common effect of bacterial extracellular proteins and polysaccharides, such as C-O, -COO-, =C = O, and aromatic Ar-H functional groups [34]. Moreover, Fe-Y7 has a higher adsorption capacity for bacteria than Y7.

3.3. Biomass Fixed by the Adsorption of Biochar

Figure 4 shows the biomass of different biochar adsorption immobilized A. faecalis JH1. From the data in the figure, it can be seen that the amount of bacteria immobilized by Fe-Y7 is the highest with 372.45 nmol P/g biochar, which confirms what was found by FTIR. The number of bacteria immobilized by adsorption was higher for Y5 than for Y3 and Y7 because Y3 and Y7 have more and stronger absorption peaks in the region of 1500–1200 cm^{-1} in FTIR, with abundant hydrophobic functional groups and therefore strong hydrophobicity, which is not favorable for their adsorption and immobilization of bacteria in the aqueous environment. Therefore, the hydrophilicity and hydrophobicity of biochar affect the amount of fixed biomass.

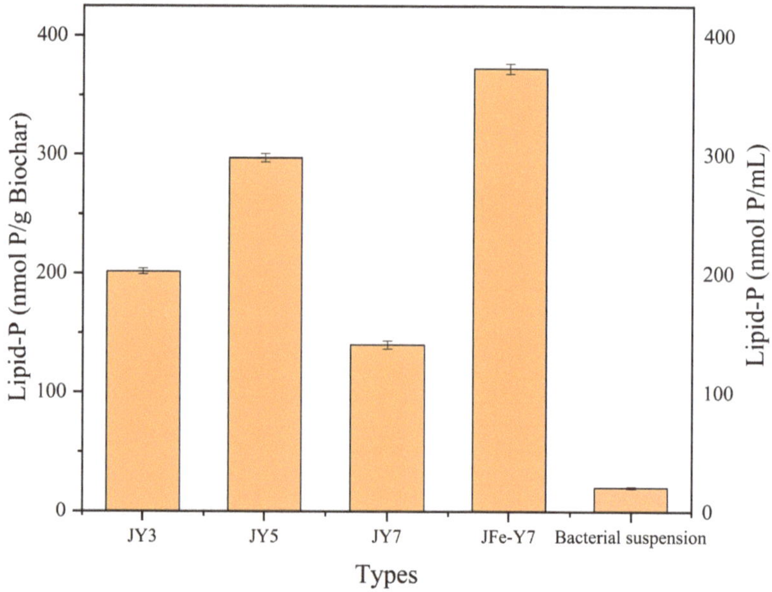

Figure 4. Biomass of strain JH1 was immobilized by biochar adsorption.

3.4. Effect of Water-Soluble Organic Carbon from Biochar

To better illustrate the process of phenol removal by biochar-immobilized bacteria, the effect of the water-soluble organic carbon (WEOC) of biochar on the bacterial degradation of phenol was investigated. Figure 5 shows the appearance of water-soluble organic carbon solutions of different biochars, which are colorless and transparent except for Y3, which is light yellow, with a darker color implying more dissolved organic carbon [35]. Generally, the absorbance values of biochar WEOC decreased significantly with increasing biochar pyrolysis temperature [36]. Figure 6a shows the effect of the WEOC of biochar on the bacterial degradation of phenol. As shown by the data at 12 h in the figure, the WEOC of biochar at different temperatures has different effects on the removal of phenol. Compared with the control experiment, WEOC in Y3 and Y7 had a facilitative effect on phenol removal. The water-soluble organic carbon of low-temperature biochar has greater DOC and abundant low molecular weight molecules [37], and these stimulate microbial activity and increase microbial abundance [36], thereby accelerating the degradation of phenol.

Figure 5. Appearance of biochars' WEOC.

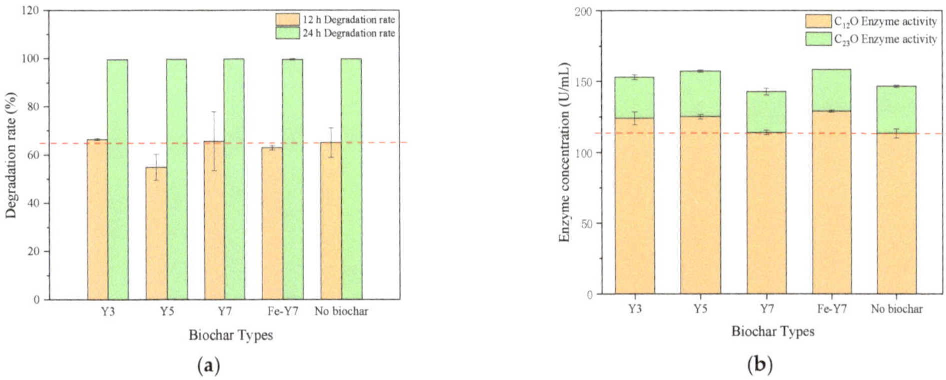

Figure 6. (**a**) Effect of biochars' WEOC on the degradation of phenol by A. faecalis JH1; (**b**) enzyme activity.

After the phenol in solution was degraded within 24 h, the enzymatic activity of the bacteria in solution was measured, as shown in Figure 6b. The results indicate that the pathway of phenol degradation by A. faecalis JH1 relies mainly on catechol 1,2-dioxygenase ($C_{12}O$) for the production of cis–cis-mucofuranic acid after the ring-opening of catechol [38], and only a small amount of phenol is assimilated by catechol 2,3-dioxygenase ($C_{23}O$) for the production of 2-hydroxymucofuranic acid semialdehyde. In addition, a comparison with the control bacterial solution revealed that the water-soluble organic carbon of all four biochars promoted the enzymatic activity of the bacteria. The results of Y3 and Y7 were consistent with the data at 12 h in Figure 6a. In contrast, Y5, which had an inhibitory effect on phenol degradation at 12 h in Figure 6b, had stronger enzyme activity measured after 24 h than that of the bacterial solution. This indicates that Y5 is susceptible to the effects of PAHs on water-soluble organic carbon in the short term. However, for Y5 and Fe-Y7, the water-soluble organic carbon under long-term use had a beneficial effect on the enzymatic activity of the bacteria.

3.5. Adsorption and Degradation Kinetics of Phenol

Figure 7 shows the adsorption degradation kinetic curves of different biochar-immobilized A. faecalis JH1. Comparing with the control biochar and bacterial suspension, the biochar-immobilized bacteria showed higher removal of phenol from the solution in different time periods, and even around 96 h, all biochar-immobilized bacterial strains JH1 were able to achieve around 99% degradation of phenol from the solution. However, the control biochar basically reached adsorption saturation at around 48 h, and only a small amount of phenol adsorption would be added with the extension of time, whereas the suspension maintained a low phenol degradation rate all the time, with only 23.4% phenol degradation at 96 h. The control biochar was also found to be a good choice for the adsorption of phenol in the suspension. Therefore, these results suggest that the removal of phenol by biochar-immobilized A. faecalis JH1 is a synergistic effect of adsorption and bacterial degradation, and that biochar itself can adsorb a large amount of phenol, and therefore increase the chance of bacterial contact with phenol and promote bacterial degradation of phenol. In addition, the large number of pores in the biochar provides an important habitat for microorganisms [26] and provides them with shelter, which reduces the toxicity of phenol to microorganisms, and at the same time, the biochar can stimulate the growth and activity of microorganisms, and microorganisms can utilize the ash of the biochar to provide mineral nutrients [28].

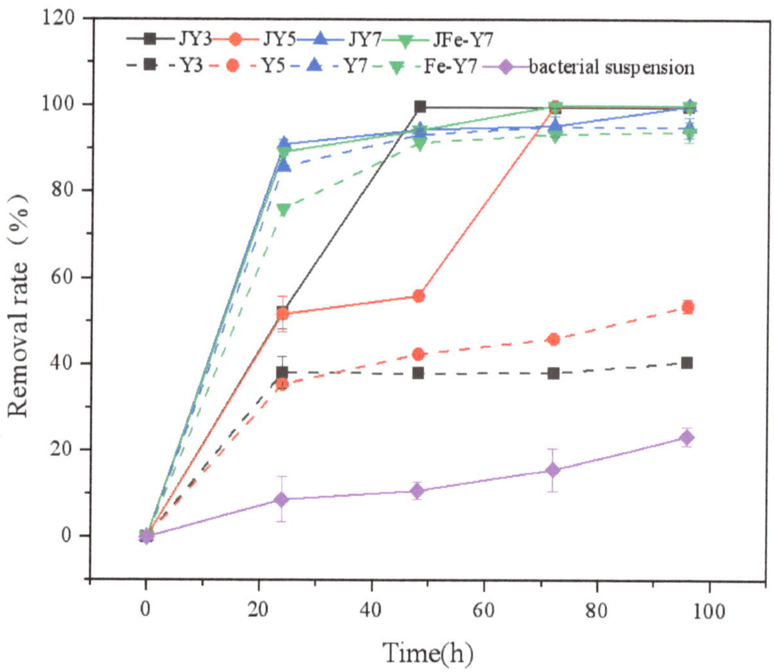

Figure 7. The adsorption and degradation kinetic curve of biochars immobilized *A. faecalis* JH1.

3.6. Effect of Temperature, pH, Initial Phenol Concentration, and the Salinity

Figure 8a–d show the effects of temperature, pH, initial concentration, and salinity of phenol on phenol removal by biochar-immobilized strain JH1 for 48 h, respectively. It can be seen from Figure 8 that JY3 shows an extremely high advantage in the removal of phenol, but the high temperature and salinity affect its removal rate of phenol. This result can indicate that the degradation of phenol by JY3 relies mainly on the degradation of suspended bacteria in solution, which is consistent with the results in WEOC. JY5 showed better phenol removal than Y5 at low temperatures, alkalinity, low phenol concentration, and low salinity, suggesting that the removal of phenol by JY5 is susceptible to environmental influences. The lower removal of phenol by JY5 under alkaline conditions may be due to the higher alkalinity of the biochar as the temperature increases [39]. Unlike JY5, JY7 and JFe-Y7 were not greatly affected by the environment, and both showed very high phenol removal rates in different environments, with bacteria still growing and remaining active in the porous structure [40]. Therefore, compared with suspended bacteria, biochar-immobilized bacteria can improve their tolerance in different environments, and they can still grow and maintain strong activity within 48 h at temperatures of 25 °C to 40 °C, pH 5~9, initial phenol concentrations of 300–500 mg/L and salinity of 3%.

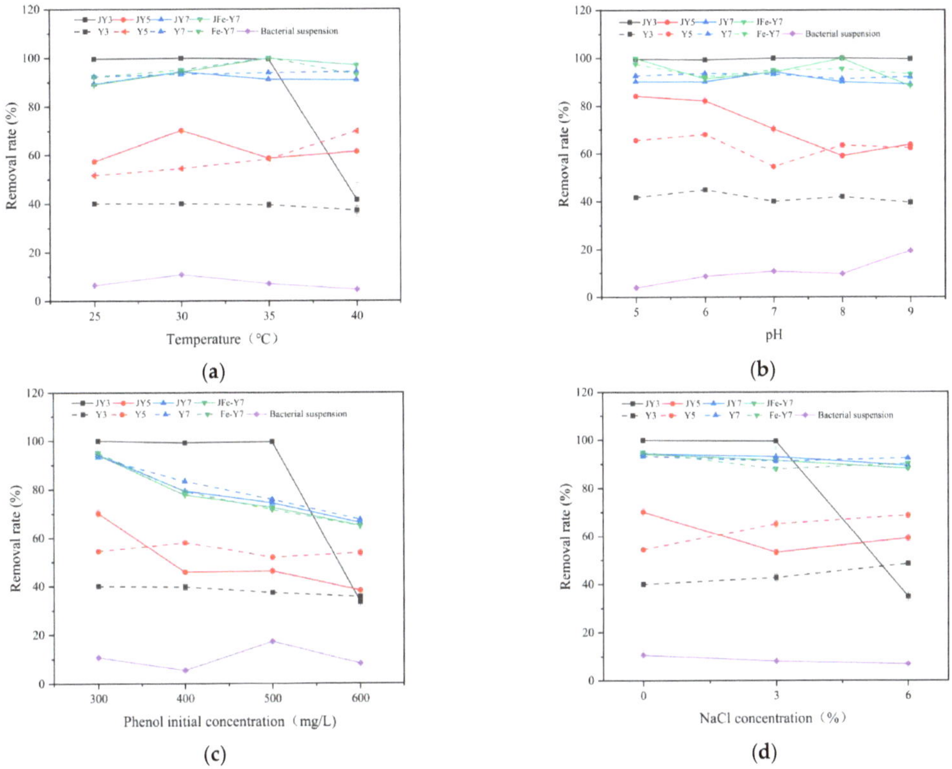

Figure 8. Effect on phenol removal from solution for 48 h. (**a**) temperature; (**b**) Ph; (**c**) initial concentration of phenol; (**d**) salinity.

3.7. Effect of Recycling Biochar-Fixed Bacteria on Phenol Adsorption

To investigate the reusability of the biochar-immobilized A. faecalis JH1 to confirm its potential for practical applications, the biochar-immobilized bacteria were subjected to recycling experiments. The eight cycles were 96, 60, 48, 48, 36, 48, 24, and 12 h. Seven of these cycles are shown in Figure 8. In the first cycle (1C), the removal of phenol relied mainly on the adsorption of biochar and the bacteria in the solution. As the number of cycles increases, the bacteria use the phenol adsorbed on the biochar as well as nutrients from the biochar itself (e.g., water-soluble organic carbon) to multiply and grow rapidly and adhere to the biochar, and more and more bacteria are adsorbed on the biochar, so that the time spent for the removal of phenol from solution by biochar immobilized bacteria will become shorter and shorter, and thus the long-term phenol removal effect depends more on the biodegradation on the biochar. After the sixth cycle, all biochar-immobilized bacteria could remove 300 mg/L phenol solution within 12 h (Figure 9).

Fe-Y7 showed the most stable removal of phenol and the fastest removal rate (Figure 10). In different cycles, JY3 and JY5 removed phenol from the solution at longer and shorter times, and the removal ability was unstable. In these seven cycles, all biochar-immobilized bacteria could remove phenol from the solution, which may be because biochar as a porous carrier provided enough space for bacterial growth and promoted the growth and activity of bacteria. These results suggest that biochar carrier-immobilized JH1 has great potential for phenol removal, is reusable, and exhibits good stability and durability, in agreement with the results of most recycling studies [18,40].

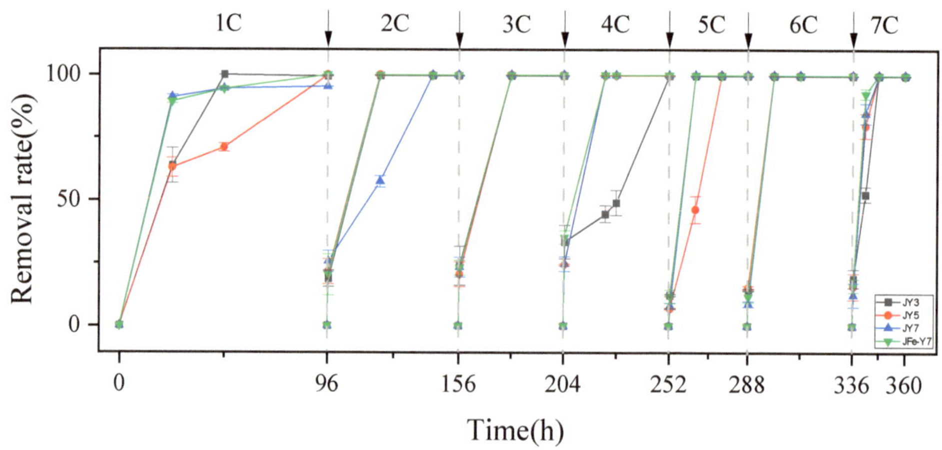

Figure 9. Biochar-immobilized bacteria for recycling.

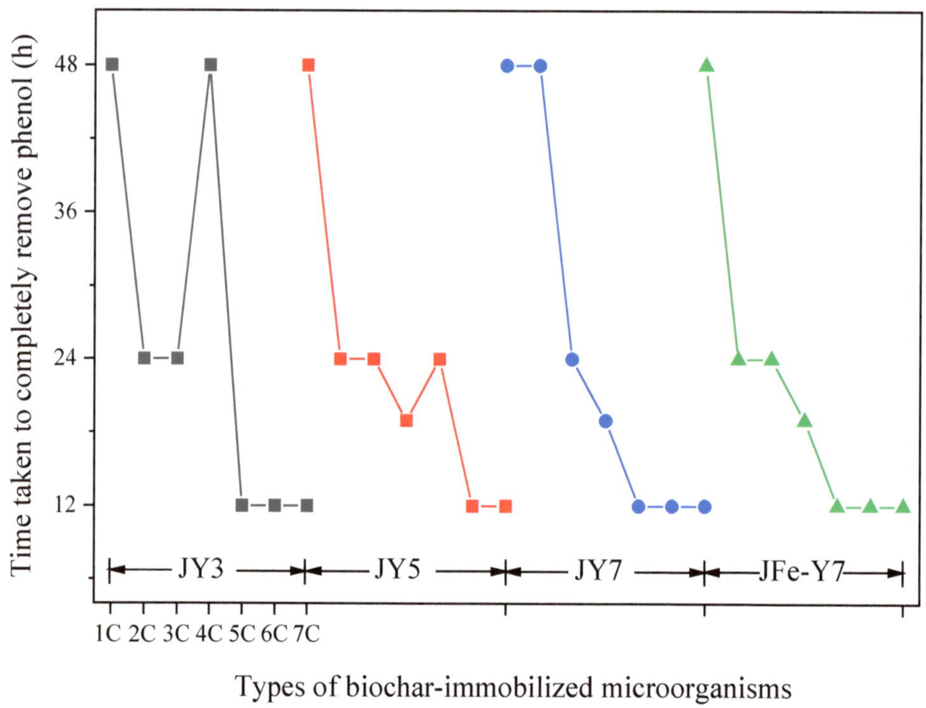

Figure 10. Time spent in each cycle to completely remove phenol from the solution.

In contrast, in the 8C cycle (Figure 11), phenol removal by JFe-Y7 was relatively rapid. The dioxygenase of A. faecalis JH1 for phenol degradation uses oxygen as an oxidant to bind two oxygen atoms to the product, while most dioxygenases require a metal cofactor, most commonly Fe(II) or Fe(III) [41]. The addition of appropriate Fe_3O_4 nanoparticles improves their enzymatic activity [42], small amounts of Fe ions are gradually released to the environment through Fe_3O_4 nanoparticles, and Fe^{2+}/Fe^{3+} in solution couples with bacterial oxidation/reduction reactions to accelerate the electron transfer rate [43,44]. Thus, the results demonstrate that Fe_3O_4 nanoparticle modification accelerates phenol removal as a result of improving the adsorption capacity of phenol or promoting bacterial growth, metabolism, and enzymatic activity.

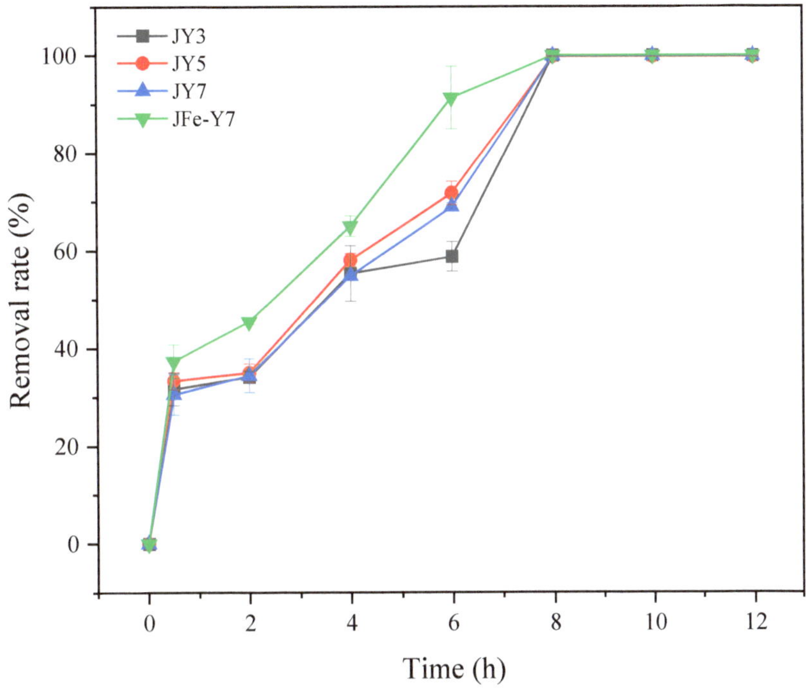

Figure 11. Phenol removal kinetics curve at 8C.

By performing further SEM analysis, Figure 12a–h shows the 2 k× and 20 k× multiples of 0C, 1C, 4C, and 8C in the cycle, respectively, and the main distribution area of A. faecalis JH1 is on the surface of the biochar and in the internal channels of the pores. At 0C, A. faecalis JH1 adhered to the biochar in relatively scattered and small numbers (Figure 12a,b). However, after one cycle, the bacteria adhering to the biochar surface started to grow slowly using the phenol adsorbed by the biochar and the water-soluble organic carbon of the biochar, splitting and multiplying around it as a single bacterium to become a multicolony, at which time there was no large number of bacteria present on the biochar surface (Figure 12c,d). However, after four cycles, a large number of bacteria were evident on the surface and in the pores of the biochar, and the bacteria multiplied and adhered to the biochar rapidly (Figure 12e,f). With repeated phenol replacement and bacterial growth, a thick and sticky biofilm was established on the biochar surface in the eighth cycle (Figure 12g,h), [42]. The biofilm thickness was in the range of 10~20 μm, and both phenol and oxygen could diffuse into the bacteria for metabolic activities [45]. The SEM images of these four biochar-immobilized A. faecalis JH1 were generally consistent, and the growth and adhesion of bacteria on biochar showed a similar variation pattern. Thus, biochar is an ideal carrier with a large specific surface area for bacterial adhesion and growth, is nontoxic and cost-effective, and additionally facilitates oxygen and substrate delivery through immobilized cells and enhances the activity of immobilized bacteria.

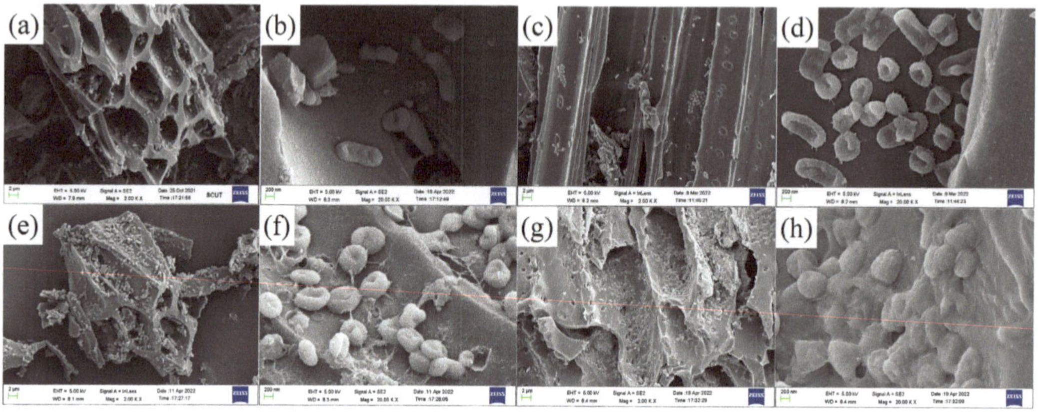

Figure 12. SEM images of different cycles. (**a**,**b**) 0C; (**c**,**d**) 1C; (**e**,**f**) 4C; (**g**,**h**) 8C.

4. Conclusions

The microstructure of *Andrographis paniculata* medicinal residue is a honeycomb with a porous structure and a large surface area. Modification by Fe_3O_4 nanoparticles increases the specific surface area during pyrolysis, and the data demonstrate the successful loading of Fe_3O_4 nanoparticles onto the surface of Fe-Y7. The adsorption of all four biochars on E. *Alcaligenes faecalis* strain JH1 relies on the combined action of bacterial extracellular proteins and polysaccharides, and Fe-Y has a higher adsorption capacity for bacteria. The biochar-immobilized bacteria were more tolerant than the suspended bacteria in different environments. During the recycling process, with the increase in the number of cycles, the bacteria will form biofilms on the biochar, and the formed biofilms make the biochar-immobilized bacteria still have a certain removal effect on phenol after several cycles, among which the removal effect of Jeffe-Y7 is relatively better. The results showed that Fe_3O_4 nanoparticle-modified biochar-immobilized microorganisms could promote the degradation of phenol with good reproducibility, and therefore, this technology has potential application in the treatment of phenol-containing wastewater. It can be used in industry to recycle phenol wastewater, which is less expensive, saves treatment costs and does not produce other environmental pollution. The removal of phenol by the biochar-immobilized A. faecalis JH1 was a dual action of adsorption-degradation. The biochar mainly relied on π–π interactions and hydrogen bonding to adsorb phenol, which was first adsorbed by the biochar before being metabolically degraded by the bacteria on the biochar.

Author Contributions: Conceptualization, J.X.; Methodology, Z.Z.; Software, Z.Z. and J.X.; Validation, M.L. and J.W.; Formal analysis, Z.Z.; Investigation, Z.Z., J.X., M.L. and J.W.; Data curation, Z.Z.; Writing—original draft, Z.Z.; Writing—review and editing, T.Z.; Supervision, T.Z.; Project administration, T.Z.; Funding acquisition, T.Z. All authors have read and agreed to the published version of the manuscript.

Funding: This study was financially supported by the National Natural Science Foundation of China Projects (U21A2003), and National Key R&D Program of China (2020YFC1808803).

Data Availability Statement: The data that supports the findings of this study are available within the article.

Conflicts of Interest: The authors declare no conflict of interest.

References

1. Panigrahy, N.; Priyadarshini, A.; Sahoo, M.M.; Verma, A.K.; Daverey, A.; Sahoo, N.K. A comprehensive review on eco-toxicity and biodegradation of phenolics: Recent progress and future outlook. *Environ. Technol. Innov.* **2022**, *27*, 102423. [CrossRef]
2. Sahoo, S.K.; Das, A.A.; Deka, D.; Naik, B.; Kumar Sahoo, N. Organic-inorganic hybrid hydroquinone bridged v-cds/hap/pd-tcpp: A novel visible light active photocatalyst for phenol degradation. *J. Mol. Liq.* **2021**, *339*, 116721. [CrossRef]

3. Darbre, P.D. The history of endocrine-disrupting chemicals. *Endocr. Metab. Res.* **2019**, *7*, 26–33. [CrossRef]
4. Martínková, L.; Kotik, M.; Marková, E.; Homolka, L. Biodegradation of phenolic compounds by basidiomycota and its phenol oxidases: A review. *Chemosphere* **2016**, *149*, 373–382. [CrossRef]
5. Kujawski, W.; Warszawski, A.; Ratajczak, W.; Porębski, T.; Capała, W.; Ostrowska, I. Removal of phenol from wastewater by different separation techniques. *Desalination* **2004**, *163*, 287–296. [CrossRef]
6. Wu, J.; Yu, H.Q. Biosorption of 2,4-dichlorophenol from aqueous solution by phanerochaete chrysosporium biomass: Isotherms, kinetics and thermodynamics. *J. Hazard. Mater.* **2006**, *137*, 498–508. [CrossRef]
7. Biswas, B.; Sarkar, B.; Rusmin, R.; Naidu, R. Bioremediation of pahs and vocs: Advances in clay mineral–microbial interaction. *Environ. Int.* **2015**, *85*, 168–181. [CrossRef]
8. Karel, S.F.; Libicki, S.B.; Robertson, C.R. The immobilization of whole cells: Engineering principles. *Chem. Eng. Sci.* **1985**, *40*, 1321–1354. [CrossRef]
9. Kim, P.; Johnson, A.M.; Essington, M.E.; Radosevich, M.; Kwon, W.-T.; Lee, S.-H.; Rials, T.G.; Labbé, N. Effect of pH on surface characteristics of switchgrass-derived biochars produced by fast pyrolysis. *Chemosphere* **2013**, *90*, 2623–2630. [CrossRef]
10. Jin, Z.; Xiao, S.; Dong, H.; Xiao, J.; Tian, R.; Chen, J.; Li, Y.; Li, L. Adsorption and catalytic degradation of organic contaminants by biochar: Overlooked role of biochar's particle size. *J. Hazard. Mater.* **2022**, *422*, 126928. [CrossRef] [PubMed]
11. Beesley, L.; Moreno-Jimenez, E.; Gomez-Eyles, J.L.; Harris, E.; Robinson, B.; Sizmur, T. A review of biochars' potential role in the remediation, revegetation and restoration of contaminated soils. *Environ. Pollut.* **2011**, *159*, 3269–3282. [CrossRef]
12. Liu, W.J.; Jiang, H.; Yu, H.Q. Development of biochar-based functional materials: Toward a sustainable platform carbon material. *Chem. Rev.* **2015**, *115*, 12251. [CrossRef]
13. Xiao, X.; Chen, B.; Chen, Z.; Zhu, L.; Schnoor, J.L. Insight into multiple and multilevel structures of biochars and their potential environmental applications: A critical review. *Environ. Sci. Technol.* **2018**, *52*, 5027–5047. [CrossRef] [PubMed]
14. Girijan, S.; Kumar, M. Immobilized biomass systems: An approach for trace organics removal from wastewater and environmental remediation—sciencedirect. *Curr Opin Environ Sci Health* **2019**, *12*, 18–29. [CrossRef]
15. Banerjee, A.; Ghoshal, A.K. Biodegradation of real petroleum wastewater by immobilized hyper phenol-tolerant strains of Bacillus cereus in a fluidized bed bioreactor. *3 Biotech* **2016**, *6*, 137. [CrossRef]
16. Manya, J.J. Pyrolysis for biochar purposes: A review to establish current knowledge gaps and research needs. *Environ. Sci. Technol.* **2012**, *46*, 7939. [CrossRef]
17. Wu, C.; Zhi, D.; Yao, B.; Zhou, Y.; Yang, Y.; Zhou, Y. Immobilization of microbes on biochar for water and soil remediation: A review. *Environ. Res.* **2022**, *212*, 113226. [CrossRef] [PubMed]
18. Zhuang, H.; Han, H.; Xu, P.; Hou, B.; Jia, S.; Wang, D.; Li, K. Biodegradation of quinoline by *streptomyces sp N01* immobilized on bamboo carbon supported Fe_3O_4 nanoparticles. *Biochem. Eng. J.* **2015**, *99*, 44–47. [CrossRef]
19. Singh, N.; Balomajumder, C. Equilibrium isotherm and kinetic studies for the simultaneous removal of phenol and cyanide by use of *S. Odorifera* (Mtcc 5700) immobilized on coconut shell activated carbon. *Appl. Water Sci.* **2017**, *7*, 3241–3255. [CrossRef]
20. Wu, Z.; Shen, J.; Li, W.; Li, J.; Xia, D.; Xu, D.; Zhang, S.; Zhu, Y. Electron self-sufficient core-shell biocl@fe-biocl nanosheets boosting Fe(III)/Fe(II) recycling and synergetic photocatalysis-fenton for enhanced degradation of phenol. *Appl. Catal. B Environ.* **2023**, *330*, 122642. [CrossRef]
21. Yang, H.J.; Yang, Z.M.; Xu, X.H.; Guo, R.B. Increasing the methane production rate of hydrogenotrophic methanogens using biochar as a biocarrier. *Bioresour. Technol.* **2020**, *302*, 122829. [CrossRef]
22. Graber, E.R.; Tsechansky, L.; Lew, B.; Cohen, E. Reducing capacity of water extracts of biochars and their solubilization of soil Mn and Fe. *Eur. J. Soil Sci.* **2014**, *65*, 162–172. [CrossRef]
23. Liu, Z.; Yang, H.; Huang, Z.; Zhou, P.; Liu, S.-J. Degradation of aniline by newly isolated, extremely aniline-tolerant *Delftia* sp. AN3. *Appl. Microbiol. Biotechnol.* **2002**, *58*, 679–682. [CrossRef]
24. Yu, J.; Tao, D.; Yu, S.; Kai, Y.; Wang, H. Biodegradation of phenol by entrapped cell of debaryomyces sp. With nano-Fe_3O_4 under hypersaline conditions. *Int. Biodeterior. Biodegrad.* **2017**, *123*, 37–45.
25. Keiluweit, M.; Nico, P.S.; Johnson, M.G.; Kleber, M. Dynamic molecular structure of plant biomass-derived black carbon (biochar). *Environ. Sci. Technol.* **2010**, *44*, 1247–1253. [CrossRef]
26. Ahmad, M.; Ok, Y.S.; Rajapaksha, A.U.; Lim, J.E.; Kim, B.Y.; Ahn, J.H.; Lee, Y.H.; Al-Wabel, M.I.; Lee, S.-E.; Lee, S.S. Lead and copper immobilization in shooting range soil using soybean stover-and pine needle-derived biochars: Chemical, microbial and spectroscopic assessments. *J. Hazard. Mater.* **2016**, *301*, 179–186. [CrossRef]
27. Jin, X.; Liu, R.; Wang, H.; Han, L.; Qiu, M.; Hu, B. Functionalized porous nanoscale Fe_3O_4 particles supported biochar from peanut shell for pb(ii) ions removal from landscape wastewater. *Environ. Sci. Pollut. Res.* **2022**, *29*, 37159–37169. [CrossRef] [PubMed]
28. Liu, L.; He, N.; Borham, A.; Zhang, S.; Xie, R.; Zhao, C.; Hu, J.; Wang, J. The effect of iron-modified biochar on phosphorus adsorption and the prospect of synergistic adsorption between biochar and iron-oxidizing bacteria: A review. *Water* **2023**, *15*, 3315. [CrossRef]
29. Cheng, C.; Lehmann, J.; Engelhard, M.H. Natural oxidation of black carbon in soils: Changes in molecular form and surface charge along a climosequence. *Geochim. Cosmochim. Acta* **2008**, *72*, 1598–1610. [CrossRef]
30. Du, J.; Sun, P.; Feng, Z.; Zhang, X.; Zhao, Y. The biosorption capacity of biochar for 4-bromodiphengl ether: Study of its kinetics, mechanism, and use as a carrier for immobilized bacteria. *Environ. Sci. Pollut. Res.* **2016**, *23*, 3770–3780. [CrossRef]

31. Liu, G.; Feng, K.; Cui, H.; Li, J.; Wang, M. Mof derived in-situ carbon- encapsulated fe3o4@c to mediate polysulfides redox for ultrastable lithium-sulfur batteries. *Chem. Eng. J.* **2019**, *381*, 122652. [CrossRef]
32. Ramos-Escobedo, G.; Pecina-Treviño, E.; Tokunaga, A.B.; Concha-Guerrero, S.; Ramos-Lico, D.; Guerra-Balderrama, R.; Orrantia-Borunda, E. Bio-collector alternative for the recovery of organic matter in flotation processes. *Fuel* **2016**, *176*, 165–172. [CrossRef]
33. Zhou, Y.; Petrova, S.P.; Edgar, K.J. Chemical synthesis of polysaccharide–protein and polysaccharide–peptide conjugates: A review. *Carbohydr. Polym.* **2021**, *274*, 118662. [CrossRef] [PubMed]
34. Yu, Y.; An, Q.; Zhou, Y.; Deng, S.; Miao, Y.; Zhao, B.; Yang, L. Highly synergistic effects on ammonium removal by the co-system of pseudomonas stutzeri xl-2 and modified walnut shell biochar. *Bioresour. Technol.* **2019**, *280*, 239–246. [CrossRef] [PubMed]
35. Xu, J.; Jian, Z.; Wang, Y.; Fang, C.; Hu, Q. Spatial–seasonal characteristics and influencing factors of dissolved organic carbon and chromophoric dissolved organic matter in poyang lake. *Environ. Earth Sci.* **2023**, *82*, 44. [CrossRef]
36. Li, M.; Liu, M.; Joseph, S.; Jiang, C.; Wu, M.; Li, Z. Change in water extractable organic carbon and microbial plfas of biochar during incubation with an acidic paddy soil. *Soil Res.* **2015**, *53*, 763–771. [CrossRef]
37. Bian, R.; Joseph, S.; Shi, W.; Li, L.; Taherymoosavi, S.; Pan, G. Biochar DOM for plant promotion but not residual biochar for metal immobilization depended on pyrolysis temperature. *Sci. Total Environ.* **2019**, *662*, 571–580. [CrossRef]
38. Jiang, Y.; Wen, J.; Bai, J.; Jia, X.; Hu, Z. Biodegradation of phenol at high initial concentration by alcaligenes faecalis. *J. Hazard. Mater.* **2007**, *147*, 672–676. [CrossRef]
39. Fidel, R.B.; Laird, D.A.; Thompson, M.L.; Lawrinenko, M. Characterization and quantification of biochar alkalinity. *Chemosphere* **2017**, *167*, 367–373. [CrossRef]
40. Liu, Y.; Gan, L.; Chen, Z.; Megharaj, M.; Naidu, R. Removal of nitrate using paracoccus sp. Yf1 immobilized on bamboo carbon. *J. Hazard. Mater.* **2012**, *229–230*, 419–425. [CrossRef]
41. Bugg, T.D.H. Dioxygenase enzymes: Catalytic mechanisms and chemical models. *Tetrahedron* **2003**, *59*, 7075–7101. [CrossRef]
42. Zeng, Q.; Xu, J.; Hou, Y.; Li, H.; Du, C.; Jiang, B.; Shi, S. Effect of Fe_3O_4 nanoparticles exposure on the treatment efficiency of phenol wastewater and community shifts in sbr system. *J. Hazard. Mater.* **2021**, *407*, 124828. [CrossRef]
43. He, S.; Feng, Y.; Ni, J.; Sun, Y.; Xue, L.; Feng, Y.; Yu, Y.; Lin, X.; Yang, L. Different responses of soil microbial metabolic activity to silver and iron oxide nanoparticles. *Chemosphere* **2016**, *147*, 195–202. [CrossRef] [PubMed]
44. He, S.; Zhong, L.; Duan, J.; Feng, Y.; Yang, B.; Yang, L. Bioremediation of wastewater by iron oxide-biochar nanocomposites loaded with photosynthetic bacteria. *Front. Microbiol.* **2017**, *8*, 823. [CrossRef] [PubMed]
45. Wang, X.; Wang, X.; Liu, M.; Bu, Y.; Zhang, J.; Chen, J.; Zhao, J. Adsorption–synergic biodegradation of diesel oil in synthetic seawater by acclimated strains immobilized on multifunctional materials. *Mar. Pollut. Bull.* **2015**, *92*, 195–200. [CrossRef]

water

Article

Over-Produced Extracellular Polymeric Substances and Activated Antioxidant Enzymes Attribute to Resistance of Pb(II) for Algal–Bacterial Granular Sludge in Municipal Wastewater Treatment

Juanru Yang [1], Yu Zhang [1] and Shulian Wang [2,3,4,*]

1 School of Civil Engineering, Architecture and Environment, Hubei University of Technology, Wuhan 430068, China
2 Key Laboratory of Health Intelligent Perception and Ecological Restoration of River and Lake, Ministry of Education, Hubei University of Technology, Wuhan 430068, China
3 Innovation Demonstration Base of Ecological Environment Geotechnical and Ecological Restoration of Rivers and Lakes, School of Civil and Environmental Engineering, Hubei University of Technology, Wuhan 430068, China
4 Hubei Cooperative Innovation Center for Industrial Fermentation, Hubei University of Technology, Wuhan 430068, China
* Correspondence: wangshulian@hbut.edu.cn

Abstract: Algal–bacterial granular sludge technology is a new type of wastewater treatment and resource regeneration process, which has received widespread attention due to its excellent nitrogen and phosphorus removal performance, and energy-saving and emission reduction effects. Although algal–bacterial granular sludge technology has achieved an ideal nutrient removal ability, some pollutants in wastewater might affect the symbiotic relationship between algae and bacteria. This study investigated the impact of coexisting Pb(II) on the symbiosis of algal–bacterial granular sludge. It was found that 2.5–10.0 mg/L of Pb(II) exposure increased the relative abundance of Pro-teobacteria. In addition, more protein in extracellular polymeric substances (EPS-PN) was secreted at 2.5 mg/L of Pb(II) exposure while EPS-PN content reduced at a rate of 5.0–10.0 mg/L of Pb(II). Under different concentrations of Pb(II), the damage degree of algal–bacterial granular sludge was exacerbated, evidenced by increased malondialdehyde (MDA) content. To cope with these adverse circumstances, the antioxidant enzyme activity of both super-oxide dismutase (SOD) and peroxidase dismutase (CAT) was boosted. With the help of these adaptive strategies, the symbiosis of algal–bacterial granular sludge was stable. Moreover, the performance of algal–bacterial granular sludge in treating COD, ammonia-N and phosphate-P was kept at above 95%. This study approved that a Pb(II) concentration less than 10.0 mg/L had little effect on the performance of algal–bacterial granular sludge in wastewater treatment. It is hoped that this study can provide useful information for an improved engineering feasibility of algal–bacterial granular sludge process.

Keywords: algal–bacterial granular sludge; divalent lead ion; extracellular polymeric substance; antioxidant enzyme; resistance

Citation: Yang, J.; Zhang, Y.; Wang, S. Over-Produced Extracellular Polymeric Substances and Activated Antioxidant Enzymes Attribute to Resistance of Pb(II) for Algal–Bacterial Granular Sludge in Municipal Wastewater Treatment. *Water* **2023**, *15*, 3833. https://doi.org/10.3390/w15213833

Academic Editor: Alejandro Gonzalez-Martinez

Received: 16 October 2023
Revised: 31 October 2023
Accepted: 1 November 2023
Published: 2 November 2023

1. Introduction

With the rapid development of industries such as electroplating, mining, fertilizer, tanning and papermaking, as well as the extensive use of products like batteries and pesticides, the direct and indirect discharge of heavy metal-containing wastewater into the environment has been increasing [1]. Heavy metals have high toxicity, are resistant to microbial degradation and can easily accumulate in organisms. They not only disrupt the ecological environment of water bodies but also pose risks of toxicity and carcinogenicity to organisms through the food chain [2,3]. Divalent lead (Pb(II)), in particular, has received much attention due to its wide application in industries, and its inherent toxicity [4–6]. As one of the nonessential metal elements with biomagnification ability, excess intake of Pb(II) in the human body damages the central nervous system and causes organ failures [7]. The concentrations of total lead in polluted water are reported to be in the range of 3–35 mg/L [8], which is significantly higher than the specified threshold of 0.1 mg/L for municipal wastewater discharge [9] and 0.01 mg/L for drinking water quality [10] in China.

The increased amount of Pb(II) discharged into the water environment has become a significant threat to the ecological environment. By causing deficiencies or altering the distribution in biological cells, Pb(II) can disrupt membrane permeability and mineral nutrient balance, affect the catalytic activity of many enzymes and thereby interfere with biological growth and development [11,12]. Research has shown that Pb(II) has affected the morphology, physiological characteristics and biochemical functions of plants, and inhibited root growth and synthesis of photosynthetic pigments [13]. In addition, the toxic effects of Pb(II) on plants weakened their antioxidant stress capacity, leading to lipid peroxidation of chloroplast membranes and even cell damage. It was found that Pb(II) inactivated enzymes involved in photosynthesis, inhibited chlorophyll synthesis and even disrupted chloroplast integrity [14]. Moreover, the chlorophyll in plants and algae was easily degraded by Pb(II) exposure in the growth environment [15].

As a new type of biotechnology for wastewater treatment, algal–bacterial granular sludge achieves the purpose of wastewater purification or remediation by combining the powerful nutrient uptake function of non-cellular structures (e.g., prokaryotic cyanobacteria, eukaryotic algae, diatoms, etc.) and the efficient degradation ability of bacteria (e.g., nitrifying bacteria, denitrifying bacteria, polyphosphate bacteria, etc.) towards pollutants [16–19]. Compared to traditional wastewater treatment technologies (e.g., activated sludge process), algal–bacterial granular sludge relies on the symbiotic relationship between algae and bacteria, where aerobic bacteria use the oxygen released by algae photosynthesis to convert organic carbon into carbon dioxide, while algae use inorganic nitrogen and phosphate for photosynthesis to synthesize intracellular components for their own growth and reproduction [20].

In theory, the synergistic cycle between algae and bacteria in algal–bacterial granular sludge can result in effectively improved removal efficiency of nutrients and reduced energy consumption and greenhouse gas emissions for actual wastewater treatment. However, up to now, the application of algal–bacterial granular sludge is still in its initial stages since some components (e.g., heavy metals, persistent organic pollutants and microplastics) in wastewater affect the symbiosis of microalgae and bacteria. Previous research has found that the coexistence of Cd(II) exerts toxic effects on microorganisms in algal–bacterial granular sludge, leading to changes in community structure and poor wastewater treatment efficiency [21]. Further studies have approved that algal–bacterial granular sludge exhibits a good adsorption capacity and can effectively treat Cr(VI)-containing wastewater through mechanisms of bio-reduction and bio-adsorption with a total removal efficiency of 89.1% [22,23]. In the symbiosis of microalgae and bacteria, algae can adsorb heavy metals, and bacteria are able to biotransform them to low toxicity forms [24]. The algal–bacterial granular sludge process has exhibited effective remediation efficiency for heavy metals and is hopeful to be widely used to treat heavy metal-containing wastewater. Currently, the heavy metal removal technology used for wastewater treatment has not yet focused on the algal–bacterial granular sludge process. Purebred algae are suitable for complex wastewater environments with cell breakage, and low concentrations of heavy metals cannot be highly selected. In addition, the actual application of the algal–bacterial granular sludge process is not widespread or accepted. Therefore, it is of vital importance to study the heavy metal-loaded algal–bacterial granular sludge process.

So far, most studies have focused on Pb(II) removal techniques [8]. Many researches have reported that filtration, precipitation, adsorption, ion exchange and electrodialysis could be applied as effective methods to remove Pb(II) in wastewater [8]. The detailed removal mechanisms have been thoroughly studied. On the other hand, the algal–bacterial granular sludge process has been expected to be a new type of biotechnology in treating municipal wastewater [18]. However, limited information regarding the effect of Pb(II) exposure on the removal efficiency of nutrients (e.g., COD, ammonia nitrogen and phosphates) and the corresponding metabolic responses is available. Further research is necessary to address this knowledge gap. The main objective of this experimental study is to investigate the influence of Pb(II) on the treatment efficiency of algal–bacterial granular wastewater, and analyze its impact mechanism regarding metabolic reactions and microbial community structure. It is hoped that this study can provide more information for future research on the engineering feasibility of algal–bacterial granular technology.

2. Materials and Methods

2.1. Synthetic Wastewater and Algal–Bacterial Granular Sludge

The synthesized wastewater used in this study was simulated domestic sewage, which was composed of 527 mg/L of NaAc ($CH_3COONa{\cdot}3H_2O$), 114.6 mg/L of NH_4Cl, 10 mg/L of KH_2PO_4, 10 mg/L of $CaCl_2$, 10 mg/L of $MgSO_4{\cdot}7H_2O$, 10 mg/L of $FeSO_4{\cdot}7H_2O$ and 1 mL/L of trace elements. The trace solution contained 10 g/L of EDTA, 150 mg/L of H_3BO_3, 100 mg/L of $MnSO_4{\cdot}H_2O$, 30 mg/L of $CuSO_4{\cdot}5H_2O$, 120 mg/L of $ZnSO_4{\cdot}7H_2O$, 60 mg/L of $Na_2MoO_4{\cdot}2H_2O$, 180 mg/L of KI and 150 mg/L of $CoCl_2{\cdot}6H_2O$.

In this study, algal–bacterial granular sludge was prepared from the sludge of a wastewater treatment plant in Wuhan, China, using aerobic granular sludge cultivated using the calcium salt method as raw materials [25]. Briefly, aerobic granules fed with the synthetic wastewater described above were exposed to a light-emitting diode (LED) light at a photosynthetic photon flux density (PPFD) of 10,000 lux under aerobic conditions. After one month of cultivation, algal–bacterial granular sludge was gradually matured with a typical green color, and an average size of 2 mm. Mature algal–bacterial granular sludge with a 5 min sludge volume index (SVI_5) of 77.25 mL/g was obtained [20,21,26]. The harvested mature algal–bacterial granular sludge was used in subsequent experiments.

2.2. Experimental Design

In this study, in a 60 mL glass serum bottle with a height/diameter of 2, 5 g of fresh algal–bacterial granular sludge was added to 30 mL of synthetic wastewater. The volatile suspended solids (VSS) concentration was 17.6 $\pm$ 0.2 g/L. A series of certain volumes of 500 mg/L of Pb(II) nitrate stock solution were added to obtain 2.5, 5.0, 7.5 and 10.0 mg/L of Pb(II) in stimulated wastewater. No additional Pb(II) was set as the control. The pH of the stimulated wastewater was kept at 6.4 $\pm$ 0.2. The experimental light intensity provided by an LED was controlled at 10,000 lux; the light cycle was 8 h of light and 16 h of dark to simulate natural lighting. Since nutrient removal efficiencies peaked at the end of the daytime, sampling was carried out at 8 h. The influent and effluent water samples were filtered through a 0.45 μm filter membrane for further analysis.

2.3. Analysis Methods

In this study, COD, ammonia-N, phosphate-P and VSS were determined according to the standard methods [27]. The pH of the solution was measured using a PHS-3E (INESA Scientific Instrument Co., Shanghai, China) pH meter. The concentration of total lead in the water samples was determined by an AAnalyst 800 atomic absorption spectrometry (Perkin Elmer, Waltham, MA, USA). Chlorophyll in algal–bacterial granular sludge was extracted using 95% ethanol according to the following procedures. Firstly, 0.2 g of fresh algal–bacterial granular sludge was placed in a 10 mL centrifuge tube. Then, 5 mL of 95% ethanol was added and the centrifuge tube was urgently wrapped in tin foil to prevent light transmission. The tube was then placed in a dark refrigerator at 4 °C for 24 h, followed with centrifugation at 4000 rpm/min for 10 min. The total chlorophyll (total Chl) content, including chlorophyll a (Chl a) and chlorophyll b (Chl b), was extracted using an acetone extraction method [28].

Extracellular polymeric substances (EPS) were extracted by a modified heat extraction method [29]. Briefly, 0.5 g of algal–bacterial granular sludge was placed into a 50 mL centrifuge tube, followed with washing with deionized water three times. Sodium chloride solution (0.05%, 10 mL) was added to suspended algal–bacterial granular sludge in the solution. The centrifuge tube containing the above solution was placed in an ultrasonic cleaning machine and sonicated at 20 KHz for 3 min and then heated at 60 °C in a water bath for 30 min. Finally, the centrifuge tube was frozen and centrifuged at 12,000× *g* for 30 min. The supernatant was collected and filtrated with a 0.45 μM membrane for the following determination. The extracted EPS was measured using a DM4000B LED fluorescence microscope (Leica, Heidelberg, Germany), with an emission wavelength of 250–600 nm and an excitation wavelength of 250–550 nm. The polysaccharide concentration was measured by the phenol–sulfuric acid method with glucose as the standard [30] and the Lowry–Folin method was used for protein determination [31].

After being frozen in liquid nitrogen, algal–bacterial granular sludge was ground to a paste using a high-speed grinder and mixed with 0.2 mol/L of phosphate buffer (pH 7.8) (1:9), followed with centrifugation (4 °C, 4000 r/min) for 10 min to obtain a crude enzyme solution from the supernatant for enzyme activity measurement. The activities of superoxide dismutase (SOD), peroxidase dismutase (CAT) and malondialdehyde (MDA) were tested by using kits purchased from Nanjing Jiancheng Bioengineering Institute (Nanjing, China).

The algal–bacterial granular sludge samples were collected after 90-day culture for microbial community analysis. DNA in samples was extracted using the E.Z.N.A.® Soil DNA Kit (Omega Bio-Tek, Norcross, GA, USA) according to the manufacturer's instructions. The quality of total DNA was checked by 0.8% agarose gel electrophoresis and further quantified. The 16S rRNA and 18S rRNA genes were amplified using 515F/907R prokaryotic primers targeting the V4–V5 region of the 16S rRNA gene and 528F/706R eukaryotic primers targeting the V4 region of the 18S rRNA gene, respectively [21]. The purified amplicons were collected in an equimolar manner on the Illumina MiSeq platform and sequenced by Meiji Biopharmaceutical Technology Co., Ltd. (Shanghai, China).

2.4. Statistical Analyses

In this study, all experiments were performed in three replicates. All measurements were repeated three times, and the data were represented as mean ± standard deviation (SD). The SPSS V19.0 (IBM, New York, NY, USA) software was used to conduct one-way analysis variance, and $p < 0.05$ stands for statistical significance.

3. Results and Discussion

3.1. Changes in Microbial Communities

The microbial community of eukaryote at species level in algal–bacterial granular sludge after 90-day culture is shown in Figure 1a. Eukaryote were mainly composed of *Fungi_unclassified*, *Ascomycota* and *Leucosporidiella*, accounting for 98.6%, 0.5% and 0.5% in the control. In the presence of Pb(II) from 2.5 to 10.0 mg/L, the relative abundance of *Fungi_unclassified* decreased from 97.8% to 61.4%, while the relative abundance of *Ascomycota* increased from 0.5% to 36.4%. In addition, the relative abundance of *Leucosporidiella* was kept at 0.9–1.7%. As reported, *Ascomycota* were highly tolerant to heavy metals [32,33]. The increased relative abundance of *Ascomycota* might be a way for microorganisms in algal–bacterial granular sludge to adapt to Pb(II). As seen in Figure 1b, prokaryotic diversity at the phylum level in microbial communities of algal–bacterial granular sludge was mainly composed of *Proteobacteria*, *Chloroflexi*, *Bacteroidota*, *Cyanobacteria* and *Firmicutes*, accounting for 51.8%, 31.1%, 2.8%, 7.8% and 1.3%, respectively. After a 90-day culture with 2.5–10.0 mg/L of Pb(II), the relative abundance of *Proteobacteria* increased from 62.0% to 77.8%. As a Pb-resistant strain, *Proteobacteria* was tolerant to Pb(II). Therefore, when algal–bacterial granular sludge was exposed to Pb(II) for a long time, the relative abundance of *Proteobacteria* increased [34].

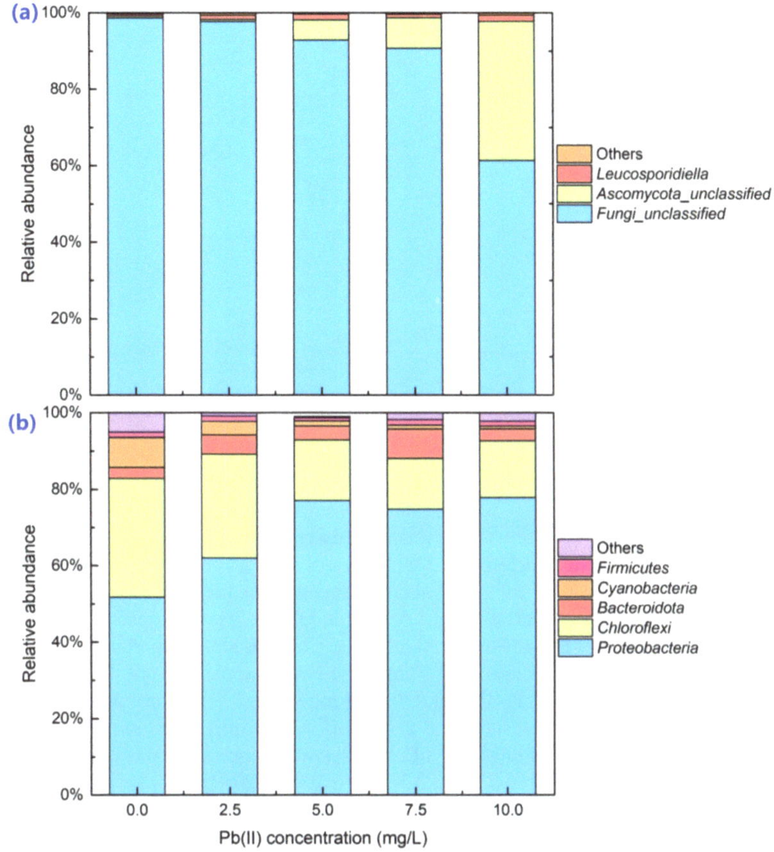

Figure 1. Distributions of eukaryotic diversity at species level (**a**) and prokaryotic diversity at phylum level (**b**) in microbial communities of algal–bacterial granular sludge after 90-day culture. The others named in the figures indicated the minority fractions that the sum of different eukaryotic species and prokaryotic phylum accounted for less than 1% of its total sequences in each sample.

The algae in the algal–bacterial granular sludge used in this study were cyanobacteria. It was found that the relative abundance of cyanobacteria in prokaryotic diversity was 7.8% in the control, which reduced to 0.7% with an increased Pb(II) concentration of 10.0 mg/L. In fact, photosynthetic pigments are one of the most commonly used indicators to determine the algae. Heavy metals can impair the biosynthesis of photosynthetic pigments in algae with varying degrees. As seen in Figure 2, insignificant chlorophyll a content was shown with a Pb(II) concentration of 0–7.5 mg/L during 60-day culture ($p > 0.05$). However, chlorophyll a content was reduced from 60–to 90 days. At 90 days, chlorophyll a content was 21.39 $\pm$ 0.48, 17.86 $\pm$ 0.22, 16.25 $\pm$ 0.14 and 13.58 $\pm$ 0.49 mg/g VSS, respectively. As for the 10.0 mg/L of Pb(II) concentration, chlorophyll a content decreased from 21.39 $\pm$ 0.48 mg/g VSS to 11.32 $\pm$ 0.41 mg/g VSS at 90 days. As the only chlorophyll in cyanobacteria, chlorophyll a was an index of cyanobacteria biomass [35]. However, when the concentration of heavy metals in the environment where cyanobacteria live was high, the growth of cyanobacteria was affected and the biomass was reduced [36]. It was reported that heavy metals may affect the electron transfer reaction of the photosystem through irreversible combination with intracellular components, which led to the obstruction of the photoreaction stage and seriously impacted photosynthesis [37].

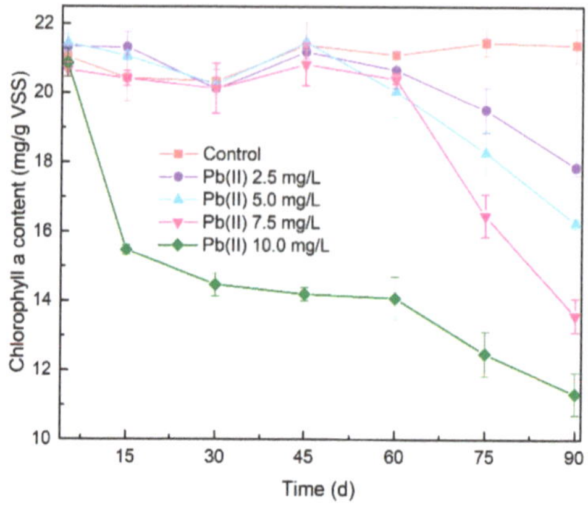

Figure 2. Chlorophyll a content in algal–bacterial granular sludge.

3.2. Extracellular Polymeric Substance Content

Extracellular polymeric substances (EPS) are macromolecular polymers secreted by microorganisms, mainly consisting of proteins, polysaccharides and a small amount of nucleic acid [38]. EPS can form a protective barrier for cells to resist the adverse effects of the external environment. The proteins in EPS play an important role in the bioremediation of heavy metals [39,40]. The changes in the content of EPS-PN and EPS-PS in algal–bacterial granular sludge is shown in Figure 3. It was found that more EPS-PN was secreted at 2.5 mg/L of Pb(II) exposure while EPS-PN content reduced at 5.0–10.0 mg/L of Pb(II). Take the 2.5 mg/L Pb(II)-exposure concentration as an example, EPS-PN content increased from 63.11 ± 2.37 mg/g VSS in the control to 68.93 ± 1.13 mg/g VSS after 30 days. At a Pb(II)-exposure concentration of 5.0–10.0 mg/L, EPS-PN content decreased from 53.69 ± 3.26 mg/g VSS to 20.01 ± 3.13 mg/g VSS after 30 days. Similar trends for EPS-PS change were observed. At a 2.5 mg/L of Pb(II)-exposure concentration after 90 days, EPS-PS content increased from 74.92 ± 4.93 mg/g VSS in the control to 94.27 ± 1.93 mg/g VSS while EPS-PN content reduced from 69.37 ± 3.86 mg/g VSS to 18.71 ± 5.79 mg/g VSS at 5.0–10.0 mg/L of Pb(II).

According to the changes in EPS-PN and EPS-PS content, low concentrations of Pb(II) (0–2.5 mg/L) promoted algal–bacterial granular sludge to secrete more EPS, while high concentrations of Pb(II) (5.0–10.0 mg/L) inhibited EPS secretion, presenting a phenomenon of low concentration promotion–high concentration inhibition. It was concluded that algal–bacterial granular sludge synthesized more EPS to cope with this adverse environment under low concentrations of Pb(II), while cell activity was inhibited to some extent under high concentrations of Pb(II).

To further verify the changes in the types and concentrations of organic substances in EPS, EPS samples from the experimental and control groups were measured using a three-dimensional fluorescence spectrometer. Figure 4 shows the fluorescence spectra of EPS from samples exposed under different Pb(II) concentrations. All the spectra showed two obvious characteristic peaks, namely peak A (Ex/Em = 290/335–338) and peak B (Ex/Em = 360/420). Peak A belonged to tryptophan protein substances and peak B was attributed to humic substances [41]. An insignificant shift in the positions of the two characteristic peaks of EPS indicated no change in the organic components in EPS of algal–bacterial granular sludge exposed to 2.5–10.0 mg/L of Pb(II). It should be noted that the fluorescence intensity of peak A increased from 19,549 in the control to 27,897, and then decreased to 10,570. The intensity of peak B decreased from 20,268 to 745 at 0–10.0 mg/L of Pb(II). Studies have shown that decreased fluorescence intensity can be attributed to fluorescence quenching [42,43]. The experimental results indicated that humic substances formed complexes with Pb(II), leading to a decrease in fluorescence intensity of peak B.

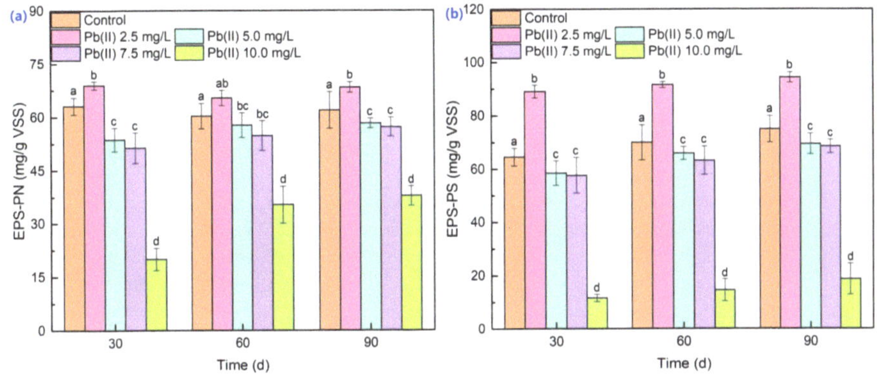

Figure 3. Content of EPS-PN (**a**) and EPS-PS (**b**) in algal–bacterial granular sludge. Different letters mean significant difference between treatments ($p < 0.05$).

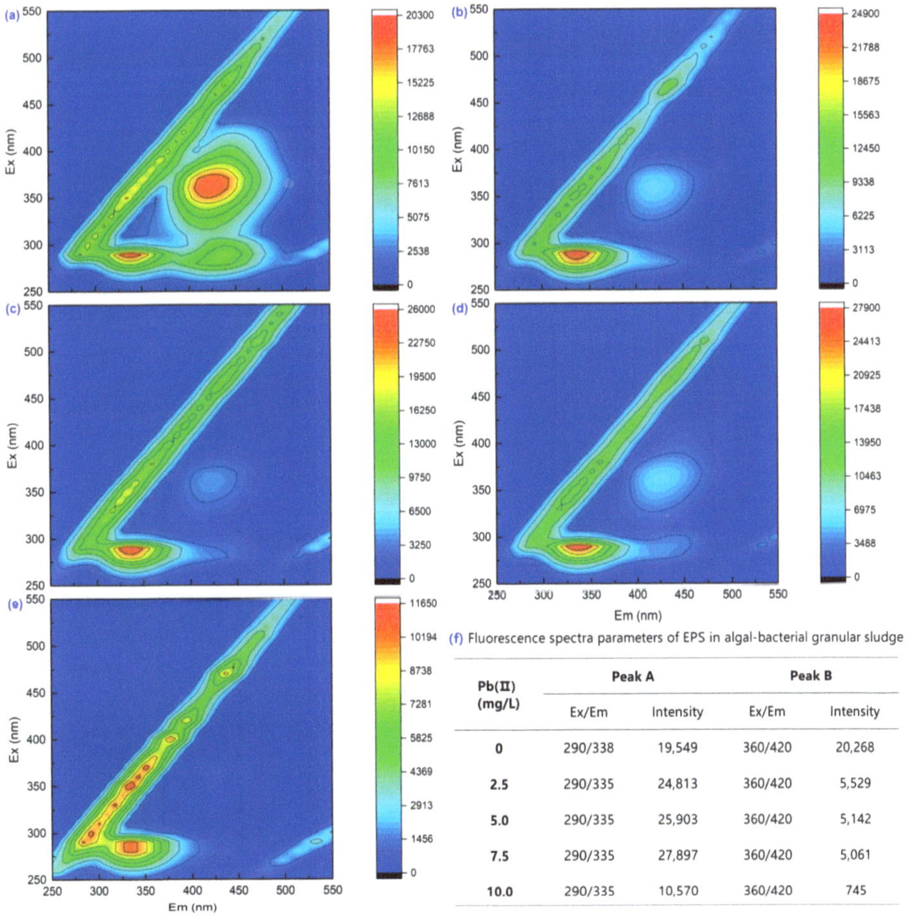

(f) Fluorescence spectra parameters of EPS in algal-bacterial granular sludge

Pb(II) (mg/L)	Peak A		Peak B	
	Ex/Em	Intensity	Ex/Em	Intensity
0	290/338	19,549	360/420	20,268
2.5	290/335	24,813	360/420	5,529
5.0	290/335	25,903	360/420	5,142
7.5	290/335	27,897	360/420	5,061
10.0	290/335	10,570	360/420	745

Figure 4. 3D-EEM spectra of EPS from algal–bacterial granular sludge after 90 day-culture in the presence of Pb(II) (**a**)—0; (**b**)—2.5 mg/L; (**c**)—5.0 mg/L; (**d**)—7.5 mg/L; (**e**)—10.0 mg/L) and the fluorescence spectra parameters (**f**).

3.3. Antioxidant Enzyme Activities and MDA Content Analysis

It was reported that cells produced active free radicals under adverse environment, including superoxide free radicals, carboxyl free radicals and hydrogen peroxide. The existence of these free radicals could disrupt the balance of intracellular oxidation, leading to permanent dysfunction and cell death [44]. In addition, the massive production of these reactive oxygen species could trigger or intensify the membrane lipid peroxidation, causing serious damage to the membrane system. The product of the membrane lipid peroxidation, MDA, could be used as an indicator to reflect the oxidative pressure faced by algae cells [45]. As seen in Figure 5, MDA content was insignificantly different at 1–7.5 mg/L of Pb(II) exposure concentration ($p > 0.05$) while increased at 10.0 mg/L of Pb(II)-exposed concentration ($p < 0.05$) at 30 days. A similar trend was obtained at 60 and 90 days. It was found that MDA content was increased from 167.48 ± 11.27 nmol/mgprot in the control to 325.47 ± 16.59, 355.68 ± 10.79, 418.23 ± 16.71 and 453.79 ± 18.53 nmol/mgprot at 2.5, 5.0, 7.5 and 10.0 mg/L of Pb(II) exposure concentration, respectively. These results indicated that the damage degree caused by Pb(II) to algal–bacterial granular sludge was exacerbated with the cultivation time and exposure concentration.

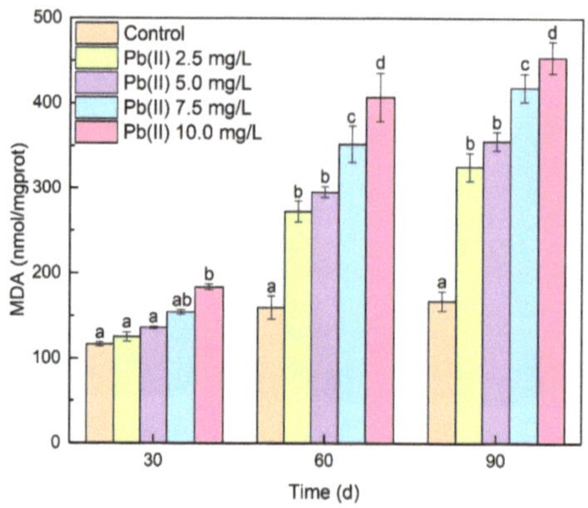

Figure 5. Antioxidant enzyme activity of MDA. Different letters mean significant difference between treatments ($p < 0.05$).

To cope with the oxidative damage caused by Pb(II) and ensure the normal metabolism of cells, several enzyme activities (e.g., SOD and CAT) could form an antioxidant enzyme system. The presence of antioxidant enzymes can effectively remove the accumulated active free radicals in organisms, ensuring the normal life activities of cells. SOD is the primary antioxidant enzyme that plays a role in scavenging free radicals in cells. Its role is to catalyze the disproportionation of free radicals and convert them into hydrogen peroxide (H_2O_2) and superoxide radicals. The accumulation of H_2O_2 further promotes the synthesis of CAT, which catalyzes H_2O_2 to decompose into H_2O and O_2 [46–48]. Figure 6 shows the antioxidant enzyme activity of SOD and CAT. At 30 days, insignificant differences in SOD activity at 0–10.0 mg/L of Pb(II) exposure concentration ($p > 0.05$) can be seen. SOD activity increased from 217.75 ± 14.57 U/mgprot in the control to 380.78 ± 17.11 U/mgprot at 10.0 mg/L of Pb(II) exposure concentration at 60 days. A similar trend for SOD activity was observed at 90 days. On the other hand, CAT activity increased with Pb(II) exposure concentration. The maximum CAT activity (i.e., 49.37 ± 2.63 U/mgprot) was obtained at 60 days with 10.0 mg/L of Pb(II). However, CAT activity decreased at 90 days. This phenomenon was similar to previous research. Algal–bacterial granular sludge effectively eliminates the free radicals produced by oxidative stress through antioxidant enzymes in the cells to maintain the metabolic balance of reactive oxygen species.

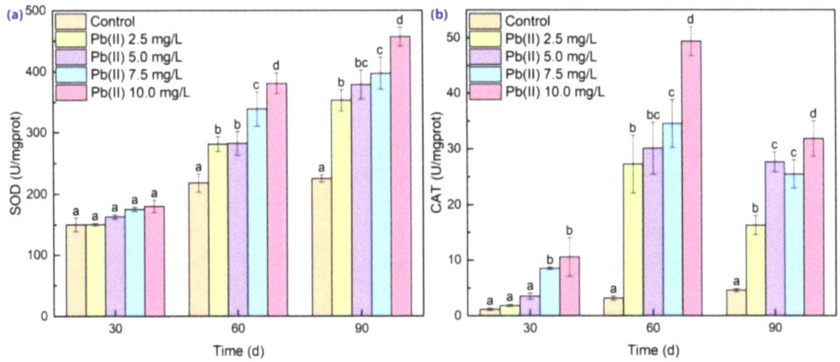

Figure 6. Antioxidant enzyme activity of SOD (**a**) and CAT (**b**). Different letters mean significant difference between treatments ($p < 0.05$).

3.4. Performance in Treating Municipal Wastewater

With different concentrations of Pb(II) exposure, insignificant differences in the removal rate of COD between the experimental groups and the control group were observed ($p > 0.05$), and both had stable removal rates of about 95%. After 90-day culture, the average COD removal rates were 95.83%, 95.51%, 95.41%, 95.09% and 95.08%, respectively, at a Pb(II) exposure concentration of 0–10.0 mg/L (Figure 7a). Moreover, algal–bacterial granular sludge maintained a high removal rate of ammonia-N, which was stable at around 97% with an insignificant difference among environmental groups ($p > 0.05$). After 90-day culture, average ammonia-N removal rates were 97.72%, 97.44%, 97.33%, 97.39% and 97.48%, respectively (Figure 7b). A similar trend was observed in the removal rate of phosphate-P, which remained around 94% ($p > 0.05$). After 90-day culture, the average phosphate-P removal rates were 94.79%, 94.06%, 94.04%, 93.55% and 93.68%, respectively (Figure 7c).

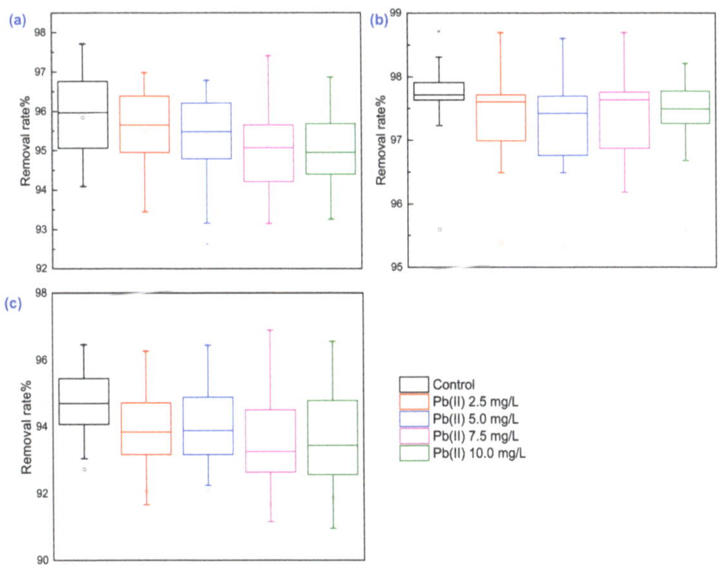

Figure 7. Removal profiles of COD (**a**), ammonia-N (**b**) and phosphate-P (**c**) across ninety days of operation. The symbols of '□' and '*' in the figures stand for average value and outside point, respectively.

It could be concluded that no inhibition of COD, ammonia-N and phosphate-P removal occurred with Pb(II) exposure less than 10.0 mg/L after 90-day culture. Unlike the presence of Cd(II) [21] and Cr(VI) [22], the synergistic effect of microalgae and bacteria could eliminate the toxic effects of Pb(II) on microorganisms and resist the adverse impact of Pb(II), evidenced by the almost unchanged nutrient removal abilities. COD, ammonia-N and phosphate-P from the control were 15, 0.80 and 0.26 mg/L, respectively, meeting the water quality standards for China, which were 75, 10 and 1 mg/L [9]. At a Pb(II) concentration of 10.0 mg/L, COD, ammonia-N and phosphate-P concentrations in the effluent water were 17, 0.88 and 0.32 mg/L, which were less than the specified threshold. Based on these results, 0–10.0 mg/L of Pb(II)-containing wastewater could be handled by the algal–bacterial granular sludge process and meet the discharge requirements. Therefore, the algal–bacterial granular sludge process could be considered as an alternative approach for treating low concentration of Pb(II)-containing wastewater.

The variations in Pb(II) removal by algal–bacterial granular sludge across 90 days of operation were obtained. It was found that 88.5%, 89.9%, 90.5% and 89.1% of Pb(II) (2.5–10.0 mg/L) was removed from the solution. In fact, the Pb(II) ions formed a stable compound, lead acetate, with sodium acetate in the feed. Lead acetate is easily soluble in water and generally generates lead hydroxide precipitates under alkaline conditions with a pH above 12. It was found that the pH of the solution varied in the range of 6.4 ± 0.2–8.5 ± 0.3, therefore Pb(II) was unlikely to be removed in the form of precipitation. The removal of Pb(II) may be attributed to electrostatic interaction with EPS, ion exchange with metal ions and bonding to functional groups [8]. Therefore, algal–bacterial granular sludge was able to effectively adsorb Pb(II) to meet the discharge requirements and could be considered as an alternative approach for treating Pb(II)-containing wastewater.

4. Conclusions

This study demonstrated the performance and the adaptive strategies of algal–bacterial granular sludge technology in wastewater treatment. Under Pb(II) stress for 90 days, algal–bacterial granular sludge secreted more EPS, which could adsorb Pb(II) through the rich negatively charged functional groups in EPS. In addition, the activities of SOD and CAT enzymes in microbial cells showed an upward trend to eliminate reactive oxygen species produced in cells. Both enzymes alleviated the oxidation burden by removing excess free radicals. In addition, the performance of algal–bacterial granular sludge maintained efficient COD, ammonia-N and phosphate-P removal capabilities. This article elaborates upon the response of algal–bacterial granular sludge to cope with Pb(II) exposure in the environment, and proposes a new and feasible approach for treating wastewater containing Pb(II).

Author Contributions: Conceptualization, J.Y. and S.W.; methodology, J.Y. and Y.Z.; validation, J.Y.; investigation, Y.Z.; data curation, J.Y. and Y.Z.; writing—original draft preparation, J.Y.; writing—review and editing, S.W. All authors have read and agreed to the published version of the manuscript.

Funding: This research received no external funding.

Data Availability Statement: Not applicable.

Conflicts of Interest: The authors declare no conflict of interest.

References

1. Shrestha, R.; Ban, S.; Devkota, S.; Sharma, S.; Joshi, R.; Tiwari, A.P.; Kim, H.Y.; Joshi, M.K. Technological trends in heavy metals removal from industrial wastewater: A review. *J. Environ. Chem. Eng.* **2021**, *9*, 105688. [CrossRef]
2. Fu, Z.; Xi, S. The effects of heavy metals on human metabolism. *Toxicol. Mech. Methods* **2020**, *30*, 167–176. [CrossRef] [PubMed]
3. Jaishankar, M.; Tseten, T.; Anbalagan, N.; Mathew, B.B.; Beeregowda, K.N. Toxicity, mechanism and health effects of some heavy metals. *Interdiscip. Toxicol.* **2014**, *7*, 60–72. [CrossRef] [PubMed]
4. Kushwaha, A.; Hans, N.; Kumar, S.; Rani, R. A critical review on speciation, mobilization and toxicity of lead in soil-microbe-plant system and bioremediation strategies. *Ecotoxicol. Environ. Saf.* **2018**, *147*, 1035–1045. [CrossRef]

5. Kalita, D.; Joshi, S.R. Study on bioremediation of Lead by exopolysaccharide producing metallophilic bacterium isolated from extreme habitat. *Biotechnol. Rep.* **2017**, *16*, 48–57. [CrossRef]
6. Das, S.; Dash, H.R.; Chakraborty, J. Genetic basis and importance of metal resistant genes in bacteria for bioremediation of contaminated environments with toxic metal pollutants. *Appl. Microbiol. Biotechnol.* **2016**, *100*, 2967–2984. [CrossRef]
7. Abbaszadeh, S.; Wan Alwi, S.R.; Webb, C.; Ghasemi, N.; Muhamad, I.I. Treatment of lead-contaminated water using activated carbon adsorbent from locally available papaya peel biowaste. *J. Clean. Prod.* **2016**, *118*, 210–222. [CrossRef]
8. Wang, Z.; Wang, H.; Nie, Q.; Ding, Y.; Lei, Z.; Zhang, Z.; Shimizu, K.; Yuan, T. Pb(II) bioremediation using fresh algal-bacterial aerobic granular sludge and its underlying mechanisms highlighting the role of extracellular polymeric substances. *J. Hazard. Mater.* **2023**, *444*, 130452. [CrossRef]
9. *GB 18918-2002*; Discharge Standard of Pollutants for Municipal Wastewater Treatment Plants. China Environment Press: Beijing, China, 2002.
10. *GB 5749-2022*; Standards for Drinking Water Quality. State Administration for Market Regulation: Beijing, China, 2022.
11. Leung, P.T.Y.; Yi, A.X.; Ip, J.C.H.; Mak, S.S.T.; Leung, K.M.Y. Photosynthetic and transcriptional responses of the marine diatom *Thalassiosira pseudonana* to the combined effect of temperature stress and copper exposure. *Mar. Pollut. Bull.* **2017**, *124*, 938–945. [CrossRef]
12. Lian, W.; Yang, L.; Joseph, S.; Shi, W.; Bian, R.; Zheng, J.; Li, L.; Shan, S.; Pan, G. Utilization of biochar produced from invasive plant species to efficiently adsorb Cd (II) and Pb (II). *Bioresour. Technol.* **2020**, *317*, 124011. [CrossRef]
13. Alkhatib, R.; Maruthavanan, J.; Ghoshroy, S.; Steiner, R.; Sterling, T.; Creamer, R. Physiological and ultrastructural effects of lead on tobacco. *Biol. Plant.* **2012**, *56*, 711–716. [CrossRef]
14. Dubey, S.; Chen, C.W.; Haldar, D.; Tambat, V.S.; Kumar, P.; Tiwari, A.; Singhania, R.R.; Dong, C.-D.; Patel, A.K. Advancement in algal bioremediation for organic, inorganic, and emerging pollutants. *Environ. Pollut.* **2023**, *317*, 120840. [CrossRef] [PubMed]
15. Wang, B.; Xiao, L.; Xu, A.; Mao, W.; Wu, Z.; Hicks, L.C.; Jiang, Y.; Xu, J. Silicon fertilization enhances the resistance of tobacco plants to combined Cd and Pb contamination: Physiological and microbial mechanisms. *Ecotoxicol. Environ. Saf.* **2023**, *255*, 114816. [CrossRef]
16. Zhang, B.; Li, W.; Guo, Y.; Zhang, Z.; Shi, W.; Cui, F.; Lens PN, L.; Tay, J.H. Microalgal-bacterial consortia: From interspecies interactions to biotechnological applications. *Renew. Sustain. Energy Rev.* **2020**, *118*, 109563. [CrossRef]
17. Ji, B.; Zhang, M.; Gu, J.; Ma, Y.; Liu, Y. A self-sustaining synergetic microalgal-bacterial granular sludge process towards energy-efficient and environmentally sustainable municipal wastewater treatment. *Water Res.* **2020**, *179*, 115884. [CrossRef] [PubMed]
18. Wang, S.; Zhu, L.; Ji, B.; Hou, H.; Ma, Y. Microalgal-bacterial granular sludge process in non-aerated municipal wastewater treatment under natural day-night conditions: Performance and microbial community. *Water* **2021**, *13*, 1479. [CrossRef]
19. Purba LD, A.; Zahra, S.A.; Yuzir, A.; Iwamoto, K.; Abdullah, N.; Shimizu, K.; Lei, Z.; Hermana, J. Algal-bacterial aerobic granular sludge for real municipal wastewater treatment: Performance, microbial community change and feasibility of lipid recovery. *J. Environ. Manag.* **2023**, *333*, 117374. [CrossRef]
20. Wang, S.; Ji, B.; Zhang, M.; Ma, Y.; Gu, J.; Liu, Y. Defensive responses of microalgal-bacterial granules to tetracycline in municipal wastewater treatment. *Bioresour. Technol.* **2020**, *312*, 123605. [CrossRef]
21. Wang, S.; Ji, B.; Cui, B.; Ma, Y.; Guo, D.; Liu, Y. Cadmium-effect on performance and symbiotic relationship of microalgal-bacterial granules. *J. Clean. Prod.* **2021**, *282*, 125383. [CrossRef]
22. Yang, X.; Zhao, Z.; Zhang, G.; Hirayama, S.; Nguyen, B.V.; Lei, Z.; Shimizu, K.; Zhang, Z. Insight into Cr(VI) biosorption onto algal-bacterial granular sludge: Cr(VI) bioreduction and its intracellular accumulation in addition to the effects of environmental factors. *J. Hazard. Mater.* **2021**, *414*, 125479. [CrossRef]
23. Yang, X.; Zhao, Z.; Yu, Y.; Shimizu, K.; Zhang, Z.; Lei, Z.; Lee, D.-J. Enhanced biosorption of Cr(VI) from synthetic wastewater using algal-bacterial aerobic granular sludge: Batch experiments, kinetics and mechanisms. *Sep. Purif. Technol.* **2020**, *251*, 117323. [CrossRef]
24. Zhao, D.; Cheah, W.Y.; Lai, S.H.; Ng, E.-P.; Khoo, K.S.; Show, P.L.; Ling, T.C. Symbiosis of microalgae and bacteria consortium for heavy metal remediation in wastewater. *J. Environ. Chem. Eng.* **2023**, *11*, 109943. [CrossRef]
25. Zhang, Y.; Dong, X.; Nuramkhaan, M.; Lei, Z.; Shimizu, K.; Zhang, Z.; Adachi, Y.; Lee, D.-J.; HwaTay, J. Rapid granulation of aerobic granular sludge: A mini review on operation strategies and comparative analysis. *Bioresour. Technol. Rep.* **2019**, *7*, 100206. [CrossRef]
26. Wang, S.; Ji, B.; Zhang, M.; Gu, J.; Ma, Y.; Liu, Y. Tetracycline-induced decoupling of symbiosis in microalgal-bacterial granular sludge. *Environ. Res.* **2021**, *197*, 111095. [CrossRef] [PubMed]
27. APHA. *Standard Methods for the Examination of Water and Wastewater*, 21st ed.; American Public Health Association: Washington, DC, USA, 2005.
28. Zhou, X.; Zhang, Y.; An, X.; De Philippis, R.; Ma, X.; Ye, C.; Chen, L. Identification of aqueous extracts from Artemisia ordosica and their allelopathic effects on desert soil algae. *Chemoecology* **2019**, *29*, 61–71. [CrossRef]
29. Zhang, W.; Cao, B.; Wang, D.; Ma, T.; Xia, H.; Yu, D. Influence of wastewater sludge treatment using combined peroxyacetic acid oxidation and inorganic coagulants re-flocculation on characteristics of extracellular polymeric substances (EPS). *Water Res.* **2016**, *88*, 728–739. [CrossRef] [PubMed]
30. Herbert, D.P.; Phipps, P.J.; Strange, R.E. Chapter III Chemical Analysis of Microbial Cells. *Methods Microbiol.* **1971**, *5*, 209–344.

31. Raunkj, R.K.; Hvitved-Jacobsen, T.; Nielsen, P.H. Measurement of pools of protein, carbohydrate and lipid in domestic wastewater. *Water Res.* **1994**, *28*, 251–262. [CrossRef]
32. Liang, X.; Chen, L.; Liu, Z.; Jin, Y.; He, M.; Zhao, Z.; Liu, C.; Niyungeko, C.; Arai, Y. Composition of microbial community in pig manure biochar-amended soils and the linkage to the heavy metals accumulation in rice at harvest. *Land Degrad. Dev.* **2018**, *29*, 2189–2198. [CrossRef]
33. Lin, Y.; Ye, Y.; Hu, Y.; Shi, H. The variation in microbial community structure under different heavy metal contamination levels in paddy soils. *Ecotoxicol. Environ. Saf.* **2019**, *180*, 557–564. [CrossRef]
34. Wang, L.; Gong, L.; Gan, D.; Li, X.; Yao, J.; Wang, L.; Qu, J.; Cong, J.; Zhang, Y. Diversity, function and assembly of the Trifolium repens L. root-associated microbiome under lead stress. *J. Hazard. Mater.* **2022**, *438*, 129510. [CrossRef]
35. Abouhend, A.S.; Milferstedt, K.; Hamelin, J.; Ansari, A.A.; Butler, C.; Carbajal-González, B.I.; Park, C. Growth progression of oxygenic photogranules and its impact on bioactivity for aeration-free wastewater treatment. *Environ. Sci. Technol.* **2020**, *54*, 486–496. [CrossRef]
36. De Souza, P.O.; Sinhor, V.; Crizel, M.G.; Pires, N.; Sanches Filho, P.J.; Picoloto, R.S.; Duarte, F.A.; Pereira CM, P.; Mesko, M.F. Bioremediation of chromium and lead in wastewater from chemistry laboratories promotes by cyanobacteria. *Bioresour. Technol. Rep.* **2022**, *19*, 101161. [CrossRef]
37. Ventrella, A.; Catucci, L.; Piletska, E.; Piletsky, S.; Agostiano, A. Interactions between heavy metals and photosynthetic materials studied by optical techniques. *Bioelectrochemistry* **2009**, *77*, 19–25. [CrossRef]
38. Tsai, B.; Chang, C.; Lee, D. Fractionation of soluble microbial products (SMP) and soluble extracellular polymeric substances (EPS) from wastewater sludge. *Environ. Technol.* **2008**, *29*, 1127–1138. [CrossRef] [PubMed]
39. Yin, Y.; Hu, Y.; Xiong, F. Sorption of Cu(II) and Cd(II) by extracellular polymeric substances (EPS) from Aspergillus fumigatus. *Int. Biodeterior. Biodegrad.* **2011**, *65*, 1012–1018. [CrossRef]
40. Wang, J.; Li, Q.; Li, M.-M.; Chen, T.-H.; Zhou, Y.-F.; Yue, Z.-B. Competitive adsorption of heavy metal by extracellular polymeric substances (EPS) extracted from sulfate reducing bacteria. *Bioresour. Technol.* **2014**, *163*, 374–376. [CrossRef] [PubMed]
41. Wen, C.; Paul, W.; AL, J.; Karl, B. Fluorescence excitation−emission matrix regional integration to quantify spectra for dissolved organic matter. *Environ. Sci. Technol.* **2003**, *37*, 5701–5710. [CrossRef]
42. Wei, D.; Wang, B.; Ngo, H.H.; Guo, W.; Han, F.; Wang, X.; Du, B.; Wei, Q. Role of extracellular polymeric substances in biosorption of dye wastewater using aerobic granular sludge. *Bioresour. Technol.* **2015**, *185*, 14–20. [CrossRef]
43. Kumari, S.; Mangwani, N.; Das, S. Interaction of Pb(II) and biofilm associated extracellular polymeric substances of a marine bacterium *Pseudomonas pseudoalcaligenes* NP103. *Spectrochim. Acta Part A Mol. Biomol. Spectrosc.* **2017**, *173*, 655–665. [CrossRef]
44. Gill, S.S.; Tuteja, N. Reactive oxygen species and antioxidant machinery in abiotic stress tolerance in crop plants. *Plant Physiol. Biochem.* **2010**, *48*, 909–930. [CrossRef] [PubMed]
45. Mo, L.; Yang, Y.; Zhao, D.; Qin, L.; Yuan, B.; Liang, N. Time-dependent toxicity and health effects mechanism of cadmium to three green algae. *Int. J. Environ. Res. Public Health* **2022**, *19*, 10974. [CrossRef] [PubMed]
46. Verma, S.; Dubey, R.S. Lead toxicity induces lipid peroxidation and alters the activities of antioxidant enzymes in growing rice plants. *Plant Sci.* **2003**, *164*, 645–655. [CrossRef]
47. Li, H.; Yao, J.; Duran, R.; Liu, J.; Min, N.; Chen, Z.; Zhu, X.; Zhao, C.; Ma, B.; Pang, W.; et al. Toxic response of the freshwater green algae *Chlorella pyrenoidosa* to combined effect of flotation reagent butyl xanthate and nickel. *Environ. Pollut.* **2021**, *286*, 117285. [CrossRef]
48. Shao, W.; Li, M.; Teng, Z.; Qiu, B.; Huo, Y.; Zhang, K. Effects of Pb(II) and Cr(VI) stress on phosphate-solubilizing bacteria (*Bacillus* sp. Strain MRP-3): Oxidative stress and bioaccumulation potential. *Int. J. Environ. Res. Public Health* **2019**, *16*, 2172. [CrossRef]

water

MDPI

Article

Adaptive Neuro-Fuzzy Inference System Modeling and Optimization of Microbial Fuel Cells for Wastewater Treatment

Mohammad Ali Abdelkareem [1,2,3], Samah Ibrahim Alshathri [4,*], Mohd Shahbudin Masdar [3,*] and Abdul Ghani Olabi [1,5]

1 Sustainable Energy and Power Systems Research Centre, RISE, University of Sharjah, Sharjah P.O. Box 27272, United Arab Emirates; mabdulkareem@sharjah.ac.ae (M.A.A.); aolabi@sharjah.ac.ae (A.G.O.)
2 Chemical Engineering Department, Faculty of Engineering, Minia University, Minia 61111, Egypt
3 Fuel Cell Institute, Universiti Kebangsaan Malaysia (UKM), Bangi 43600, Malaysia
4 Department of Information Technology, College of Computer and Information Sciences, Princess Nourah bint Abdulrahman University, P.O. Box 84428, Riyadh 11671, Saudi Arabia
5 Department of Mechanical Engineering and Design, School of Engineering and Applied Science, Aston University, Aston Triangle, Birmingham B4 7ET, UK
* Correspondence: sealshathry@pnu.edu.sa (S.I.A.); shahbud@ukm.edu.my (M.S.M.)

Abstract: Due to their toxicity, Cr(VI) levels are subject to strict legislation and regulations in various industries and environmental contexts. Effective treatment technologies are also being developed to decrease the negative impacts on human health and the environment by removing Cr(VI) from water sources and wastewater. As a result, it would be interesting to model and optimize the Cr(VI) removal processes, especially those under neutral pH circumstances. Microbial fuel cells (MFCs) have the capacity to remove Cr(VI), but additional research is needed to enhance their usability, increase their efficacy, and address issues like scalability and maintaining stable operation. In this research work, ANFIS modeling and artificial ecosystem optimization (AEO) were used to maximize Cr(VI) removal efficiency and the power density of MFC. First, based on measured data, an ANFIS model is developed to simulate the MFC performance in terms of the Cu(II)/Cr(VI) ratio, substrate (sodium acetate) concentration (g/L), and external resistance Ω. Then, using artificial ecosystem optimization (AEO), the optimal values of these operating parameters, i.e., Cu(II)/Cr(VI) ratio, substrate concentration, and external resistance, are identified, corresponding to maximum Cr(VI) removal efficiency and power density. In the ANFIS modeling stage of power density, the coefficient-of-determination is enhanced to 0.9981 compared with 0.992 (by ANOVA), and the RMSE is decreased to 0.4863 compared with 16.486 (by ANOVA). This shows that the modeling phase was effective. In sum, the integration between ANFIS and AEO increased the power density and Cr(VI) removal efficiency by 19.14% and 15.14%, respectively, compared to the measured data.

Keywords: microbial fuel cell; artificial ecosystem optimization; ANFIS modeling

Citation: Abdelkareem, M.A.; Alshathri, S.I.; Masdar, M.S.; Olabi, A.G. Adaptive Neuro-Fuzzy Inference System Modeling and Optimization of Microbial Fuel Cells for Wastewater Treatment. *Water* **2023**, *15*, 3564. https://doi.org/10.3390/w15203564

Academic Editors: Yung-Tse Hung, Tsuyoshi Imai, Rehab O. Abdel Rahman and Issam A. Al-Khatib

Received: 25 August 2023
Revised: 28 September 2023
Accepted: 2 October 2023
Published: 12 October 2023

1. Introduction

A conceptual framework known as the water–energy–food (WEF) nexus acknowledges the interdependencies and connections between water, energy, and food systems [1,2]. It emphasizes the necessity for integrated and sustainable management techniques and draws attention to the complex relationships and trade-offs among various areas [3,4]. Water is necessary for many human activities, such as residential use, industry, and agriculture. It is necessary to produce energy, process food, and irrigate crops. Water resources are limited and must contend with issues like scarcity, pollution, and rivalry from other industries. Food must be produced, processed, transported, and distributed. It is essential for the distribution, treatment, and delivery of water. Fossil fuels, hydropower, biomass, and renewable energy sources have varying effects on water resources, and energy production can exacerbate water pollution and stress [5]. Agriculture depends on energy inputs for machinery, transportation, processing, and water for irrigation. Food production can significantly influence water resources, including water use, fertilizer and pesticide contamination, and land degradation. Agricultural production and food security can both be impacted by water availability and quality. The WEF nexus paradigm acknowledges the possibility of cascading impacts from changes or disruptions in one sector to others. For instance, the availability of food and agricultural productivity may need to be improved by securing water resources. Crop yields, energy output, and water availability are all impacted by climate change. The WEF nexus demands integrated planning, policy coordination, and decision-making that considers the synergies and trade-offs across these interconnected systems to be managed effectively and sustainably [6,7]. It entails locating opportunities for resource optimization, expanding food production's water and energy efficiency, supporting renewable energy sources, lowering food waste, and boosting climate change resilience. When addressing issues related to water, energy, and food, the WEF nexus method aids policymakers, academics, and practitioners in adopting a holistic perspective to achieve more resilient and sustainable development pathways.

Although traditional wastewater treatment requires massive energy [8,9] for proper treatment before safe discharge to the environment, wastewater contains biomass energy that can provide considerable energy if properly used [10,11]. Using microorganisms and the bio-electrochemical process of microbial metabolism, microbial fuel cells (MFCs) convert organic matter into electrical energy [12,13]. A microbial fuel cell primarily transforms organic substrates into electrical energy while treating organic waste or wastewater [14,15]. MFCs are vital for generating renewable energy. The fuel cell's anode can capture the released electrons when microbes degrade organic matter and transport them to the cathode, creating an electrical current. MFCs may be used as sustainable power sources in distant or off-grid places [16]. Microbial fuel cells may be essential in the treatment of wastewater. The anode chamber's microorganisms break down organic debris as they consume it, detoxifying the wastewater. The MFC procedure can assist in removing organic contaminants and lowering the wastewater's COD. MFCs may find use in bioremediation, which entails eliminating or degrading contaminants from the environment. Utilizing microorganisms' metabolic processes, MFCs can speed up the breakdown of some pollutants or change them into less dangerous forms [17,18]. MFCs can be a power source for low-power sensors or monitoring equipment. Thanks to their ability to produce power from organic matter, they can supply sensors used for environmental monitoring or remote sensing applications with sustainable and self-sufficient energy. In disciplines including microbiology, bioelectrochemistry, and renewable energy, MFCs are valuable tools in research. They serve as a platform for research into electrochemical and microbiological processes and improve energy conversion effectiveness. Even though microbial fuel cells have shown promise in several applications, more study and development are still required to increase their effectiveness, scalability, and viability for commercial use.

MFCs can perform better thanks to artificial intelligence (AI) approaches since they can help optimize, control, and make decisions [19–21]. AI has been used to improve MFC performance [22]. Particle swarm optimization and genetic algorithms are two examples of AI techniques that can improve MFC systems' design and configuration. AI can assist in identifying the best electrode materials, reactor layouts, and operating conditions to maximize power output and efficiency by considering various characteristics and restrictions. AI-based solutions are capable of enabling real-time monitoring and control of MFC functioning. To gain insights into the performance of MFCs, machine learning techniques can be used to analyze sensor data, microbial activity, and other pertinent characteristics. Using this data, operational parameters can be changed, power generation can be optimized, and system problems can be avoided. AI can help create performance-based predictive models for MFC [23]. Machine learning algorithms can analyze historical data to grasp the intricate correlations between operational parameters and power output. Then, these models may be applied to forecast MFC performance under various scenarios, facilitate decision-making, and enhance system performance. By automating some jobs and decision-making procedures, AI can assist in streamlining the overall MFC process. Continuous data analysis, adaptive operational condition adjustment, and resource allocation optimization are all capabilities of AI algorithms. This can increase power generation, maximize energy efficiency, and simplify MFC operation. It is crucial to remember that AI approaches in MFCs are still a developing topic, and work is being undertaken to fully realize their potential. The particular MFC system, the data that are accessible, and the optimization or control goals will determine the individual AI methodologies and approaches used. The performance, effectiveness, and sustainability of microbial fuel cells could be improved by using AI technology in research and development.

Both fuzzy sets and fuzzy logic are used in the computational methods of ANFIS (adaptive neuro-fuzzy inference system) and fuzzy logic. ANFIS and fuzzy logic, however, differ significantly in a few crucial ways [24,25]. Fuzzy logic is a mathematical framework for representing and manipulating imprecise or uncertain information using linguistic variables and fuzzy rules [26]. It is based on the fuzzy sets theory, in which membership functions define the degree of membership of an element in a set. A hybrid computational model that blends fuzzy logic and neural networks is called ANFIS. ANFIS uses a neural network topology to represent a fuzzy inference system, and a learning algorithm is used to update the membership function and fuzzy rule parameters. Human experts often define fuzzy rules and membership functions based on their domain expertise and intuition. To modify the fuzzy inference system's parameters, ANFIS uses a learning technique often based on gradient descent or backpropagation. Thanks to this learning process, ANFIS may automatically modify the fuzzy rules and membership functions based on training data.

The removal of Cr(VI) depends on an MFC systems working circumstances, including substrate type and concentration [27,28], external resistance [29], and others. According to Koók et al. [29], external resistance impacts how extracellular electrons are transferred from the bacteria that produce them to the electrode surface. The substrate concentration influences the number of electrons created at the anode during the oxidation process, which correlates to the number of electron moves from the anode to the cathode, and thus significantly impacts electricity production [28]. Therefore, this research aims to improve the MFC's power density and Cr(VI) removal efficiency by utilizing artificial ecosystem optimization (AEO) and ANFIS modeling. To simulate the MFC in terms of the Cu(II)/Cr(VI) ratio, substrate (sodium acetate) concentration (g/L), and external resistance, an ANFIS model is first created based on observed data. In order to achieve the highest Cr(VI) removal efficiency and power density, the ideal values of the Cu(II)/Cr(VI) ratio, substrate concentration, and external resistance are then determined using AEO.

2. Microbial Fuel Cells

Microbial fuel cells (MFCs) have a generally established working mechanism. The microbial metabolism of organic matter, which entails several crucial phases and activities, is the basis upon which MFCs operate [26]. A biofilm or anode electrode coated with microorganisms, usually bacteria, is found in the anode chamber of an MFC. These bacteria can oxidize organic material found in substrates or effluents. Microbes break down organic matter as they metabolize it through various biochemical processes, such as oxidation or fermentation [30]. Electrons and protons are liberated from the organic molecules during this process. The anode electrode, which serves as the electron acceptor, receives the emitted electrons from the microbial metabolism.

The microorganisms on the anode surface create direct or mediated electron transfer channels to aid in this electron transfer process. The anode chamber becomes more acidic due to the simultaneous release of protons (H+) created during the oxidation of organic materials into the solution. These protons move towards the cathode chamber through the electrolyte. A membrane or ion exchange separator separates the anode and cathode chambers, which also permits proton migration. As a result, the anode and cathode solutions cannot be mixed directly. A second electrode serves as the electron acceptor in the cathode chamber, typically for oxygen reduction. Water is created when oxygen from the cathode and protons and electrons from the anode chamber come together. Electrons produced at the anode travel via an external circuit to produce an electric current that can be used to power electronics or recharge batteries, among other valuable applications. A continuous flow of electrons from the anode to the cathode is made possible by the external circuit's flow of electrons, which completes the electrochemical circuit. The microbial metabolism and the voltage differential between the anode and cathode electrodes propel this electron flow. Figure 1 shows schematic diagram of the main components of the MFC. The microbial oxidation of organic materials at the anode, which produces electrons and protons, is the main component of the microbial fuel cell mechanism [31]. It is important to note that different MFCs may use different microbial species, substrate compositions, electrode materials, and system configurations, which might affect performance and efficiency. Microbial fuel cells' processes and operating parameters are now being further understood and optimized for better performance and broader applications.

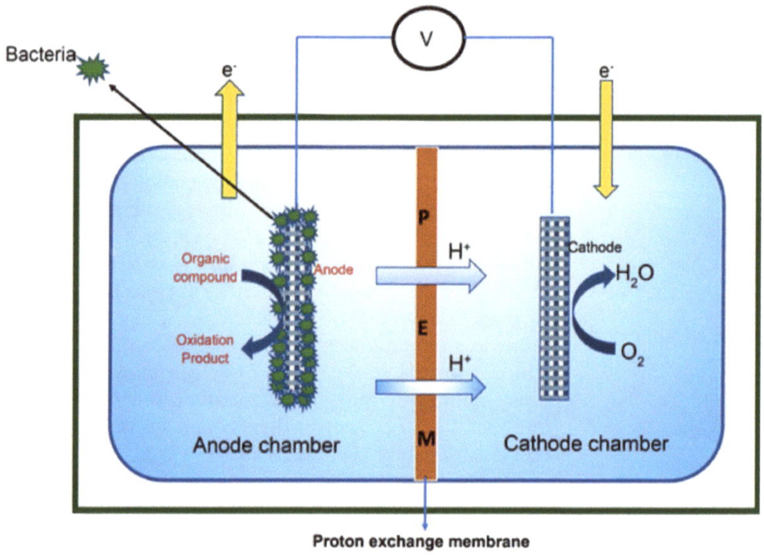

Figure 1. Operation of MFC.

Given the potential risks to both human health and the environment, treating wastewater containing Cr(VI) is garnering more attention. Because Cr(VI) is highly hazardous, carcinogenic, and mutagenic, it must be removed correctly to keep it from entering the environment. Cr(VI) concentrations in wastewater produced by industrial processes such as electroplating, leather tanning, welding, and dyeing can range from 5.7 to 87 mg/L (wastewater from electroplating) [32]. If these concentrations are not appropriately handled, they can present serious dangers. On the other hand, Cr(III) is less harmful to organisms and tends to precipitate as $Cr(OH)_3$. Wastewater containing Cr(VI) must be treated to lessen its adverse effects on ecosystems and human health. Chemical precipitation, adsorption, ion exchange, membrane filtration, and biological therapy are some of the treatment techniques that can be used. Each approach has pros and cons, and the choice is made based on the baseline Cr(VI) concentration, the required level of treatment efficacy, cost, and infrastructure accessibility. Cr(VI) concentrations can be effectively brought down to safe levels by effective treatment techniques, allowing for regulatory compliance and reducing environmental contamination. To ensure the preservation of the environment and human health, applying effective treatment technologies and adopting sound management practices for Cr(VI) is essential. Additionally, continuous monitoring and adherence to environmental standards are crucial to stopping the discharge of Cr(VI) into ecosystems and water bodies.

An MFC system's ability to remove Cr(VI) is influenced by its operational circumstances, including substrate type and concentration, external resistance, and others [33]. The external resistance impacts the extracellular electron transfer from the bacteria that produce extracellular electrons to the electrode surface. The number of electron suppliers, or the substrate concentration, determines how many electrons are produced at the anode during the oxidation process, which significantly impacts how much power is produced. The kind and concentration of the substrate used in the MFC are essential for removing Cr(VI) and producing electricity. The substrate supplies the microorganisms in the anode chamber with the carbon they need for their metabolic activity. The bacterial oxidation of organic material causes the release of electrons that help produce power. Up to a certain optimum point, the substrate concentration influences electron donor availability, and a higher concentration may result in increased electricity production and improved Cr(VI) elimination. However, extremely high substrate concentrations may result in unwanted microbial metabolic byproducts or substrate inhibition. The MFC's exterior resistance impacts how extracellular electrons are transported from the bacteria that produce them to the electrode surface. The resistance, which is frequently managed by an external load or resistor, affects the circuit's electron flow and regulates the amount of current produced. The external resistance can be adjusted to achieve the best balance between power production and Cr(VI) removal efficiency. Depending on the specific goals of the MFC operation, adjusting the external resistance might help maximize power output or prioritize the removal of Cr(VI).

Therefore, to effectively remove Cr(VI) and generate energy, an MFC system's working parameters must be optimized, considering substrate concentration, external resistance, and the Cu(II)/Cr(VI) ratio. The MFC's performance can be optimized by adjusting these parameters, making it more suitable for specific applications and ensuring efficient Cr(VI) remediation while producing sustainable energy.

In the current research, the considered data were obtained from [32]. These measurements were obtained using a two-tank microbial fuel cell (MFC). Each tank had a 700 mL functional capacity. The anode was made of carbon cloth, measuring 12 cm^2. Meanwhile, the cylindrical cathode was made of conductive carbon black combined with 10.24% wt polyvinyl alcohol (PVA). External resistors of 12 mm in length were used to link the anode and cathode. More details can be found in [32]. The number of data points is 19. Three input parameters are used as follows. The lower and upper limits for Cu(II)/Cr(VI) ratio are 0.33 and 1.672, respectively. The lower and upper percentages of the substrate concentration are 0.244 and 1.756. The minimum and maximum external resistance values are 244 Ω and 1000 Ω, respectively. Under these conditions, the power density (PD) of the MFC system ranged from 0.45 mW/ m^2 to 36.76 mW/m^2, and the Cr(VI) removal (RE) ranged from 30% to 75%, suggesting that the PD and Cr(VI) RE of the MFC system were affected by the Cu(II)/Cr(VI) ratio, substrate concentration, and external resistance. A modeling approach may successfully handle small amounts of nonlinear data. ANFIS combines the advantages of fuzzy logic and neural networks to create a hybrid model capable of capturing nonlinear relationships; therefore, it is used in the current study.

3. ANFIS Model of MFCs

Fuzzy offers a versatile and user-friendly framework for processing imprecise or uncertain information. A rich representation of information and reasoning is possible thanks to its ability to handle complicated fuzzy rules and membership functions [34]. ANFIS introduces neural networks' processing power and learning capacity, enabling them to learn and optimize the fuzzy inference system's parameters automatically. When the fuzzy rule base or membership functions need to be modified based on data, or when they are not well defined, ANFIS can be especially helpful. Fuzzy is widely employed in many areas, including expert systems, decision-making, pattern recognition, and control systems.

ANFIS has found use in tasks including system identification, modeling, prediction, and control, where both fuzzy logic and neural networks are advantageous. In conclusion, ANFIS and fuzzy logic use fuzzy sets and fuzzy logic concepts. Still, ANFIS blends fuzzy logic with neural networks and includes learning algorithms to change the fuzzy inference system's settings. Unlike classical fuzzy logic, which depends on predefined rules and membership functions, this hybrid method enables ANFIS to learn from data and optimize its performance [35].

Numerous steps are usually involved in the modeling process for the ANFIS. Preparing and prepping the input and output data for ANFIS modeling is the initial stage of the procedure. Data normalization, cleansing, and set division into training and testing sets are all included in this. ANFIS begins by creating a set of fuzzy rules based on the input-output information. The intricacy of the problem and the quantity of input variables determine the number of rules and the structure of those rules. Typically, clustering methods or grid partitioning techniques are used to construct the fuzzy rules. The membership function parameters connected to each input variable in the created fuzzy rules are determined in the following step. The least squares approach and gradient descent algorithms are frequently used parameter estimation techniques. The fuzzy inference system is built at this stage using the generated fuzzy rules and the estimated membership function parameters. The fuzzy inference system integrates the input variables and their corresponding fuzzy sets to make fuzzy inferences using fuzzy rules. To modify the fuzzy inference system's parameters, ANFIS employs a hybrid learning technique. The fuzzy rules' associated weights and membership function parameters are updated in this phase. The learning algorithm, such as the backpropagation method, often uses the training data to reduce the error between the expected and actual output values. Following the learning phase, the ANFIS model is assessed and validated using the testing data. Metrics like mean squared error (MSE), root mean squared error (RMSE), or correlation coefficients are used to evaluate the model's performance to gauge its accuracy and generalizability. The ANFIS model can make predictions or carry out system control operations based on new input data once it has been trained, validated, and optimized. It is crucial to remember that the specific implementation and variations in ANFIS modeling can change depending on the issue domain, the data that is accessible, and the particular application needs. Although the stages above offer a broad framework for ANFIS modeling, specifics may change based on the model architecture and learning methods.

The output of the ANFIS model is mapped to crisp form during the defuzzification phase. The map of input–output is formulated by IF–THEN rules. An example of the fuzzy rule is presented below.

$$\text{IF } x \text{ is } A_1 \text{ and } y \text{ is } B_1 \text{ then } f_1 = g_1(x, y)$$

$$\text{IF } x \text{ is } A_2 \text{ and } y \text{ is } B_2 \text{ then } f_2 = g_2(x, y)$$

where, the A_1 and B_1 are the MFs of the two inputs x and y.

The final output f is estimated as following.

$$f = \tilde{\omega}_1 f_1 + \tilde{\omega}_2 f_2 \quad \text{(Output Layer)}$$

where $\tilde{\omega}_1$ and $\tilde{\omega}_2$ are the firing strength of rules.

Evaluating

$$\tilde{\omega}_1 g_1(x, y) \text{ and } \tilde{\omega}_2 g_2(x, y) \quad \text{(Defuzzification Layer)}$$

$$\tilde{\omega}_1 = \frac{\omega_1}{\omega_1 + \omega_2} \text{ and } \tilde{\omega}_2 = \frac{\omega_2}{\omega_1 + \omega_2} \quad \text{(N Layer)}$$

where ω_1 and ω_2 are the weights.

$$\omega_1 = \mu_{A_1} * \mu_{B_1} \text{and } \omega_2 = \mu_{A_2} * \mu_{B_2} \quad (\pi \text{ Layer})$$

μ_{A_1}, μ_{A_2}, μ_{B_1} and μ_{B_2} are the MF values of the two inputs (Fuzzification Layer)

4. Parameter Identification by an Artificial Ecosystem Optimizer

An innovative method for resolving optimization issues is called artificial ecosystem-based optimization (AEO). It uses the principles of adaptability, cooperation, and rivalry seen in biological systems to tackle challenging optimization issues. It takes inspiration from natural ecosystems. AEO assembles a population of individual solution candidates, or "organisms," that represent potential answers to the optimization issue to form an artificial ecosystem. These organisms interact and change over time due to numerous factors, including competition, mutation, reproduction, and selection. The following steps are commonly included in the optimization process in AEO: A population of initial solution candidates is randomly generated within the search space. The fitness or objective function value is computed to evaluate each potential solution. Reproduction, crossover, and mutation are just a few of the mechanisms the solution candidates use to interact with one another and imitate the principles found in natural ecosystems. This enables the sharing of genetic data and the research of various search space locations [36]. Candidates for the solution are chosen depending on how likely they are to survive and reproduce in the following generation. Positive traits or qualities are transmitted to succeeding generations. A termination criterion, such as reaching a present number of generations or arriving at a good solution, must be satisfied before the optimization process can be completed. Similar to other methods like genetic algorithms, particle swarm optimization, or ant colony optimization, AEO is regarded as a metaheuristic or nature-inspired optimization methodology. Simulating the dynamics and interconnections of natural ecosystems provides a new way to approach the solution of optimization problems. It is crucial to remember that the precise application and variants of AEO can change based on the issue at hand, the goal of the optimization, and the particular algorithms or methods employed inside the framework. To find a solution, AEO typically uses the following guidelines. The production operator can be modeled as follows [36].

$$\begin{array}{l} x_1^{t+1} = (1-a)x_n^t + ax_{rand}^t \\ a = (1 - \frac{t}{t_{Max}})r_1;\ x_{rand} = r(ub - lb) + lb \end{array}$$

where n presents the population size, r_1 and r are random factors, and x_{rand} is a random position produced in the search space. The consumption operator can be modeled as

$$\begin{array}{l} x_i^{t+1} = x_i^t + C(x_i^t - x_1^t),\ i \in [2,..n];\ if\ rand < 1/3 \\ \left\{ \begin{array}{l} x_i^{t+1} = x_i^t + C(x_i^t - x_j^t),\ i \in [3,..n] \\ j = randi([\ 2 \quad i-1\ ]) \end{array} \right.;\ if\ 1/3 < rand < 2/3 \\ \left\{ \begin{array}{l} x_i^{t+1} = x_i^t + C(r_2(x_i^t - x_j^t) \\ \qquad +(1-r_2)(x_i^t - x_j^t)), \quad i \in [3,..n] \\ j = randi([\ 2 \quad i-1\ ]) \end{array} \right.;\ if\ rand > 2/3 \end{array}$$

The decomposition operator can be modeled as

$$\begin{array}{c} x_i^{t+1} = x_n^t + D(ex_n^t - hx_i^t),\quad i \in [1,..n] \\ D = 3u,\quad u \sim N(0,1) \\ e = r_3.randi([1 \quad 2]) - 1 \\ h = 2r_3 - 1 \end{array}$$

The AEO flowchart is presented in Figure 2. As explained in Figure 2, in the consumption stage, the consumer feeds on another consumer with a lower energy level or on a producer. For the consumer classes, carnivores, herbivores, and omnivores.

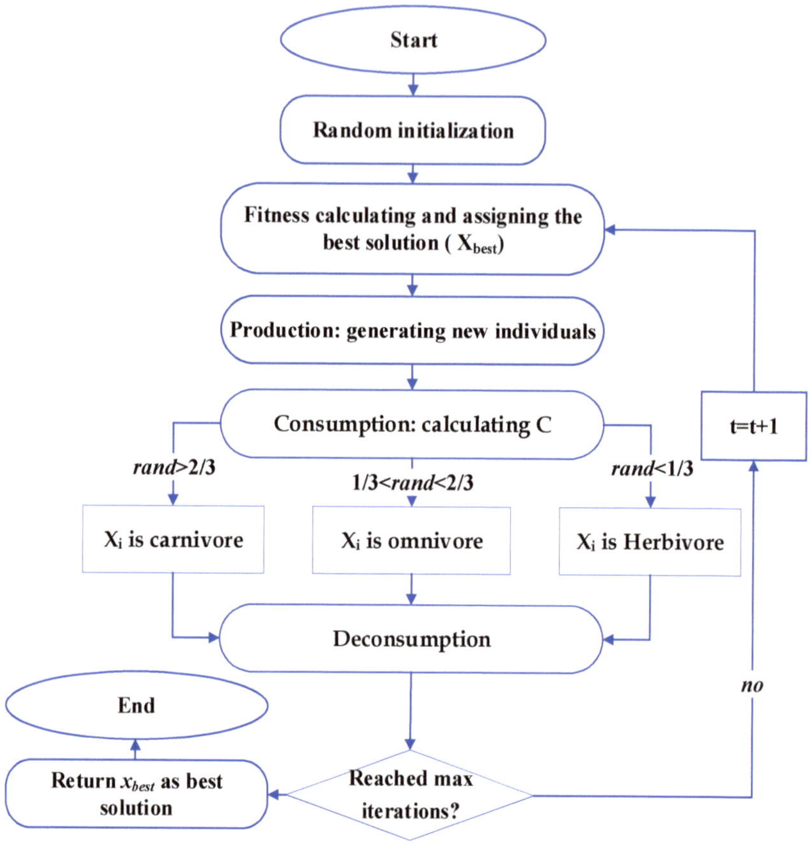

Figure 2. AEO flowchart.

The primary purpose of the optimization phase is to acquire the optimal values of Cu(II)/Cr(VI) ratio, substrate concentration, and external resistance that generate the maximum Cr(VI) removal efficiency and power density. Therefore, during the optimization process, the Cu(II)/Cr(VI) ratio, substrate concentration, and external resistance are assigned as decision variables, whereas the Cr(VI) removal efficiency and power density are the objective function that had to be most significant. The problem statement of the current optimization procedure may be written as

$$x = \underset{x \in R}{\operatorname{argmax}}(y)$$

where x is the set of input variables, and y is the output variable.

5. Results and Discussion

5.1. Modeling Phase

The ANFIS model was developed using 19 experiments. The data are divided into two groups: training and testing. Fifteen points comprise the first component, which is used to train the model; the remaining points are used to test the model. The hybrid training method employs LSE for the forward path and backpropagation for the backward direction. The system's rules, which number 15, were constructed using the SC. These models were subsequently trained until a reduced RMSE was attained. Table 1 displays the statistical metrics that the ANFIS model produced.

Table 1. Statistical metrics of ANFIS models.

RMSE			Coefficient of Determination (R^2)		
Train	**Test**	**All**	**Train**	**Test**	**All**
Model of power density					
9.64×10^{-6}	1.0598	0.4863	1.0	0.9864	0.9981
Model of Cr(VI) removal efficiency					
2.63×10^{-5}	2.1275	0.9761	1.0	0.9991	0.9963

Note: All means the total data containing both training and testing data sets.

Regarding Table 1, the RMSE values for the training and testing data sets for the ANFIS model of the power density are 9.64×10^{-6} and 1.0598, respectively. For training and testing, the coefficients of determination are 1.0 and 0.9864, respectively. The coefficient of determination is enhanced from 0.992 (by ANOVA) to 0.9981 (by ANFIS), and the RMSE is decreased from 16.486 (by ANOVA) to 0.4863 (by ANFIS). For the training and testing data sets, the RMSE values for the ANFIS model of Cr(VI) removal efficiency are 2.63×10^{-5} and 2.1275, respectively. For training and testing, the coefficients of determination are 1.0 and 0.9991, respectively. The coefficient of determination is enhanced from 0.951 (by ANOVA) to 0.9963 (by ANFIS), and the RMSE is decreased from 22.60 (by ANOVA) to 0.9761 (by ANFIS). This shows that the fuzzy modeling phase was effective. Figure 3 depicts the 3-input, single-output fuzzy model architecture, and Figure 4 depicts the general contours of the Gaussian form of MFs.

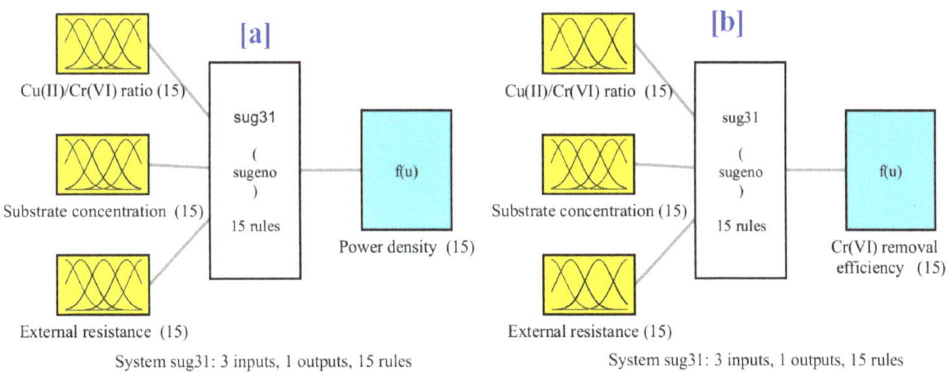

Figure 3. Configuration of ANFIS models [**a**] power density and [**b**] Cr(VI) removal efficiency.

Figure 5 provides a three-dimensional spatial representation, including the contour plots of the input-output functions for every combination of inputs. The color scale ranges from dark red, representing the maximum output value, to dark blue, which indicates the minimum output value. There is an enhancement in PD and Cr(VI) RE with an increase in substrate concentration. However, a decrease in PD and Cr(VI) RE is observed when external resistance deviates from an optimal range, either by increasing too much or decreasing substantially. This can be attributed to the fact that excessively high external resistance raises the obstacle to electron transfer from anode to cathode, lowering the current and making electricity generation and Cr(VI) removal less favorable. On the other hand, if the external resistance is too low, it results in an inadequate potential difference between the anode and cathode, impairing electron transfer and destabilizing the MFC system. Hence, an optimal level of external resistance is conducive to power generation in the MFC system and the removal of Cr(VI). Among the factors, substrate concentration exerts the most significant influence on PD; therefore, an adequate amount of substrate is preferable. On the other hand, a shortage of substrate greatly diminishes the PD. Consequently, addressing both external resistance and substrate concentration in evaluating the PD is imperative.

Figure 5 also elucidates that augmenting the substrate concentration and the ratio of Cu(II) to Cr(VI) is beneficial for the PD. This is associated with the rise in electron production owing to the availability of organic material in the anode compartment for microbial degradation.

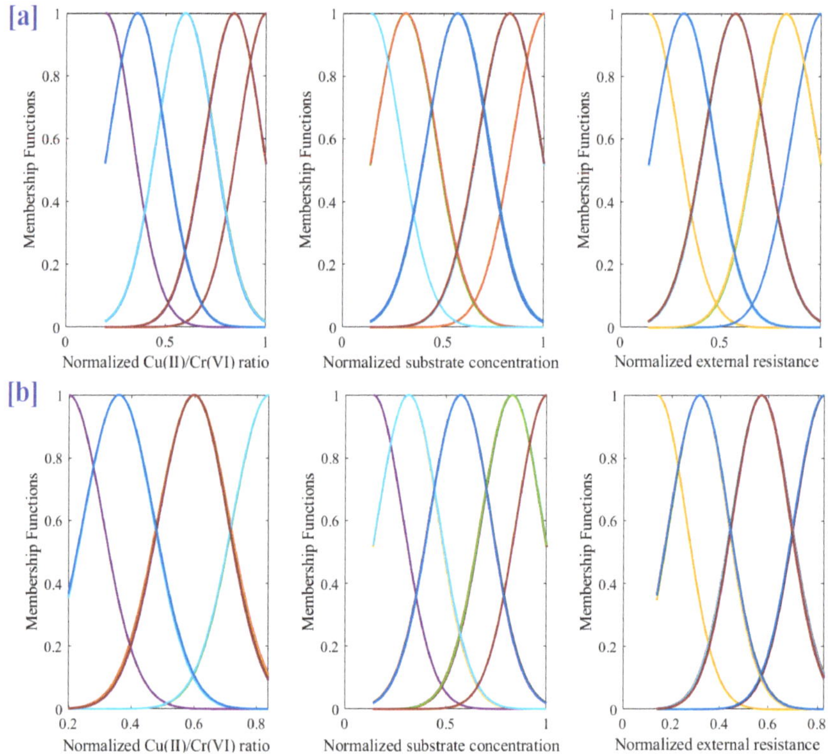

Figure 4. Inputs' MFs of ANFIS models: [**a**] power density and [**b**] Cr(VI) removal efficiency.

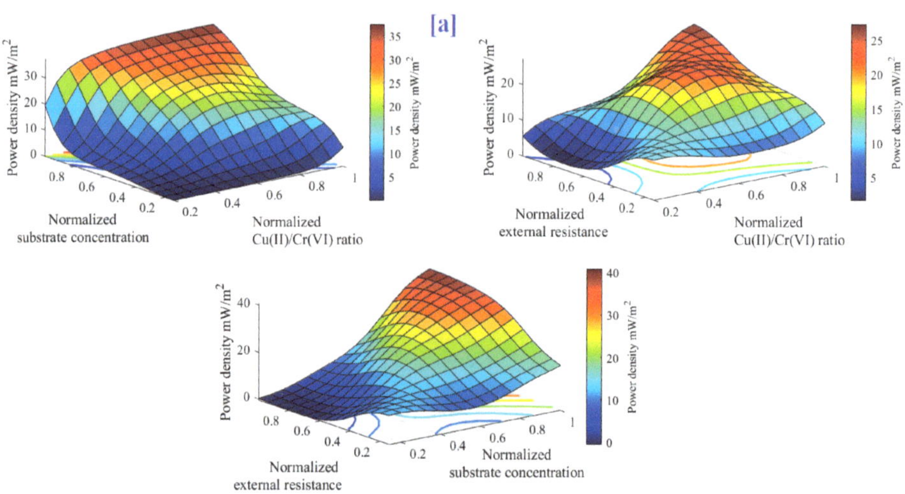

Figure 5. *Cont.*

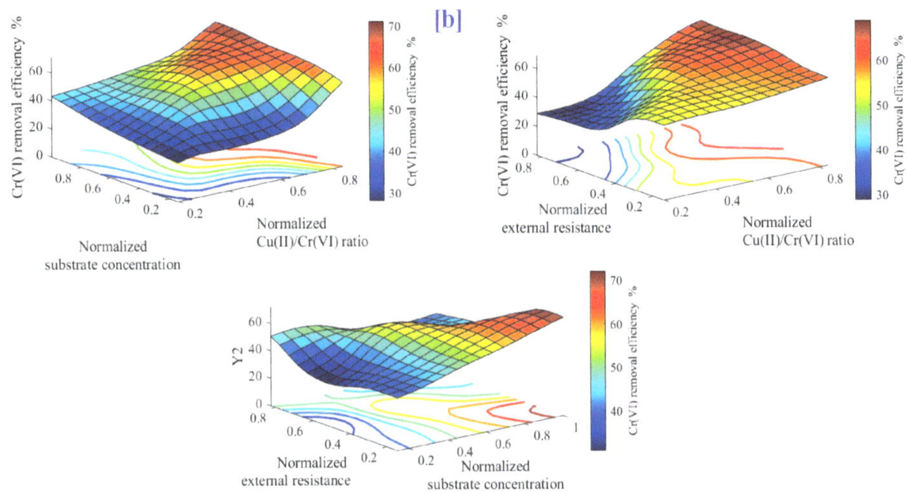

Figure 5. Three-dimensional plot of controlling parameters: [**a**] power density and [**b**] Cr(VI) removal efficiency.

Figure 6 shows the predicted versus measured data of the ANFIS model of power density and Cr(VI) removal efficiency. It is clear that the estimated and measured values fit each other well. The image presented in Figure 7 also shows the predictions' plots around the line of 100% accuracy for both the training and testing phases.

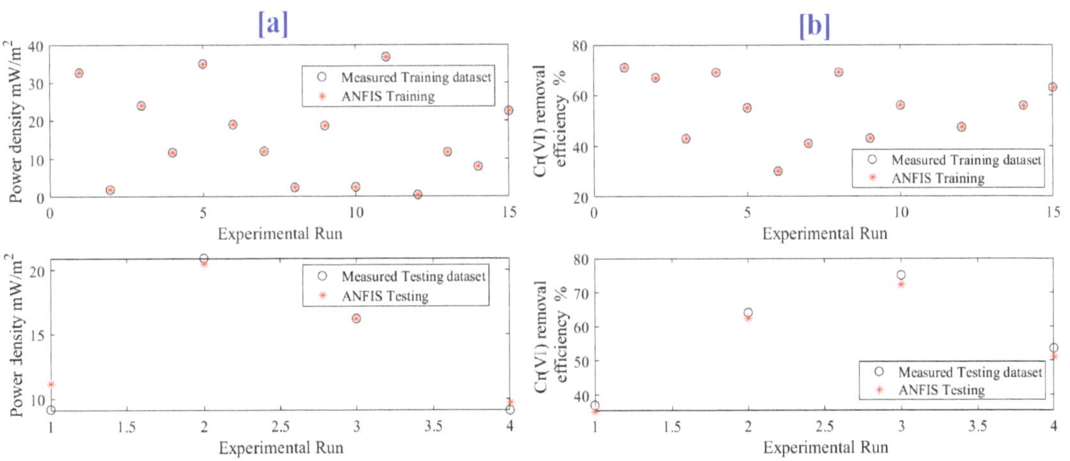

Figure 6. Predicted versus measured data of the ANFIS model: [**a**] power density and [**b**] Cr(VI) removal efficiency.

5.2. Optimization Phase

This section aims to identify the optimal levels of Cu(II)/Cr(VI) ratio, substrate concentration, and external resistance corresponding to maximum Cr(VI) removal efficiency and power density. As a result, AEO has been used to predict the best values for three regulating parameters after building trustworthy ANFIS models. The top experimental input and output parameters and the AEO are shown in Table 2. The integration between ANFIS and AEO increased the power density from 32.7 mW/m^2 to 38.96 mW/m^2 (by 19.14%) compared to measured data. In addition, boosting the Cr(VI) removal efficiency from 71% to 81.75% (by 15.14%) compared to measured data. Under this condition, the optimal values are 1.672, 1.756 (g/L), and 1404.8 Ω, respectively, for the Cu(II)/Cr(VI) ratio, substrate concentration, and external resistance. Figure 8 shows the particle convergence of objective function, normalized Cu(II)/Cr(VI) ratio, normalized substrate concentration, and normalized external resistance. The figure demonstrated that all particles converged to the optimal value after 35 iterations.

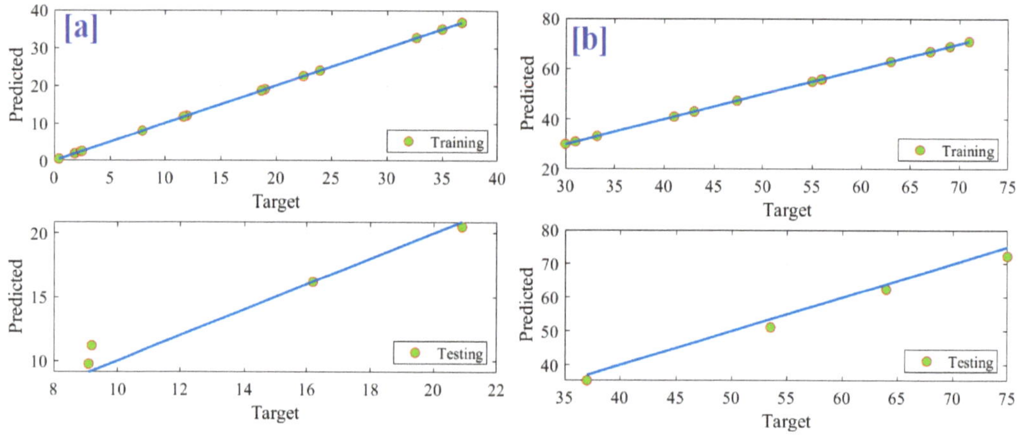

Figure 7. Prediction accuracy of the ANFIS model: **[a]** power density and **[b]** Cr(VI) removal efficiency.

Table 2. Best parameters by measured, ANOVA, and proposed methods.

	Cu(II)/Cr(VI) Ratio	Substrate Concentration	External Resistance	Power Density, mW/m^2	Cr(VI) Removal Efficiency	Change in Power Density	Change in Cr(VI) Removal Efficiency
Measured [32]	1.4	1.45 (g/L)	1000 Ω	32.7	71%	0.0	0.0
ANOVA [32]	1.65	1.36	1360	33.84	71%	3.48%	0.0
ANFIS and AEO	1.672	1.756 (g/L)	1404.8 Ω	38.96	81.75%	19.14%	15.14%

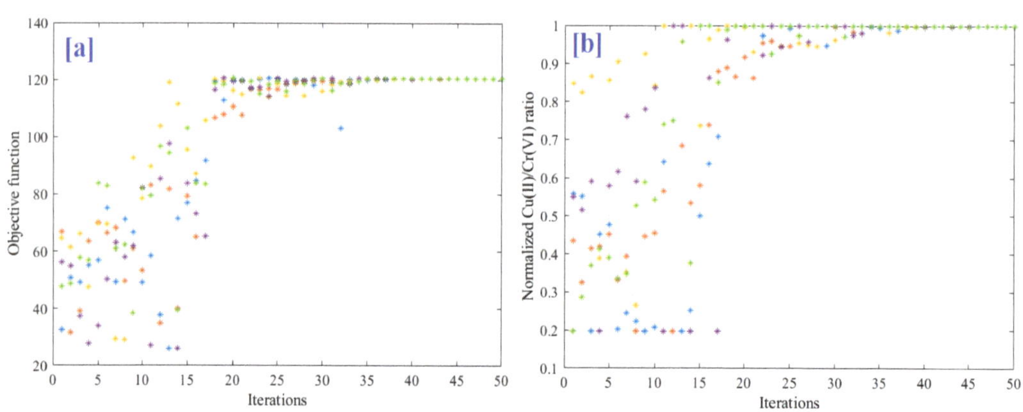

Figure 8. *Cont.*

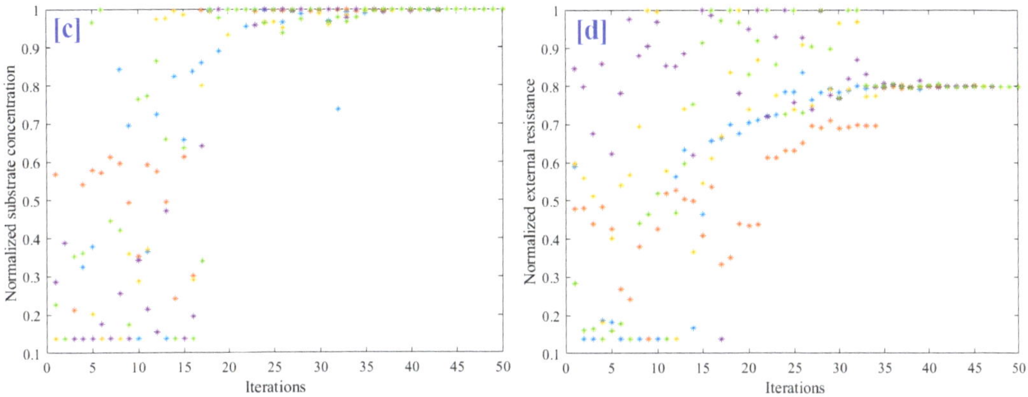

Figure 8. Particles convergence during parameter identification: [**a**] cost function, [**b**] normalized Cu(II)/Cr(VI) ratio, [**c**] normalized substrate concentration, and [**d**] normalized external resistance.

6. Conclusions

Boosting Cr(VI) removal efficiency and power density of microbial fuel cells using ANFIS and artificial ecosystem optimization (AEO) is the main target of this research. The proposed methodology contains two phases: modeling and parameter identification by artificial ecosystem optimization (AEO). First, based on measured data, an ANFIS model is developed to simulate the MFC in terms of the Cu(II)/Cr(VI) ratio, substrate (sodium acetate) concentration (g/L), and external resistance Ω. Then, using AEO, the optimal values of the Cu(II)/Cr(VI) ratio, substrate concentration, and external resistance are identified, corresponding to maximum Cr(VI) removal efficiency and highest power density. For the training and testing data sets, the RMSE values for the ANFIS model of Cr(VI) removal efficiency are 2.63×10^{-5} and 2.1275, respectively. For training and testing, the coefficients of determination are 1.0 and 0.9991, respectively. The coefficient of determination is enhanced from 0.951 (by ANOVA) to 0.9963 (by ANFIS), and the RMSE is decreased from 22.60 (by ANOVA) to 0.9761 (by ANFIS). This shows that the fuzzy modeling phase was effective. Finally, the integration between ANFIS and AEO increased the power density from 32.7 mW/m^2 to 38.96 mW/m^2, by 19.14%, compared to measured data. In addition, it boosted the Cr(VI) removal efficiency from 71% to 81.75%, by 15.14%, compared to measured data. Under this condition, the optimal values are 1.672, 1.756 (g/L), and 1404.8 Ω, respectively, for the Cu(II)/Cr(VI) ratio, substrate concentration, and external resistance. The obtained results are not verified experimentally; therefore, they could be used as a basis for further investigations using more parameters than those investigated in the current research.

Author Contributions: Conceptualization, M.A.A. and M.S.M.; Methodology, S.I.A. and M.S.M.; Software, S.I.A. and M.S.M.; Validation, S.I.A.; Formal analysis, M.A.A. and A.G.O.; Investigation, M.A.A. and A.G.O.; Resources, A.G.O.; Writing—original draft, M.A.A., S.I.A., M.S.M. and A.G.O.; Writing—review & editing, M.A.A., S.I.A., M.S.M. and A.G.O.; Supervision A.G.O. All authors have read and agreed to the published version of the manuscript.

Funding: This research was funded by Princess Nourah bint Abdulrahman University Researchers Supporting Project number (PNURSP2023R197), Princess Nourah bint Abdulrahman University, Riyadh, Saudi Arabia.

Data Availability Statement: Not applicable.

Conflicts of Interest: The authors declare no conflict of interest.

References

1. Lazaro, L.L.B.; Bellezoni, R.A.; Puppim de Oliveira, J.A.; Jacobi, P.R.; Giatti, L.L. Ten Years of Research on the Water-Energy-Food Nexus: An Analysis of Topics Evolution. *Front. Water* **2022**, *4*, 859891. [CrossRef]
2. Zhang, C.; Chen, X.; Li, Y.; Ding, W.; Fu, G. Water-energy-food nexus: Concepts, questions and methodologies. *J. Clean. Prod.* **2018**, *195*, 625–639. [CrossRef]
3. Abulibdeh, A.; Zaidan, E. Managing the water-energy-food nexus on an integrated geographical scale. *Environ. Dev.* **2020**, *33*, 100498. [CrossRef]
4. James Rubinsin, N.; Daud, W.R.W.; Kamarudin, S.K.; Masdar, M.S.; Rosli, M.I.; Samsatli, S.; Tapia, J.F.D.; Wan Ab Karim Ghani, W.A.; Hasan, A.; Lim, K.L. Modelling and optimisation of oil palm biomass value chains and the environment–food–energy–water nexus in peninsular Malaysia. *Biomass Bioenergy* **2021**, *144*, 105912. [CrossRef]
5. Jin, Y.; Behrens, P.; Tukker, A.; Scherer, L. Water use of electricity technologies: A global meta-analysis. *Renew. Sustain. Energy Rev.* **2019**, *115*, 109391. [CrossRef]
6. Bazzana, D.; Comincioli, N.; El Khoury, C.; Nardi, F.; Vergalli, S. WEF Nexus Policy Review of Four Mediterranean Countries. *Land* **2023**, *12*, 473. [CrossRef]
7. Albrecht, T.R.; Crootof, A.; Scott, C.A. The Water-Energy-Food Nexus: A systematic review of methods for nexus assessment. *Environ. Res. Lett.* **2018**, *13*, 043002. [CrossRef]
8. Longo, S.; d'Antoni, B.M.; Bongards, M.; Chaparro, A.; Cronrath, A.; Fatone, F.; Lema, J.M.; Mauricio-Iglesias, M.; Soares, A.; Hospido, A. Monitoring and diagnosis of energy consumption in wastewater treatment plants. A state of the art and proposals for improvement. *Appl. Energy* **2016**, *179*, 1251–1268. [CrossRef]
9. Yu, Y.; Zou, Z.; Wang, S. Statistical regression modeling for energy consumption in wastewater treatment. *J. Environ. Sci.* **2019**, *75*, 201–208. [CrossRef]
10. Ibrahim, N.; Kamarudin, S.K.; Minggu, L.J. Production of electricity from ethanol and ammonium based wastewater via photo-electrochemical process. *Int. J. Hydrogen Energy* **2017**, *42*, 9051–9062. [CrossRef]
11. Kollmann, R.; Neugebauer, G.; Kretschmer, F.; Truger, B.; Kindermann, H.; Stoeglehner, G.; Ertl, T.; Narodoslawsky, M. Renewable energy from wastewater—Practical aspects of integrating a wastewater treatment plant into local energy supply concepts. *J. Clean. Prod.* **2017**, *155*, 119–129. [CrossRef]
12. Rusli, S.F.N.; Daud, S.M.; Abu Bakar, M.H.; Loh, K.S.; Masdar, M.S. Biotic Cathode of Graphite Fibre Brush for Improved Application in Microbial Fuel Cells. *Molecules* **2022**, *27*, 1045. [CrossRef] [PubMed]
13. Rusli, S.F.N.; Abu Bakar, M.H.; Loh, K.S.; Mastar, M.S. Review of high-performance biocathode using stainless steel and carbon-based materials in Microbial Fuel Cell for electricity and water treatment. *Int. J. Hydrogen Energy* **2019**, *44*, 30772–30787. [CrossRef]
14. Tsekouras, G.J.; Deligianni, P.M.; Kanellos, F.D.; Kontargyri, V.T.; Kontaxis, P.A.; Manousakis, N.M.; Elias, C.N. Microbial Fuel Cell for Wastewater Treatment as Power Plant in Smart Grids: Utopia or Reality? *Front. Energy Res.* **2022**, *10*, 843768. [CrossRef]
15. Sayed, E.T.; Alawadhi, H.; Olabi, A.G.; Jamal, A.; Almahdi, M.S.; Khalid, J.; Abdelkareem, M.A. Electrophoretic deposition of graphene oxide on carbon brush as bioanode for microbial fuel cell operated with real wastewater. *Int. J. Hydrogen Energy* **2021**, *46*, 5975–5983. [CrossRef]
16. Kurniawan, T.A.; Othman, M.H.D.; Liang, X.; Ayub, M.; Goh, H.H.; Kusworo, T.D.; Mohyuddin, A.; Chew, K.W. Microbial Fuel Cells (MFC): A Potential Game-Changer in Renewable Energy Development. *Sustainability* **2022**, *14*, 16847. [CrossRef]
17. Bala, S.; Garg, D.; Thirumalesh, B.V.; Sharma, M.; Sridhar, K.; Inbaraj, B.S.; Tripathi, M. Recent Strategies for Bioremediation of Emerging Pollutants: A Review for a Green and Sustainable Environment. *Toxics* **2022**, *10*, 484. [CrossRef] [PubMed]
18. Roy, H.; Rahman, T.U.; Tasnim, N.; Arju, J.; Rafid, M.M.; Islam, M.R.; Pervez, M.N.; Cai, Y.; Naddeo, V.; Islam, M.S. Microbial Fuel Cell Construction Features and Application for Sustainable Wastewater Treatment. *Membranes* **2023**, *13*, 490. [CrossRef] [PubMed]
19. Rezk, H.; Olabi, A.G.; Abdelkareem, M.A.; Sayed, E.T. Artificial intelligence as a novel tool for enhancing the performance of urine fed microbial fuel cell as an emerging approach for simultaneous power generation and wastewater treatment. *J. Taiwan Inst. Chem. Eng.* **2023**, *148*, 104726. [CrossRef]
20. Rezk, H.; Olabi, A.G.; Abdelkareem, M.A.; Maghrabie, H.M.; Sayed, E.T. Fuzzy Modelling and Optimization of Yeast-MFC for Simultaneous Wastewater Treatment and Electrical Energy Production. *Sustainability* **2023**, *15*, 1878. [CrossRef]
21. Sayed, E.T.; Rezk, H.; Abdelkareem, M.A.; Olabi, A.G. Artificial neural network based modelling and optimization of microalgae microbial fuel cell. *Int. J. Hydrogen Energy* **2023**. [CrossRef]
22. Abdollahfard, Y.; Sedighi, M.; Ghasemi, M. A New Approach for Improving Microbial Fuel Cell Performance Using Artificial Intelligence. *Sustainability* **2023**, *15*, 1312. [CrossRef]
23. Garg, A.; Vijayaraghavan, V.; Mahapatra, S.S.; Tai, K.; Wong, C.H. Performance evaluation of microbial fuel cell by artificial intelligence methods. *Expert Syst. Appl.* **2014**, *41 Pt 1*, 1389–1399. [CrossRef]
24. Arslankaya, S. Comparison of performances of fuzzy logic and adaptive neuro-fuzzy inference system (ANFIS) for estimating employee labor loss. *J. Eng. Res.* **2023**, 100107. [CrossRef]
25. Zaghloul, M.S.; Hamza, R.A.; Iorhemen, O.T.; Tay, J.H. Comparison of adaptive neuro-fuzzy inference systems (ANFIS) and support vector regression (SVR) for data-driven modelling of aerobic granular sludge reactors. *J. Environ. Chem. Eng.* **2020**, *8*, 103742. [CrossRef]

26. Li, M.; Zhou, M.; Tian, X.; Tan, C.; McDaniel, C.T.; Hassett, D.J.; Gu, T. Microbial fuel cell (MFC) power performance improvement through enhanced microbial electrogenicity. *Biotechnol. Adv.* **2018**, *36*, 1316–1327. [CrossRef] [PubMed]
27. Nourbakhsh, F.; Pazouki, M.; Mohsennia, M. Simultaneous Investigation of Three Effective Parameters of Substrate, Microorganism Type and Reactor Design on Power Generation in a Dual-Chamber Microbial Fuel Cells. *Iran. J. Biotechnol.* **2020**, *18*, e2292.
28. Ramya, M.; Senthil Kumar, P. A review on recent advancements in bioenergy production using microbial fuel cells. *Chemosphere* **2022**, *288*, 132512. [CrossRef]
29. Koók, L.; Nemestóthy, N.; Bélafi-Bakó, K.; Bakonyi, P. The influential role of external electrical load in microbial fuel cells and related improvement strategies: A review. *Bioelectrochemistry* **2021**, *140*, 107749. [CrossRef]
30. Sharma, R.; Garg, P.; Kumar, P.; Bhatia, S.K.; Kulshrestha, S. Microbial Fermentation and Its Role in Quality Improvement of Fermented Foods. *Fermentation* **2020**, *6*, 106. [CrossRef]
31. Ucar, D.; Zhang, Y.; Angelidaki, I. An Overview of Electron Acceptors in Microbial Fuel Cells. *Front. Microbiol.* **2017**, *8*, 643. [CrossRef] [PubMed]
32. Lin, C.-W.; Chung, Y.-P.; Liu, S.-H.; Chen, W.T.; Zhu, T.-J. Optimizing the parameters of microbial fuel cells using response surface methodology to increase Cr(VI) removal efficiency and power production. *Process Saf. Environ. Prot.* **2023**, *172*, 369–378. [CrossRef]
33. Hidayat, A.R.P.; Widyanto, A.R.; Asranudin, A.; Ediati, R.; Sulistiono, D.O.; Putro, H.S.; Sugiarso, D.; Prasetyoko, D.; Purnomo, A.S.; Bahruji, H.; et al. Recent development of double chamber microbial fuel cell for hexavalent chromium waste removal. *J. Environ. Chem. Eng.* **2022**, *10*, 107505. [CrossRef]
34. Pedrycz, W. Interfaces of fuzzy models: A study in fuzzy information processing. *Inf. Sci.* **1996**, *90*, 231–280. [CrossRef]
35. Buragohain, M.; Mahanta, C. A novel approach for ANFIS modelling based on full factorial design. *Appl. Soft Comput.* **2008**, *8*, 609–625. [CrossRef]
36. Zhao, W.; Wang, L.; Zhang, Z. Artificial ecosystem-based optimization: A novel nature-inspired meta-heuristic algorithm. *Neural Comput. Appl.* **2020**, *32*, 9383–9425. [CrossRef]

MDPI AG
Grosspeteranlage 5
4052 Basel
Switzerland
Tel.: +41 61 683 77 34

Water Editorial Office
E-mail: water@mdpi.com
www.mdpi.com/journal/water

www.ingramcontent.com/pod-product-compliance
Lightning Source LLC
Chambersburg PA
CBHW041353210926
51850CB00013B/3067
* 9 7 8 3 7 2 5 8 3 4 7 6 1 *